Reinhard Selten (Ed.)

Game Equilibrium Models I

Evolution and Game Dynamics

With Contributions by

I. Eshel, J. W. Friedman, R. Gardner, P. Hammerstein
P. F. Hoekstra, Y. Iwasa, D. Messick, M. Morris
H. J. Poethke, R. Selten, A. Shmida, F. J. Weissing

With 41 Figures

Springer-Verlag Berlin Heidelberg GmbH

Professor Dr. Reinhard Selten
Institut für Gesellschaft- und
Wirtschaftswissenschaften
der Universität Bonn
Wirtschaftstheoretische Abteilung I
Adenauerallee 24-42
D-5300 Bonn 1, FRG

ISBN 978-3-642-08108-8   ISBN 978-3-662-02674-8 (eBook)
DOI 10.1007/978-3-662-02674-8

Preface to the Series "Game Equilibrium Models"

The four volumes of the series "Game Equilibrium Models" are the result of a research year at the Center for Interdisciplinary Research of the University of Bielefeld, Germany. The German name of this center is Zentrum für interdisziplinäre Forschung, but everybody who is familiar with this unique institution refers to it by the official acronym *ZiF*.

In the time from October 1, 1987, to September 30, 1988, the ZiF was the home of the interdisciplinary research group which produced the papers in the four volumes of this series. Participants coming from many parts of the world lived in the guest apartments of the ZiF for the whole time or part of it and worked on a common project. The name of the project was "Game Theory in the Behavioral Sciences". It occurred to me only later that "Game Equilibrium Models" - the title of the series - even more appropriately indicates the unifying theme of the research group.

Among the participants were economists, biologists, mathematicians, political scientists, psychologists and a philosopher. A lively interaction resulted from the mix of disciplines. The common methodological basis of non-cooperative theory was the shared culture which facilitated communication across disciplines. The intense exchange of ideas coming from different fields had a profound influence on the thinking of many among the participants.

It was not easy to find a coherent way to group the papers into the four volumes and to find appropriate titles for the books. These and other difficult decisions have been made by an editorial committee consisting of Wulf Albers, Rudolf Avenhaus, Eric van Damme, Werner Güth, Peter Hammerstein, Ronald Harstad, Franz Weissing, and myself.

In the behalf of the whole research group I want to thank all those who helped to make the research year possible. We owe special thanks to the staff of the ZiF and in particular to Mrs. Lilo Jegerlehner for her technical help in the preparation of the four volumes.

Finally, I want to express my gratitude to all those who assisted me in the organizational and editorial work, especially to Franz Weissing whose efforts were indispensable.

Bielefeld/Bonn, January 1991   Reinhard Selten

Contents

Ilan Eshel
GAME THEORY AND POPULATION DYNAMICS IN COMPLEX GENETICAL SYSTEMS:
THE ROLE OF SEX IN SHORT TERM AND IN LONG TERM EVOLUTION

Franz J. Weissing
EVOLUTIONARY STABILITY AND DYNAMIC STABILITY IN A CLASS OF EVOLUTIONARY
NORMAL FORM GAMES

Reinhard Selten
ANTICIPATORY LEARNING IN TWO-PERSON GAMES

Rolf F. Hoekstra, Yoh Iwasa and Franz J. Weissing
THE ORIGIN OF ISOGAMOUS SEXUAL DIFFERENTIATION

Contributors

Ilan Eshel, Faculty of Mathematical Sciences, Tel Aviv University Raymond and Beverly Sackler, Ramat Aviv, 69978 Tel Aviv, Israel

James W. Friedman, Department of Economics, University of North Carolina, Gardner Hall 01719, Chapel Hill, NC 27514, USA

Roy Gardner, Department of Economics, Indiana University, Ballantine Hall, Bloomington, IN 47405, USA

Peter Hammerstein, Max-Planck-Institut für Verhaltensphysiologie, Abteilung Wickler, D-8131 Seewiesen, FRG

Rolf Hoekstra, Universität of Wageningen, Department of Genetics, Dreijenlaan 2, 6703-HA Wageningen, The Netherlands

Yoh Iwasa, Kyushu University, Department of Biology, Faculty of Science 33, Fukkoka 812, Japan

David Messick, Department of Psychology, University of California Santa Barbara, Santa Barbara, CA 93106, USA

Molly Morris, Department of Biology, Indiana University, Bloomington, IN 47405, USA

Hans-Joachim Poethke, Universität Mainz, Institut für Zoologie, AG Populationsbiologie, Postfach 3980, D-6500 Main, FRG

Reinhard Selten, Institut für Gesellschafts- und Wirtschaftswissenschaften der Universität Bonn, Wirtschaftstheoretische Abteilung I, Adenauerallee 24-42, D-5300 Bonn 1, FRG

Avi Shmida, Department of Botany, The Hebrew University, Givat Ram, Jerusalem 91904, Israel

Franz J.Weissing, Rijksuniversiteit Groningen, Biologisch Centrum, Kerklaan 30, 9751-NN Haren, The Netherlands

Introduction to the Series "Game Equilibrium Models"

Game equilibrium models are descriptions of interactive decision situations by game in extensive or normal form. The analysis of such models is based on the equilibriu point concept, often refined by additional requirements like subgame perfectness. Th series consists of four volumes:

I: Evolution and Game Dynamics
II: Methods, Morals and Markets
III: Strategic Bargaining
IV: Social and Political Interaction.

The game equilibrium models presented in these books deal with a wide variety o topics. Just one example from each of the volumes may provide an illustration: Eg trading in hermaphrodite fish (*Friedman and Hammerstein* in Volume I), the social or ganization of irrigation systems (*Weissing and Ostrom* in Volume II), wage bargainin (*Haller* in Volume III), and beheading games in mediaeval literature (*O'Neill* i Volume IV).

Non-cooperative game theory is a useful research tool not only in economics an the social sciences, but also in biology. Game theory has been created as a theory o conflict and cooperation among rational individuals. For a long time strong rational ity assumptions seemed to be indispensable at the foundations of game theory. In thi respect, biological applications have changed our perspectives. Game equilibrium ma be reached as the result of processes of natural selection without any rational de liberation. Therefore, game equilibrium models can contribute to the explanation o behavioral and structural features of animals and plants.

The interpretation of game equilibrium as the result of evolution raises the question of dynamic stability with respect to evolutionary processes. Similar prob lems also arise in theories of game learning. The first volume contains three papers on game dynamics. Two of them are concerned with the dynamic foundations of evol utionary game theory and the third one explores stability in a model of anticipatory learning. The remaining papers in the first volume present evolutionary game equilib rium models ranging from abstract investigations of phenomena like bluffing or group based altruism to the examination of concrete systems observed in nature like "competition avoidance in a dragonfly mating system". Not only theoretical clarifica tions of the foundations of evolutionary game theory and related research can be found in **Evolution and Game Dynamics**, but also exciting new biological applications.

The title of the second volume, **Methods, Morals, and Markets**, points to several areas of research which attract the interest mainly of economists, but also of pol itical scientists, mathematicians and philosophers. The first paper is a sophisti cated mathematical contribution which applies new tools to basic questions of non-co operative game theory. The word "method" mainly refers to this paper, but to some extent also to the next three contributions, which discuss basic conceptual problems in the interpretation of game equilibrium. Two papers relate to the philosophical notion of the social contract and its exploration with the help of game theoretical models. This work in concerned with "morals", a theme which is also touched by a paper on irrigation institutions. The remaining four papers of the second volume explore game equilibrium models of markets; two of these contributions are experimental and compare theoretical solutions with experimental data.

The third volume on **Strategic Bargaining** collects ten papers on game equilibrium models of bargaining. All these papers look at bargaining situations as non-cooperative games. Unlike in cooperative game theory, cooperation is not taken for granted, but explained as an outcome of equilibrium analysis. General models of two-person and n-person bargaining are explored, sometimes enriched by institutional detail like the availability of long-term contracts. Other papers explore bargaining in special contexts like wage negotiations. Two contributions concern spatial games; one of these contributions is experimental.

The exploration of strategic models of bargaining is an active field of research which attracts the attention of many game theorists and economists. The ten papers in the third volume contribute to the progress in this field.

The fourth volume on **Social and Political Interaction** mainly presents game equilibrium models in the area of political science. Three of the papers concern topics in other fields: the distribution of foreign language skills, altruism as a social dilemma (an experimental paper) and beheading games in mediaeval literature. Five contributions to the area of international relations deal with game theoretical models of the balance of power, of alliance formation, and of an issue in armament policy. An investigation of inspection problems like those arising in connection with the non-proliferation treaty also touches the area of international relations. Other papers on problems of political science deal with the game theoretical resolution of the Condorcet paradox by equilibrium selection, the modelling of political pressure exerted by firms on the government and the draft resistance problem.

The main emphasis is on biology in Volume I, on economics in Volumes II and III, and on political science in Volume IV. This is the result of an attempt to group the great variety of papers resulting from a year long interdisciplinary research project in a reasonably coherent way. However, not only biologists, but also economists and a psychologist have contributed to Volume I. Similarly, not only economists and mathematicians, but also political scientists as well as a biologist and a psychologist are among the authors of Volumes II and III. All four volumes are the result of the cooperation of researchers from many disciplines united by a common interest in game equilibrium models within and beyond the borders of their fields.

Bielefeld/Bonn, January 1991   Reinhard Selten

Introduction to Volume I: Evolution and Game Dynamics

Game theory has been introduced by von Neumann and Morgenstern (1944) as a tool for the analysis of problems arising in economics and social science. Strong assumptions on the rationality of the players seemed to be indispensable at the basis of game theory. Therefore, it came as a surprise that after the pioneering paper of Maynard Smith and Price (1973) more and more biologists began to apply game theory.

It is a tenet of the Darwinian view of evolution that natural selection leads to a tendency towards optimization. The maximization of fitness has proved to be a powerful principle in the explanation of empirical phenomena. Roughly speaking, fitness is the expected number of offspring in the next generation. The principle of fitness maximization makes it possible to understand morphological and behavioral traits of animals and plants as the results of optimal adaptation to selective forces.

In retrospect it may seem to be only a small step from non-interactive optimization approaches to applications of game theory in biology. As soon as problems of social interactions among members of the same species are considered, game theoretical modelling becomes almost unavoidable. This consequence of the principle fitness maximization may now seem to be obvious, but nevertheless its discovery by Maynard Smith and Price (1973) was a great conceptual innovation. It is often very difficult to reach conclusions which later seem to be obvious.

Even among biologists one sometimes finds a misunderstood Darwinism according to which evolution tends to optimize the welfare of the species. The principle of fitness maximization does not imply anything like this. Selection forces work on the level of the individual. The motor of evolution is competition among the members of the same species. Therefore, evolutionary reasoning must be based on methodological individualism.

Evolutionary game theory is non-cooperative game theory. The methodological individualism of evolutionary theory seems to leave little room for cooperative game theory. Even more important in this respect is a focus on strategies rather than payoffs. Biologists want to explain morphological and behavioral traits. Therefore, they need game models in which traits appear as features of strategies. Non-cooperative games in normal form or extensive form meet this requirement.

As far as basic concepts are concerned, evolutionary game theory restricts its attention to symmetric games. This is understandable in view of the interpretation of a strategy as a genotype. Maynard Smith and Price (1973) introduced the central concept of an "evolutionarily stable strategy" applicable to symmetric normal form games. An evolutionarily stable strategy is a symmetric equilibrium strategy with an additional stability property concerning alternative best replies to itself: Against any alternative best reply the evolutionarily stable strategy is more successful than this alternative best reply.

Maynard Smith and Price (1973) justified the concept of an evolutionarily stable strategy by a heuristic argument which requires that a mutant entering a monomorphic population is driven out again. The word "monomorphic" indicates the use of only one strategy by all members of the population. The mutant is assumed to play a different strategy.

In biological applications equilibrium is not interpreted as the outcome of rational deliberation, but as the result of a dynamic process of natural selection. It is therefore necessary to explore the question whether such processes converge to evolutionarily stable strategies. Most of the literature on this problem is based on a simple selection process called "replicator dynamics". The paper by *Franz Weissing*, the second one in this volume, is a thorough investigation of the consequences of the replicator dynamics for a class of 3 x 3-games.

Stable dynamic equilibria of the replicator dynamics are not necessarily evolutionarily stable strategies, but they are always symmetric equilibrium strategies. As far as the replicator dynamics is concerned, one can speak of a good agreement of dynamic theory and non-cooperative game theory. However, serious difficulties arise in more complex genetical systems involving sexual reproduction and recombination. This is the subject matter of the paper by *Ilan Eshel*, the first one in this volume. The difficulties do not only concern game theory, but much more fundamentally the principle of fitness maximization.

Even without any game interaction it can happen in complex genetical systems that the relative frequency of the fittest genotypes declines and approaches a minimum. Such cases of a tendency towards fitness minimization are by no means degenerate or exceptional. Ilan Eshel argues that we must distinguish between two processes of evolution: a relatively short run process of adaptation of genotype frequences without mutation and a more long run process of gene substitution by mutation. The short run process does not necessarily optimize, but stability with respect to the long run process implies game theoretical equilibrium. This conclusion is of fundamental importance for evolutionary game theory.

Animal behavior and even more so human behavior is not only influenced by natural selection, but also by processes of individual learning. Processes of learning are different from processes of natural selection. One cannot expect that the same stability criteria apply in both cases. The third paper in this volume explores a learning process involving an element of anticipation which stabilizes some mixed strategy equilibria which would not be stable without it.

Rolf Hoekstra, Yoh Iwasa and Franz Weissing investigate the origin of isogamous sexual differentiation. The authors argue that the evolution of anisogamy, the size difference between female and male gametes is preceded by the evolution of isogamous mating types. Previous attempts to model the phenomenon remained unsatisfactory. The paper presents a class of game models which is analysed both game theoretically and dynamically. The results are discussed in the light of biological facts reported in the literature.

The paper by *Roy Gardner and Molly Morris* considers the problem of bluffing in the context of an extensive game model suggested by empirical observations. The paper does not only throw light on the biological phenomenon, but also on theoretical questions arising with respect to evolutionary stability in extensive games.

The paper by *Selten and Shmida* presents a game equilibrium model of an ecological system involving one pollinator and two flower species. It provides a possible answer to the question why flowers offer resources to pollinators. The analysis is an integrated approach to pollinator foraging behavior and flower competition.

The game theoretically intriguing phenomenon of egg trade observed in a species of hermaphrodite fish is analysed in the paper by *James W. Friedman and Peter Hammerstein*. The model presented in this paper succeeds to be close to the empirical background without becoming too complex to be analysed.

The paper on competition avoidance in a dragonfly making system written by *Joachim Poethke and Franz Weissing* is based on extensive field observations. It is an excellent example for the power of imaginative game theoretical reasoning to exhibit and to explain regularities in field data. Empirical work of this type has the potential to become more and more important in the exploration of nature.

The last paper in this volume has been written by one of the few psychologists in our research group. *David Messick* explores the evolution of group altruism and shows how evolutionary thinking can contribute to the understanding of human behavioral tendencies.

This volume is the first one of four in the series on game equilibrium models. It is devoted to evolution and game dynamics, topics which are of interest not only to biologists but also to behavioral science in general. Evolutionary game theory has introduced a new approach to non-cooperative game theory. Game equilibrium is viewed as the result of a dynamic process and not as the product of rational deliberation. Other fields can be expected to profit from the development of a similar approach, even if a direct and unmodified transfer of biological game theory may not be the best route to take.

Bielefeld/Bonn, December 1990 Reinhard Selten

References

Maynard Smith, J. and G.R. Price. 1973. The Logic of Animal Conflict. Nature **246**: 15-18.

von Neumann, J. and O. Morgenstern. 1944. Theory of Games and Economic Behavior. Princeton, New Jersey: The Princeton University Press.

GAME THEORY AND POPULATION DYNAMICS IN COMPLEX GENETICAL SYSTEMS: THE ROLE OF SEX IN SHORT TERM AND IN LONG TERM EVOLUTION

by

Ilan Eshel

Abstract

The article maintains three major points:
(1) In sexual populations with recombination, there is a qualitative difference between one process of natural selection which manifests itself in terms of changes of frequencies of genotypes already present in the population (say, short term selection) and another process which manifests itself in terms of selective gene substitutions (say, long term selection).
(2) It is only the process of long term selection (due to gene substitutions) and not the process of short term selection (due to changes in genotype frequencies) that, quite generally, leads to the individual optimization of certain evolutionarily relevant payment functions and thus guarantees the stabilization of ESS population strategies.
(3) These findings, based on theoretical analysis of the well studied genetic structure of sexual reproduction, fit into the main bulk of current theories about the evolution of the sexual system of reproduction in a changing environment. They can, in turn, throw some new light on the role of sex in preventing fast adaptation to short—term, not persisting environmental changes and, at the same time, allow (and, as I claim, even facilitate) slow adaptation to long—term, persisting environmental changes.

1. Game Theory and the Dynamics of Evolutionary Changes in Complex Sexual Systems — The Problem

An indispensable theoretical strategy in the struggle of modern biology to cope with the complexity of patterns, either behavioral, physiological or biochemical, exhibited by natural organisms, is to explain such patterns on the basis of their adaptive value to the organism exhibiting them. At the core of such a program lies the crucial, though sometimes tacit, assumption that natural selection is bound to operate towards the optimization of certain biological traits, at least on the level of the individual (though not necessarily on the level of the population, e.g. Williams 1966). Tacitly, this assumption became a synonym to Darwinism. Indeed, when one keeps in mind the idea of individual optimization in conflicts within a population, population game theory and the concept of ESS has become indispensable tools to analyze the outcome of a conflict.

In considering, however, the applicability of population game theory to the actual dynamics of evolutionary changes in sexual diploid populations, the question is what relevant payment function, if any, tends to be maximized (at least locally and on the individual level) by the combined effects of the mating system (e.g., even random mating), Mendelian segregation, recombination (when more than one locus is involved) and selection pressure. Thus, in the simplest case of a two strategy viability population game in a one locus, random mating (infinite) population, changes in the genotype frequencies are proved to always determine either convergence of the population strategy to an ESS of the genetically feasible strategies or oscillation around it. Moreover, with some upper bound on the intensity of selection (measured in terms of the ratio between the extreme values of the payment function), convergence to the viability–ESS is the only possibility (Eshel 1982, Lessard 1984; see also the discussion in Maynard Smith 1982).

Unfortunately, these results remain true for only very special cases when more than one locus is involved. Moreover, as it has been first pointed out by Moran (1964), quite counterintuitively at first sight, in a random mating diploid population with viability determined by two loci or more, natural selection does not guarantee an increase in the average viability of the population even under the condition of a fixed environment. Since then, this apparent contradiction to the Darwin–Fisher–Wright expectation has been shown to occur in most multilocus nonadditive fitness regimes (Ewens 1968, Lewontin 1971, Karlin 1975). As a result, even in a constant environment and even when only individual viability is involved, it cannot be generally true that natural selection operates to produce an optimal population strategy. It is, therefore, not surprising that in a population–game structure (see Maynard Smith and Price 1973), when more than one locus is responsible for the individual strategy, genetic equilibria are most unlikely to determine an ESS distribution of phenotypes even when such a distribution is feasible (Lessard 1984), this in contrast to the case of one locus.

The situation becomes even worse if viability is replaced by any other 'natural' payment function, say inclusive fitness (Hamilton 1964), expected number of grandoffspring (e.g. Fisher 1930) or even fertility. Thus, it has been shown be Hadeler and Liberman (1975) that average fertility is not generally maximized even by natural selection at one locus; and it has been shown by Cavalli Sforza and Feldman (1978) and by Uyenoyama and Feldman (1981) that even inclusive fitness is not monotonically increasing under one locus selection pressure (but it is locally maximized at a one locus fixation). Furthermore, none of these is maximized by the multilocus dynamics (indeed, the viability–selection counterexample is sufficient for this because it can be represented as equivalent to special cases of either fertility selection or inclusive fitness selection). All the same it was shown by Karlin and Lessard (1986) that selection on the sex ratio, when determined by two loci, is not likely to lead to a stable equilibrium that maximizes the expected number of grandoffspring, contrary to the prediction made by Fisher (1930) and to the result achieved by Eshel and Feldman (1982) for a one locus case.

I think that these seemingly counter–intuitive and rather disturbing findings should not be overlooked on the basis of simple minded common sense. It is true that they are deduced from an analytically complicated structure which is sometimes hard to follow by simple intuitive arguments, but so is the very process of reproduction in a complex multilocus sexual system and, indeed, one should keep in mind that it may be even harder to explain, on the basis of simple minded common sense, the very evolution of this complex structure of reproduction, say the evolution of sex itself. But we do, in fact, reproduce sexually.

During the last decade, many attempts have been made to explain the evolution of sex on the basis of exact genetic models, e.g. Maynard Smith (1975, 1978), Williams and Mitton (1973), Hamilton (1980, 1982), Hamilton et al. (1981), Weinshall (1986), Weinshall and Eshel (1987), Bell and Maynard Smith (1987), Bernstein et al. (1984, 1985), Maynard Smith (1987). With the exception of the last one, all these attempts converge on the assumption of some sort of irregularity of the environment as a necessary prerequisite for the establishment of sexual reproduction in a population. Later in this work (see Section 5) I will try to maintain that the same structure that allows us to explain the evolution of sex in an irregular environment is most likely to explain as well the seemingly 'nonadaptive' failure of the multilocus system to stabilize individually optimal strategies in sexual populations due to selective change in genotype frequencies.

But maybe a more disturbing aspect of this analytically established but intuitively unwelcome result has to do with its apparent theoretical implication on the most crucial component of Darwinism, say the causal bond between natural selection and adaptation. Does it mean that near optimality (or, as well, mutual optimality, say ESS) is not likely to be observed when complex features in sexual populations are concerned? Such theoretical prediction would indeed stand in contrast with many observations of both complex and amazingly close to optimal traits like flight–ability, eyesight or programmed nest–building, to mention a few traits which are rather found in sexual populations and undoubtedly involve many recombining genes. Another property of these traits, however, is that they are all likely to evolve throughout a long process of 'trial and error', say by the successive replacement of genes rather than by changes in the relative frequencies within a given set of genotypes. And, as we see in the next two sections (see Section 2 for viability selection, Section 3 for other modes of selection), contrary to a widely accepted though tacit assumption, this long term process of selected gene substitution is qualitatively different from the extensively studied process of change in genotype frequencies.

We should, therefore, start by a clear distinction between the two processes, that of *short term selection* (namely due to changes in genotype frequencies) and *long term selection* (namely due to selected gene substitution). On the basis of a bulk of quite recent theoretical findings we see that, contrary to short term selection, the hitherto less studied process of long term selection does lead, in fact, to the establishment of individually optimal and mutually optimal strategies, say ESS.

Moreover, in Section 4 we see how evolutionarily relevant payment functions, to be optimized by the process of long term selection in multilocus sexual populations, can be analytically deduced from the system (i.e. from the genetic structure and the selection forces operating on it) rather than assumed on pure intuitive arguments. Finally, I try to explain the qualitative difference between the two processes of selection on the basis of the very role of sex in evolution as it appears to follow from modern quantitative theories about the evolution of sex.

2. Selective Gene Substitution and Long Term Convergence to an ESS in a Viability Population Game

Let us start from the simplest case of pure viability selection operating on a large diploid population with random mating an no family structure and let us assume the most general two locus system of genotype determination of the individual phenotype (or strategy). More specifically, let $A_1,...,A_n$ be the alleles present on one locus and $B_1,...,B_m$ the ones present on the other locus. Let $0 \leq r \leq \frac{1}{2}$ be the rate of recombination between the two loci. The genotypes are A_iB_k/A_jB_l where $i,j = 1,2,...,n;\ k,l = 1,2,...,m$.

Let us assume that individuals of different genotypes differ only in their viability and let ω_{ijkl} be the viability of the genotype A_iB_k/A_jB_l, $\omega_{ijkl} = \omega_{jikl} = \omega_{ijlk} = \omega_{jilk} \geq 0$. Finally, let p_{ik} be the relative frequency of the chromosome A_iB_k passed to newborn offspring in the population. With random mating and Hardy Weinberg law for combination of gametes after recombination, the average viability in the population is, therefore:

$$W = \sum_{ijkl} p_{ik}p_{jl}\omega_{ijkl} . \qquad (2.1)$$

By straightforward calculation one gets the frequency of the chromosome A_iB_k, $(i = 1,...,n;\ k = 1,...m)$ after random mating, recombination and selection:

$$p'_{ik} = \frac{1}{W} \cdot \sum_{jl} [(1-r)p_{ik}p_{jl} + rp_{il}p_{jk}] \cdot \omega_{ijkl} . \qquad (2.2)$$

As we know to be the general case in multilocus systems, $W = W(p)$ is not a Liapunov function of the transformation and it is possible that $W(p') < W(p)$. Moreover, a stable equilibrium of the system may not (and quite often does not) maximize the average viablity of the population, even locally (e.g. Karlin 1975). As having been noted by Moran (1964) this means that as the genotype frequencies are changing due to random mating, recombination, Mendelian segregation and natural selection, the average viability of a population near equilibrium may monotonously drop down from one generation to the next, tending from above to a lower bound which is only materialized at the equilibrium itself.

10

But it can be shown (Eshel and Feldman 1984; see also Liberman 1988 for generalization of this result to any number of loci) that the situation is different if we start from a population which is already in an internally stable equilibrium (i.e. stable in respect to changes in the frequencies of those genotypes already in the population) and ask about successful invasion of such a population by a new, random mutant. As it turns out, one can solve this problem quite generally without resorting to any explicit information about the specific distribution of genotypes at equilibrium (an information which, as known to students of multilocus systems, is rather impossible to obtain analytically in general). Actually, employing (2.2) one only needs the equilibrium condition:

$$p_{ik} = \frac{1}{W} \cdot \sum_{jl} [(1-r)p_{ik}p_{jl} + rp_{il}p_{jk}] \cdot \omega_{ijkl} \, , \tag{2.3}$$

$i = 1,...,n; \ k = 1,...,m.$

Without loss of generality we may assume now that the new mutant allele is introduced into the first locus. This allele, say A_{n+1}, may appear in combination with each of the allels at the other locus, thus let ϵ_k $(k = 1,...,m)$ be the frequency of the mutant chromosome $A_{n+1}B_k$. Let $\epsilon = \sum_k \epsilon_k > 0$ be a small number. Finally we take over the standard assumption that the frequencies of those chromosomes A_iB_k already present at the original equilibrium are not changed by more than the order of ϵ as a result of the first invasion of the mutant. Employing (2.2) for the dynamics of the new $(n+1,m)$ allele system and ignoring terms of the order of ϵ^2 we get

$$\epsilon'_k = \frac{1}{W} \cdot \sum_{jl} [(1-r)\epsilon_k p_{jl} + r\epsilon_l p_{jk}] \cdot \omega_{n+1,jkl} \, , \tag{2.4}$$

$j = 1,...,n; \ k,l = 1,...,m;$ where W is the average equilibrium viability, defined in (2.1). The Perron–Frobenius theorem implies that the transformation $\epsilon \to \epsilon'$ being determined by (2.4) has a unique normalized positive leading right eigenvector $V = (V_1,...,V_m), \sum_k V_k = 1,$ corresponding to the (positive) leading eigenvalue $\lambda > 0$. The mutant will, indeed, be established in the population if $\lambda > 1$ (and only if $\lambda \geq 1$).

It can be shown (Eshel and Feldman 1984, Liberman 1988) that $\lambda > 1$ if and only if

$$\sum_{jkl} V_k p_{jl} \omega_{n+1,jkl} > W. \tag{2.5}$$

Equivalently, let ϵ be in the direction of the right eigenvector V, then the mutant will become established in the population if

$$W_{n+1} := \frac{1}{\epsilon} \cdot \sum_{jkl} \epsilon_k p_{jl} \omega_{n+1,jkl} > W. \tag{2.6}$$

But the lefthand side of (2.6) is the average viability of a random mutant in the population. Moreover, the average viability of the population, with ϵ deviation from the equilibrium, is $(1-\epsilon)W + \epsilon W_{n+1}$. Hence we get as a corollary (Eshel and Feldman 1984):

A new mutant, being introduced into the population will be successfully established if and only if it initially increases the average viability of the population, at least in the direction of the leading right eigenvector.

Note that the relative frequencies of the mutant chromosomes tend to the components of the (normalized) leading right eigenvector, if converging, or else, even if diverging, they get as close to these components as we wish before leaving the ϵ–vicinity of the equilibrium, provided we start with a small enough deviation. Hence one can conclude that starting from any two locus viability equilibrium, a new mutation will successfully enter the population if and only if it initially increases the average viability at least in the direction of the main eigenvector of the population. (See for comparison Taylor 1985.)

Furthermore, assume now a population game structure of viability selection, i.e., let each individual in the population choose one of the N strategies $\alpha_1,...,\alpha_N$ and let the probability that an individual of genotype A_iB_k/A_jB_l chooses the strategy α_ν be $S_{ijkl}(\nu)$. Indeed, $S_{ijkl}(\nu) \geq 0$ and $\Sigma \, S_{ijkl}(\nu) = 1$ for all $i,j = 1,...,n;$ $k,l = 1,...,m.$ Let the relative frequency of the chromosome A_iB_k among newborn offspring be p_{ik}, as before. Then the population strategy $S = (S_1,...S_N)$ will be

$$S_\nu = \underset{ijkl}{\Sigma} \, p_{ik}p_{jl}S_{ijkl}(\nu), \quad \nu = 1,...,N. \tag{2.7}$$

Assume now that the individuals in the population meet at random and the payment function of an individual playing α_ν against an opponent playing α_μ is $m_{\nu\mu}$, payment being measured in terms of an additive component of viability. We get a general two locus, linear frequency dependent viability system in which the viability of the genotype A_iB_k/A_jB_l is given by

$$\omega_{ijkl} = \omega_{ijkl}(p) = \underset{\mu,\nu}{\Sigma} \, S_{ijkl}(\nu) \cdot S_\mu \cdot m_{\nu\mu} \tag{2.8}$$

where the $S_\mu = S_\mu(p)$ are given by (2.7).

Consider first the simplest case of a 2x2 population game ($N = 2$). In Section 1, we have seen (e.g. Lessard 1984) that, concerning changes in two–locus genotype frequencies, an ESS of the population game is likely to be unstable and natural selection, combined with random mating and recombination, can render the population strategy further apart from the ESS. If, on the other hand, the population is already in a stable equilibrium, determining a population strategy close to an ESS, one can show with a similar technique that a new mutation will be established in the population if and only if it initially renders the population strategy closer to the ESS, at least when the relative frequencies of the mutant–chromosomes approach the direction of the main eigenvector. No mutation can successfully invade a population which already determines an ESS strategy. This last result is true, moreover, for any (N strategy) population game. If the population is only close to an ESS, a new mutation will successfully invade it if it initially shifts the population strategy into a strategy–cone in the direction of the ESS. Also these results (Eshel and Feldman 1984) have recently been generalized by Liberman (1988) to any number of loci.

3. Other Relevant Payment Functions and Evolutionary Genetic Stability (EGS)

A similar approach can lead to the demonstration of a long–term maximization of other payment functions, depending on the specific (short term) forces of selection operating on the population. Thus, the general n–locus result for initial increase under viability selection has been also generalized by Liberman (1988) to the most general case of sex–dependent viability. In this case, the condition for initial increase of a new mutant at any of the n involved loci is

$$\frac{1}{2} \cdot \left[\frac{W_f'}{W_f} + \frac{W_m'}{W_m} \right] > 1, \tag{3.1}$$

where W_f and W_m are the population–average viabilities of females and males respectively at (internal) equilibrium, W_f' and W_m' are the average viabilities of (heterozygote) mutant females and males where the distribution of mutant chromosomes is in the direction of the main eigenvector. In the case of a trade–off restriction $W_f = h(W_m)$ between male and female viabilities (e.g. where phenotypic features which are advantageous for one sex are disadvantageous for the other), one gets stability to any new mutation (at any locus), affecting W_m (and thus $W_f = h(W_m)$) if and only if the function

$$F(W_m, X) = \frac{1}{2} \left[\frac{X}{W_m} + \frac{h(X)}{h(W_m)} \right] \tag{3.2}$$

is maximized for $X = W_m$ (where it obtains the value 1). In all other cases, natural selection will operate in favor of those mutants which (by changing $W_m' = X$) increase the value of $F(W_m, X)$, thus long term natural selection operates for individual increase of the payment function F.

But it can readily be shown (Liberman et al. 1989) that F is the expected number of grand–offspring, males and females altogether, descending from an adult (heterozygote) mutant, either male or female. The result being obtained therefore follows and generalizes the argument of Fisher (1930) for the individual maximization of grandoffspring number by shifting the sex ratio in the direction of 1:1 (see also Eshel 1974 for detail).

Thus, it can be shown that in the most general, one locus autosomal system of sex determi–nation, a new mutation without pleiotropic effects (i.e., a mutation that affects only the sex ratio) will successfully be established in the population if and only if it initially renders the population sex ratio closer to 1:1 (Eshel and Feldman 1982). In this case, it was further proved by Karlin and Lessard (1983) that the sex ratio corresponding to the newly established equilibrium will be closer to it than the old one. Yet they have maintained (Karlin and Lessard 1986, see also Liberman et al. 1989) that in a two locus system of sex determination, changes in genotype frequencies are not likely to lead to an even sex ratio, even when feasible. Instead, any sex ratio can be stably maintained, depending on the parameters of the model.

As in the case of viability selection, however, a different result is obtained if we ask about the long term process of evolution, namely selective allele substitution. As can be shown for the most general two locus autosomal system of sex determination, a new mutation will be established in the population when introduced into an internally stable equilibrium (i.e. a genotype frequency equilibrium) if and only if it initially renders the sex ratio closer to 1:1, at least in the direction of the main eigenvector. As being shown by Liberman et al. (1989), this result readily follows from the general sex–dependent viability model (Liberman 1988) by just formulating the model of n–locus sex determination and then formally interpreting the parameters of individual sex determination as parameters of sex–dependent viability selection with the special restriction $W_f = h(W_m) = 1 - W_m$. (For first observation of the general analytic equivalence of the two models, see Karlin and Lessard 1986). Moreover, in the same way it is shown that even with pleiotropic effects of sex determination, although the sex ratio of 1:1 is not necessarily obtained, Fisher's criterion of grandoffspring maximization still holds for initial increase at any locus.

Following the terminology of Hamilton (1967), one can say that the even sex ratio (namely the ESS of the population game with individual maximization of grandoffspring number) and (as demonstrated in the previous section) the ESS of a viability population game are *unbeatable* population strategies, namely: If adopted by almost the entire population, this population must be stable against any new non–pleiotropic mutation, affecting the individual behavior in respect to this strategy (see also Maynard Smith and Price 1973). Yet, in both these cases we can prove more. It is therefore worthwhile to introduce a somehow stronger concept of long term stability (Eshel and Feldman 1982).

Definition:
A strategy α has the property of *Evolutionary Genetic Stability (EGS)* within the genetic structure G if
1. Any genetic equilibrium in G which determines the population strategy α is stable against any mutation that changes the population strategy (i.e., α is an unbeatable strategy in G).
2. Any genetic equilibrium in G which determines a population strategy $\beta \neq \alpha$, where β is close to α, is unstable in face of mutations that initially render the population strategy closer to α. It is stable in face of mutations that initially render the population further apart from α.

While the concept of dynamic stability is relevant to short term evolution (i.e. change in genotype frequencies due to a direct selection pressure) the concept of Evolutionary Genetic Stability is relevant to long term evolution (i.e. selective gene substitution).

We see that an ESS of a viability population game has the property of Evolutionary Genetic Stability in a one locus as well as in a multilocus diploid system with random mating. This may justify the game–theory approach to viability selection, when long–term evolution is concerned. In the same way, the even sex ratio of 1:1 (which is the ESS for the payment function of the number of grandoffspring) has the property of EGS for a one locus as well as for the two locus autosomal system of sex determination (but it does not have the EGS property in respect to systems which involve sex–linked modifiers, e.g., Hamilton 1967, Eshel and Feldman 1982, Eshel 1984). In the same way one can show that the Mendelian (even) rate of segregation is EGS in respect to a non–sexlinked modifier of meiotic drive (Eshel 1985) but not in respect to sex–linked modifiers (Liberman 1976). And a zero rate of mutation and recombination is an EGS for two locus systems in a fixed environment (Liberman and Feldman 1986).

Finally, following certain assumptions about environmental fluctuations which lead to the evolution of sexual reproduction (Weinshall 1986), one can ask about the exact conditions for the establishment of a new mutation in a modifier locus that regulates the rate of (maybe partial) sexual reproduction. In this case, an EGS rate of sexual reproduction is shown to exist and it can be calculated, depending on the parameters of the process. We will return to this last example in Section 5.

In all these examples, a mutually 'optimal' (EGS) strategy is selected in the population only throughout the long process of 'trial and error' due to selective gene substitution.

4. Is it Possible to Infer a (Long Term) Game Structure From the (Short Term) Population Dynamics?

A serious difficulty, concerning the application of classic population game theory to biological conflicts of complicated social structure concerns the necessary asssumption about the specific payment function which is supposed to be individually optimized by natural selection in that specific situation. In the previous sections we have attempted to justify the assumption of individual optimization of specific, rather simple payment functions in complex (multilocus) genetic structures, when long term selection due to gene substitution is concerned. In some cases of biological interest, however, it is very hard to choose a 'natural' candidate for such a justification since the very nature of a 'descent' candidate as the inclusive fitness is rather controversial. One example is the case of worker–queen conflict on the nest sex ratio in a social haplodiploid hymenoptera population. Another is the parent–offspring conflict over the sex ratio in a diploid population, when the cost of rearing a male offspring is different from the cost of rearing a female one. In such cases, it is possible to objectively infer the relevant payment function (if there is one) which is individually optimized by natural selection from the local dynamics of the population, rather than to speculate about it on the basis of intuitive, maybe plausible arguments.

We concentrate on the first example. (For a similar analysis of the second example, the reader is referred to Eshel and Sansone (1989).) As it has been suggested by Trivers and Hare (1976), the fact that (in the case of single insemination) a mother in a haplodiploid population is equally related to her male and to her female offspring while a daughter (including a worker) in such a population is only half as strongly related to her male brothers as to her female sisters leads to a conflict between the mother–queen and her worker–daughters about the sex ratio among reproductive offspring of the nest. Basing their argument on the assumption of maximization of the expected number of identical–by–descent genes, passed to the third generation, Trivers and Hare have concluded that while the queen's preferred sex ratio is 1:1, the workers preferred ratio is 1:3. The expected number of identical–by–descent genes passed to the third generation was regarded by Trivers and Hare as the inclusive fitness.

Assuming, further, that worker–queen disagreement on the sex ratio can reduce the entire reproductive success of the nest, a game–structure of the conflict naturally emerges. This is the case, for example, if a worker eats some male eggs (e.g. Trivers and Hare 1976) or if they manipulate the proportion of differently treated reproductive and non–reproductive (worker) females (e.g. Bulmer and Taylor 1981, Bulmer 1981); or else, if the queen, in turn, lays unfertile eggs in a cell, allocated by workers to a fertile egg (e.g. Matessi and Eshel 1989). A crucial question, however, concerns the relevant payment function to be maximized by workers and queen. More specifically, since male offspring carry half as many genes as female offspring and since (unlike in a diploid population) not all individuals born in the population have both mother and father (males don't have a father), it is not clear what the 'reproductive value' is (in terms of genes, passed to further generations) of a male, relative to that of a female. This has been proved, furthermore, to depend on the sex ratio at a given time (Oster et al. 1977). Hence, so is the 'value', for future generations of a gene, carried by a male or by a female offspring.

In order to avoid these difficulties, however, one can employ an exact two locus model (Matessi and Eshel 1989) – one locus affecting the queen's behaviour, another affecting workers' behaviour with any positive rate of recombination between them. Starting from a fixation equilibrium, a mutation can appear in either the first locus, the second one or in both. It can appear in a male or in a female individual, hence one should analyse a six–dimensional model of change in genotype frequencies. However, as long as the rate of recombination is not too small (i.e., if it is higher than the maximal effect of a single mutation), a spectral analysis of the matrix of linear approximation for rare mutations of the six–dimensional dynamics leads to two (out of six) candidates for leading eigenvalues and the condition for stability (in face of the specific pair of mutations) reduces to two separate conditions, namely that each of the two candidates for a leading eigenvalue should be smaller than one in absolute value.

More specifically, denote by $\varphi(a,b)$ the total success of a nest in which the queen's strategy is a and the (average) worker's strategy is b. Denote by $F(a,b)$ the frequency of females among the reproductives of such a nest. Let α and β be the common strategy of queens and workers in the population respectively and let α' and β' be the relevant strategies of mutant queen (being heterozygous at the A locus) and a worker (being heterozygous at the B locus) respectively. Then, by the laborious work of solving for the linear approximation of the six–dimensional transform–ation of genotype frequencies, one can straightforwardly calculate the two candidates for the leading eigenvalues as:

$$\lambda_1 = \tfrac{1}{2} \cdot \left[\, H_1(\alpha,\beta;\alpha') + \sqrt{H_1(\alpha,\beta;\alpha')^2 + 4 \cdot H_2(\alpha,\beta;\alpha')} \, \right] , \tag{4.1}$$

where

$$H_1(\alpha,\beta;\alpha') := \frac{\varphi(\alpha',\beta)F(\alpha',\beta)}{2 \cdot \varphi(\alpha,\beta)F(\alpha,\beta)} \; ; \quad H_2(\alpha,\beta;\alpha') := \frac{\varphi(\alpha',\beta)[1-F(\alpha',\beta)]}{2 \cdot \varphi(\alpha,\beta)[1-F(\alpha,\beta)]} \tag{4.2}$$

and

$$\lambda_2 = \tfrac{1}{2} \cdot \left[\, G_1(\alpha,\beta;\beta') + \sqrt{G_1(\alpha,\beta;\beta')^2 + 4 \cdot G_2(\alpha,\beta;\beta')} \, \right] , \tag{4.3}$$

where

$$G_1(\alpha,\beta;\beta') := \frac{\varphi(\alpha,\bar\beta)F(\alpha,\bar\beta)}{2 \cdot \varphi(\alpha,\beta)F(\alpha,\beta)} \; ; \quad G_2(\alpha,\beta;\beta') := \frac{\varphi(\alpha,\beta') \cdot \varphi(\alpha,\bar\beta) \cdot F(\alpha,\beta') \cdot [1-F(\alpha,\bar\beta)]}{2 \cdot \varphi(\alpha,\beta)^2 \cdot F(\alpha,\beta) \cdot [1-F(\alpha,\beta)]}$$

$$\text{and} \quad \bar\beta := \frac{\beta + \beta'}{2} . \tag{4.4}$$

(Note that the rate of recombination $r > 0$ dissappears from the terms (4.1) and (4.3). It does affect, however, two of the other four eigenvalues.) Indeed a population being fixed on the pair of strategies (α,β) will be stable against a specific pair of mutations, determining (as heterozygotes) queen and worker strategies α' and β' respectively if $\lambda_1 < 1$ and $\lambda_2 < 1$ (only if $\lambda_1 \leq 1$ and $\lambda_2 \leq 1$). By straightforward calculation one can show that this is equivalent to:

$$H(\alpha,\beta;\alpha') := H_1(\alpha,\beta;\alpha') + H_2(\alpha,\beta;\alpha') < 1 \tag{4.5}$$

and

$$G(\alpha,\beta;\beta') := G_1(\alpha,\beta;\beta') + G_2(\alpha,\beta;\beta') < 1 . \tag{4.6}$$

Thus a population fixed on a pair of strategies (α,β) will be stable against any new mutation at either locus if both inequalities (4.5) and (4.6) will hold for any $\alpha' \neq \alpha$ and $\beta' \neq \beta$. But by inserting $\alpha' = \alpha$ and $\beta' = \beta$, these inequalities indeed turn into equalities ($H(\alpha,\beta;\alpha) = 1$, $G(\alpha,\beta;\beta) = 1$). Hence, a necessary and sufficient condition for (α,β) being an *unbeatable* pair of strategies (i.e., being stable against any new mutation) is that $H(\alpha,\beta;\alpha')$ will obtain its maximum at $\alpha = \alpha'$ and $G(\alpha,\beta;\beta')$ will obtain its maximum at $\beta = \beta'$. But by stating this condition, however, we formally define an asymmetric population game with $H(\alpha,\beta;\alpha')$ and $G(\alpha,\beta;\beta')$ being the payment functions of a player choosing the strategy α' in the first role (i.e., as a queen) and choosing β' in the second role (i.e., as a worker) in a population playing (α,β). (See Maynard Smith and Parker 1976; Selten 1980, 1983).

Let us now consider the payment functions H and G which emerged from the analysis of two locus stability. In $H(\alpha,\beta;\alpha')$, which is given by

$$H(\alpha,\beta;\alpha') = \frac{\varphi(\alpha',\beta)}{\varphi(\alpha,\beta)} \cdot \frac{1}{2} \cdot \left[\frac{F(\alpha',\beta)}{F(\alpha,\beta)} + \frac{1-F(\alpha',\beta)}{1-F(\alpha,\beta)} \right] , \tag{4.7}$$

one can easily recognize the inclusive fitness of the mother, as interpreted by Trivers and Hare, namely the expected number of her own genes passed to all grandoffspring, males and females altogether. Indeed, $\varphi(\alpha',\beta)/\varphi(\alpha,\beta)$ is the expected number of all the reproducing offspring (relative to the population average), and $\frac{1}{4}\cdot 1/F(\alpha,\beta)$ and $\frac{1}{4}\cdot 1/[1-F(\alpha,\beta)]$ are the (relative) expected numbers of her genes, passed to the third generation by a female or by a male offspring respectively. Thus

$$\frac{1}{2}\cdot\left[\frac{F(\alpha',\beta)}{F(\alpha,\beta)} + \frac{1-F(\alpha',\beta)}{1-F(\alpha,\beta)}\right]$$

is the relative expected number of a mother's random reproductive offspring and, therefore, $H(\alpha,\beta;\alpha')$ is the expected number of her genes passed to (either male or female) grandoffspring by either male or female offspring.

Yet, the inclusive fitness of a mixed–nest worker, as interpreted by Trivers and Hare, namely the number of genes, identical by descent to hers, passed to the next generation by either male or female reproductive sibs is (by the same argument):

$$\widetilde{G}(\alpha,\beta;\beta') := \frac{\varphi(\alpha,\overline{\beta})}{\varphi(\alpha,\beta)} \cdot \frac{1}{4}\cdot\left[\,3\cdot\frac{F(\alpha,\overline{\beta})}{F(\alpha,\beta)} + \frac{1-F(\alpha,\overline{\beta})}{1-F(\alpha,\beta)}\,\right], \tag{4.8}$$

a payment function which looks quite different from the workers' payment function $G(\alpha,\beta;\beta')$ emerging from the analysis and determined in (4.6):

$$G(\alpha,\beta;\beta') = \frac{\varphi(\alpha,\overline{\beta})F(\alpha,\overline{\beta})}{2\cdot\varphi(\alpha,\beta)F(\alpha,\beta)} + \frac{\varphi(\alpha,\beta')\cdot\varphi(\alpha,\overline{\beta})\cdot F(\alpha,\beta')\cdot[1-F(\alpha,\overline{\beta})]}{2\cdot\varphi(\alpha,\beta)^2\cdot F(\alpha,\beta)\cdot[1-F(\alpha,\beta)]}. \tag{4.9}$$

Note, however, that without the additional effect on the nest productivity assumed here, say with full power to the workers (as guessed by Trivers and Hare), both G and $\widetilde{G}$ are maximized for the sex ratio 1:3 as predicted by Trivers and Hare (1976) and further argued by Oster et al. (1977). Moreover, under quite plausible assumptions, $\widetilde{G}(\alpha,\beta;\beta')$, though looking quite differently from $G(\alpha,\beta;\beta')$, leads to the same ESS when combined with an opponent's payment function $H(\alpha,\beta;\alpha')$ (Matessi and Eshel 1989). In fact, $\widetilde{G}(\alpha,\beta;\beta')$ can also be interpreted as a sort of workers' inclusive fitness, namely the number of her identical–by–descent genes passed by reproductive sisters to reproductive female offspring plus the number of such genes passed to female grandoffspring through male siblings (Matessi and Eshel 1989). Maybe more important is the fact that for mutation of small effect, the expression $\widetilde{G}(\alpha,\beta;\beta')$ can be recognized as the inclusive fitness, calculated with the regression coefficients of relatedness, as being suggested by Uyenoyama and Bengtsson (1981, 1982). A somewhat different result is achieved when mutation can occur at very tightly linked loci. In that case, the ESS condition for the population game $(H,\widetilde{G})$ remains necessary but not sufficient for (α,β) to be an unbeatable pair of strategies.

The main point of this section is that even when the feature of the unbeatable pair of strategies (α,β) can be and, for better conceptual understanding, are interpreted in terms of an ESS of a population game with plausible payment functions, there is an objective way, emerging from the analysis of the exact model, to determine these payment functions.

5. The Role of Sex in Short and in Long Term Evolution

In this section, I will try to establish the hypthesis that maintaining a qualitative difference between short term and long term response of a population to selection pressure is the main role of sexual reproduction and that this difference is manifested in all sorts of hindering effects concerning a short term response (i.e., through changes in genotype frequencies) to a short term non—persisting selection pressure, but not concerning (sometimes even facilitating) long term response (i.e., through selective gene substitution) to a persistent selection pressure.

Speaking about the role of sex in evolution one is generally concerned with one of the following two questions:
1. How does sexual reproduction affect natural selection?
2. How does natural selection, operating on the level of the individual, enable the evolution of sex?

Although the participants of this workshop are interested mainly in the first question, I find it hard and sometimes misleading to deal with one of these questions separately while ignoring the other one.

One effect of sex which was not fully comprehended till the appearance of William's book 'Sex and Evolution' (1975; see also Eshel and Feldman 1970; Eshel 1971, 1972) is its tendency to slow down and even counterbalance natural selection (or, more specifically, what we refer here as to short term selection). This is true even at the first level of one locus re—coupling of alleles, without recombination.

In order to understand this, let us consider a one—locus polymorphism with any number of alleles. Except for the trivial case of non—selection, we know that in such a polymorphic population there must be genotypes which are more viable than the average and others which are less so (e.g., in a two allele polymorphism, the necessarily more viable heterozygotes coexist with the less viable homozygotes). Let p_i be the frequency of the allele A_i among adults, $i = 1,...,n$. We know that the proportion of the genotype $A_i A_j$ among newborn offspring is $P_{ij} = 2p_i p_j$ if $i \neq j$ and $P_{ij} = p_i^2$ if $i = j$. Now, if the viability of $A_i A_j$ is ω_{ij} $(i,j = 1,...,n)$ and if the average viability in the population is $W = \Sigma \, p_i p_j \omega_{ij}$, then the proportion of $A_i A_j$ among adults (after selection) will be $\bar{P}_{ij} = P_{ij}\omega_{ij}/W$. This obviously means that $\bar{P}_{ij} > P_{ij}$ if $\omega_{ij} > W$ and vice versa if $\omega_{ij} < W$. But then, random mating with re—coupling of alleles will render the proportion of $A_i A_j$ back to P_{ij}.

In other words: *Re—coupling of gametes due to the Hardy—Weinberg law will always increase the proportions of those combinations which are less fit than the average and decrease the proportion of those which are more fit than the average.*

With somehow more difficulties, this result can be shown to be valid for any distribution of alleles, not necessarily at equilibrium. The non—trivial part of Fisher—Kingman's fundamental law of natural selection (Fisher 1930, Kingman 1961) can therefore be stated as follows:

In a one–locus random mating diploid population under viability selection, the effect of sexual reproduction, which (except in fixation) is always to decrease the average fitness of a population, is never sufficient to overbalance the effect of selection, which is always to increase the average fitness. They are perfectly balanced only at equilibrium.

A natural question is, indeed, why not reproduce asexually, thereby to increase the average viability of the offspring? This question becomes more serious if one takes into consideration the cost of sex. First, the twofold (or close to it) cost of meiosis (e.g., Maynard Smith 1971, 1978; Williams 1975; Uyenoyama 1984; Feldman and Christiansen 1984), then the cost of courtship, sexual attraction and the chance of not finding a mate. Indeed, what remains to be considered is the effect of recombination which for a long time has been assumed to enhance the rate of evolution by ever providing new combinations as a material for natural selection (Fisher 1930; Muller 1932, 1958; Crow and Kimura 1965, 1969). A missing link in this argument is that recombination may not only create successful combinations of maybe not that successful single–locus mutations, it can as well break down such successful combinations. And, unfortunately, it can be shown (Eshel and Feldman 1970; see also Williams 1975, Levin 1988) that the rate by which recombination destroys successful combinations of alleles is faster than the rate by which it builds them up. It is often claimed, though, that this phenonemon is true for infinite but not for finite populations (e.g. Crow 1988 and references therein) but this is not quite so. The truth is that the argument, though first being proved for infinite populations, holds for finite populations as well. The first appearance of a successful combination of alleles (in fact of any combination of alleles) is in finite populations indeed faster if recombination creates (and destroys) new combinations in any generation. But a single 'first appearance' of any combination of alleles is of limited evolutionary importance if this combination, successful as it may be, is soon to be destroyed by recombination. Thus it appears to me that, at least qualitatively, recombination has the same hindering effect on natural selection as re–coupling of gametes. More careful quantitative study shows, moreover, that unlike the one–locus effect of re–coupling of alleles alone, its combined effect with recombination (the only combination which produces the real effect of sex) may be strong enough as to overbalance the effect of natural selection, not only to slow it, thus leading, as we have already seen, to the violation of Fisher's fundamental law of natural selection. As being concluded by Bernstein et al. (1985; quoted also by Crow 1988): "The traditional view of the consequence of sex for evolution is that sex speeds up adaptation by promoting the spread of favorable mutants and elimination of deleterious mutation. However, we argue here that the opposite is true; that sex acts as a constraint to adaptation." Or, as stated by Williams (1988): "Levin's discussion (see Levin 1988) of fitness reduction by recombination in bacteria says, in effect, that Eshel and Feldman's unwelcome conclusion has wide applicability."

It is therefore not that surprising that, even without the appalling 1:2 cost of meiosis, natural selection when operating in a fixed environment leads to the reduction of either outbreeding (Karlin 1968) or recombination (Fisher 1930, Feldman 1972, Feldman et al. 1980, Feldman and Liberman 1986, Liberman and Feldman 1986).

The situation is different, however, if short term environmental uncertainty is the main source of mortality in the population. With various plausible assumptions about the mode of the environmental change one can readily show that a too fast adaptation of an asexual population to a non—persistent environmental change can easily drive such a population to total extinction. More important to the theory of evolution is the attempt to explain the evolution of sex on the basis of quantitative models, demonstrating an increase in the frequency of sexually reproducing (or more sexually reproducing) individuals within a mixed partly sexual population exposed to selection pressure of a short—term environmental uncertainty. Such uncertainties may be of either spatial character (Williams and Mitton 1973, Cohen, unpublished) or of a temporal one (Maynard Smith 1971, 1975, 1978, Charlesworth 1976, Bell and Maynard Smith 1987; see also Sturtevant and Mather 1938). As it appears, the most plausible candidate for a source of permanent environmental changes leading to the evolution of sexual reproduction is the interminable host—parasite struggle and coevolution (Levin 1975, Charlesworth 1976, Glesner and Tilman 1978, Jaenike 1978, Hamilton 1980, 1982, Hamilton et al. 1981, Bremermann 1980, 1985, Bell 1982, Toby 1982, Rice 1983, Weinshall 1986, Eshel and Weinshall 1987, Weinshall and Eshel 1987, Seger and Hamilton 1988). On the basis of both empirical observation and theoretical findings (e.g. Selten 1980, 1983; Eshel and Akin 1983) it appears that such host—parasite coevolution is most likely to result in some sort of permanent cyclings. These may occur simultaneously, but not necessarily synchronized, at the level of different loci of the same host, concerning different parasites (Eshel and Hamilton 1984). The effect of such a host—parasite cycling may not be dramatic or even conspicuous from the point of view of long—term, fossil—recorded evolution nor may it have any substantial effect on the morphology or behaviour of individuals in the host population. Yet it may impose a high enough toll on the generation—to—generation survival of the host population as to ensure the extinction of non—sexual clones.

For example, in the simplest to explain model, suggested by Weinshall (1986), it is shown that if the population is repeatedly infested by a cycle of parasite types (at least three parasite types for a diploid population) and if, for any parasite type α_i there is a single allele A_i which makes its carrier immune to it, any non—sexual clone will be wiped out within one cycle of the parasite. This is so because after an attack by α_1 only the genotypes A_1A_1, A_1A_2 and A_1A_3 will survive. Of those, only A_1A_2 is immune to α_2 and will, thus, survive its attack. Thus no genotype carrying A_3 will remain to successfully face the attack of α_3. As for sexual individuals in the same population, it can be shown that not only some of them will survive any parasite attack but, with some further (and rather plausible) assumptions, the toll of death imposed by the parasite on the sexual subpopulation will not be intolerable (say, less than 50% each generation).

The reader is referred to all other models mentioned above (including some criticism of Weinshall's model in Williams 1988). In all of them the main role of sex (re—coupling of alleles in Weinshall's model as well as in Cohen's; mainly recombination in all other models) is just to prevent fast response to strong, short—term selection forces which are likely to change their direction and, thus, endanger the too fastly adapted non—sexual part of the population.

It should be mentioned, however, that at the present stage no single theory about the evolution of sex is well—established on a quantitative level as not to leave open questions, some of them quite crucial (see, for example, Crow 1988, Williams 1988) and, indeed, some alternative explanations cannot yet be precluded. One is the old theory of sex as a short term means to 'cover' deleterious mutations (Muller 1932; see also Feldman and Balkan 1972, Feldman et al. 1980). Another, rather new alternative approach concerns the role of recombination in repairing DNA sequences (Bernstein 1983, Bernstein et al. 1984, 1985; see also Gould and Lewontin 1979). The reader is also referred to an interesting attempt of Ettinger (1986) to explain the evolution of meiosis on the basis of its possible role in purging parasitic DNA.

I do maintain, however, that the well established though long unwelcome finding that sexual reproduction, via all its aspects, slows down short—term selection (i.e. due to changes in genotype frequencies) and keeps genetic variance at the expense of optimization, stands quite in agreement with the basic prediction of the main bulk of theories, attempting to explain the evolution of sex on the basis of individual selection in a changing environment. From the viewpoint of this bulk of theories, the seemingly non—adaptive features of multilocus sexual reproduction is, indeed, adaptive in the sense of insurance against unpredictable, short term selection forces. As it has been suggested by Thompson (1976; see also Eshel 1971, 1972), the main role of sex is to slow rather to accelerate the evolutionary response to a changing environment, and thus to act as a rein to prevent the population from chasing after transitory environmental changes. This suggestion is criticized by Crow (1988) on the basis of historical records which appear to show that asexual species change more slowly in evolutionary time. This argument can, however, be turned upside down by stating the counter—hypothesis that those species which were less exposed to fast environmental changes (and, therefore, changed more slowly) remained asexual.

Yet, as I have tried to maintain in this work, the effect of sexual reproduction to slow down evolutionary changes has to do only with the short—term aspects of natural selection. Those aspects concerning cycling changes in genotype frequencies may be crucial in coping with short—term environmental changes as parasite pressure but, as I have tried to maintain in the previous sections, they are most unlikely to leave their prints in fossil records. On the other hand, sexual reproduction may not virtually affect the slower, long—term process of adaptation due to selective gene substitutions. Moreover, allowing for the short—term 'insurance' effect of sex, it might well be that sexual reproduction enables a higher rate of mutations and, maybe, other patterns of structural flexibility, facilitating long—term evolution (for a special continuous type model predicting this possibility see Eshel 1971, 1972).

6. Summary

We conclude that, contrary to the long time widely accepted assumption of population biologists, the well–studied process of changes in frequencies within a given set of genotypes may not well represent the appropriate dynamics which is responsible for the long term evolution of either morphological or behavioral patterns in natural populations, and it does not lead to the mutually optimizing structure of a population game with any payment function, except for oversimplifying, singular (though easy to analyse) situations.

It is the long term process of selective gene substitution that is likely to be responsible for the long term, fossil–recorded morphological change of populations as well as for the establishment of the genetical basis for behavioral changes. This process is qualitatively different from the short term process of changes in genotype frequencies. Unlike the latter, it is likely to lead to individual optimization and, in case of frequency dependent selection, to the convergence to ESS solutions of the population game.

As we see, these theoretical findings stand in agreement with the role of sex according to the bulk of present theories about the evolution of sexual reproduction due to selection pressure in a permanently changing environment.

Discussion

Cohen: Could you explain to us what you mean by long term and short term evolution? I mean in terms of standard time measure, say thousands of years or millions of years?

Eshel: I am afraid I cannot. In fact, we have no good theoretical measure for the rate of evolution, especially when it involves selective gene substitution. And in fact, we have no empirical estimation on the rate and effect–distribution of beneficial mutations. The distinction I have suggested is just qualitative. I have tried to show that there must be a qualitative difference between the short–term, well–studied process of change in genotype frequencies and the long–term process of selective allele substitution. I have maintained that it is only the long–term process in which the concept of ESS and local optimization can be defended. And I have tried to show that this difference should not be that surprising as it was considered in the sixties when counterexamples to Fisher's fundamental law began to be established. They are in good agreement with the present main stream of theories about the evolution of sex and its role in natural selection.

Cohen: I still have the feeling that natural populations can adapt themselves to a new environmental situation in such a short time that cannot possibly be regarded as a long–term process. Take, for example, the fast adaptation of insects to insecticides.

Eshel: This has admittedly happened in quite a short time in terms of years but the process was, most likely, based on selection of new mutations occuring at one or few loci, rather than on a change in genotype frequencies. Indeed, insect populations are large enough and their generation time is short enough to make long–term evolution (in the qualitative sense suggested above) quite observable to the human eye.

Cohen: This may be true but we can observe many other examples in which a natural population has undergone a very fast adaptation to human–induced environmental changes. And at least some of these examples are most likely expressing fast changes in genotype frequencies. Maybe the most well–known example is that of the color–adaptation of the moth Biston betularia and related species to a black background at the time of the industrial revolution and back to white when coal was widely replaced by oil and laws for the preservation of the environment were applied.

Eshel: To quote Dick Lewontin, the fact that, for more than fifty years, this example and virtually only this one, is found in any textbook of evolution may put it as an exception that proves the rule. The fact is that lack of melanin is, often, governed by one gene and can, thus, be directly selected in agreement with the fundamental law of natural selection. Still, even in this example, people tend to miss two non–trivial points: The first is the relative inefficiency of natural selection, even when operating on one locus, in sexual, Mendelian populations. The second is the importance of this 'inefficiency' for the chance of offspring survival. Indeed, if these moths were reproducing asexually, the white variant would disappear completely and its recuperation would have to wait much longer for new mutations to occur. But my main point is that in the case of more complicated features (and concerning animal conflict we generally deal with such features) natural selection is likely to become even less efficient in the short run, thus not even 'inefficiently' leading to optimization.

Weissing: Do we really have a good general theory about natural selection in the most general multilocus system? I mean an asymptotic one concerning very large numbers of loci, each with a small effect.

Eshel: At this point, I am afraid, not.

Weissing: In this case we cannot preclude the possibility that, with an increase in the number of loci, Fisher's fundamental law tends again to become a good approximation, an assumption that is implicitly made in many models of quantitative inheritance.

Eshel: This is a working assumption, all right, which is backed only by one's willingness to cope with the unwelcome finding about the hindering effect of sex; and it is tacitly based on the postulate of negligible epistasis. In many of the examples we are involved in, epistasis is an essential component of the non–genetic model and, thus, cannot be ignored. But still, on a hard analytical basis, I cannot preclude this possibility in some cases.

Hoekstra: In this case one still may not need the assumption of environmental fluctuations because sexual systems should behave very much like asexual ones.

Eshel: Except for the 1:2 cost of meiosis. In fact, I find it hard to believe that the sexual system has evolved to behave 'descently' like the asexual one, except for its being half as efficient.

Hoekstra: There is a point, indeed. Unless there is another explanation.

Eshel: That is true.

Cohen: Still, I am not sure of how really exceptional the example of Biston betularia is. In fact, there is the overwhelming evidence of semi–artificial selection to which newly domesticated populations have reacted in a very rapid behavioral and morphological change.

Eshel: Indeed, there is no question about population changes as a result of selection pressure. The question is not even of adaptation but of optimization, which is absolutely different.

Weissing: It is true, though, that a breakthrough in artificial selection has occured only when people started to combine it with inbreeding, thus when they started to remove, in essence, the effect of sex.

Eshel: This is a very interesting remark. I never thought about it in this connotation but indeed it stands in agreement with the main thesis that short term selection is only likely to lead to an optimal peak (in respect to selection) when the effects of sex are removed. Thank you.

Acknowledgement

I wish to thank Professors Reinhard Selten, Peter Hammerstein, Dan Cohen, Bill Hamilton, John Maynard Smith, Peter Taylor, James Crow and Franz Weissing for helpful remarks on this manuscript. I wish to thank the Center for Interdisciplinary Research in Bielefeld, Germany, for its hospitality while I was doing this research.

References

Bell, G. (1982): The Masterpiece of Nature: The Evolution and Genetics of Sexuality. Berkeley: University of California Press.

Bell, G. and J. Maynard Smith (1987): Short–term selection for recombination among mutually antagonistic species. Nature **328**: 66–68.

Bernstein, H. (1983): Recombinational repair may be an important function of sexual repro–duction. BioScience **33**: 326–331.

Bernstein, H., H.C. Byerly, F.A. Hopf and R.E. Michod (1984): Origin of sex. J. theor. Biol. **110**: 323–351.

Bernstein, H., H.C. Byerly, F.A. Hopf and R.E. Michod (1985): Genetic damage, mutation and the evolution of sex. Science **229**: 1277–1281.

Bremermann, H.J. (1980): Sex and polymorphism as strategies in host–pathogen interaction. J. theor. Biol. **87**: 671–702.

Bremermann, H.J. (1985): The adaptive significance of sexuality. Experientia **41**: 1245–1254.

Bulmer, M.G. (1981): Worker–queen conflict in annual social hymenoptera. J. theor. Biol. 72: 701–727.

Bulmer, M.G. and P.D. Taylor (1981): Worker–queen conflict and sex ratio in social hymenoptera. Heredity 47: 197–207.

Cavalli–Sforza, L.L. and M.W. Feldman (1978): Darwinian selection and altruism. Theor. Pop. Biol. 14: 268–280.

Charlesworth, B. (1976): Recombination modification in a fluctuating environment. Genetics 83: 181–195.

Crow, J.F. (1988): The importance of recombination. In: R.E. Michod and B.R. Levin (Eds.): The Evolution of Sex: An Examination of Current Ideas. Sunderland, MS: Sinauer Ass., pp. 56–73.

Crow, J.F. and M. Kimura (1965): Evolution in sexual and asexual populations. Amer. Natur. 99: 439–450.

Crow, J.F. and M. Kimura (1969): Evolution in sexual and asexual populations: A reply. Amer. Natur. 103: 89–91.

Eshel, I. (1970): On the evolution in a population with an infinite number of types. Theor. Pop. Biol. 2: 209–236.

Eshel, I. (1971): Evolution processes with continuity of types. Adv. Appl. Prob. 4: 475–507.

Eshel, I. (1972): Evolution in diploid populations with continuity of gametic types. Adv. Appl. Prob. 5: 55–65.

Eshel, I. (1974): Selection on the sex ratio and the evolution of sex–determination. Heredity 34: 351–361.

Eshel, I. (1982): Evolutionarily stable strategies and viability selection in Mendelian populations. Theor. Pop. Biol. 22: 204–217.

Eshel, I. (1984): On the evolution of an inner conflict. J. theor. Biol. 108: 65–76.

Eshel, I. (1985): On the evolutionary genetic stability of Mendelian segregation and the role of free recombination. Amer. Natur. 125: 412–420.

Eshel, I. and E. Akin (1983): Coevolutionary instability of mixed Nash solutions. J. Math. Biol. 18: 123–133.

Eshel, I. and M.W. Feldman (1970): On the evolutionary effect of recombination. Theor. Pop. Biol. 1: 88–100.

Eshel, I. and M.W. Feldman (1982): On the evolutionary genetic stability of the sex ratio. Theor. Pop. Biol. 21: 430–439.

Eshel, I. and M.W. Feldman (1984): Initial increase of new mutants and some continuity properties of ESS in two locus systems. Amer. Natur. 124: 631–640.

Eshel, I. and W.D. Hamilton (1984): Parent–offspring correlation in fitness under fluctuating selection. Proc. Roy. Soc. Lond. B 222: 1–14.

Eshel, I. and E. Sansone (1989): Parent offspring conflict over the sex ratio in a diploid population with different investment in male and in female offspring. Theor. Pop. Biol. (in press).

Eshel, I. and D. Weinshall (1987): Sexual reproduction and viability of future offspring. Amer. Natur. 130: 775–787.

Ettinger, L. (1986): Meiosis: A selection stage preserving the genome's pattern of organization. Evol. Theor. **8**: 17–26.

Ewens, W.J. (1968): A genetic model having complex linkage behavior. Theor. Appl. Genet. **38**: 140–143.

Feldman, M.W. (1972): Selection for linkage modifications. I. Random mating populations. Theor. Pop. Biol. **3**: 324–346.

Feldman, M.W. and B. Balkau (1972): Some results in the theory of three gene loci. In: T.N.E. Greville (Ed.): Population Dynamics. New York: Academic Press.

Feldman, M.W. and F.B. Christiansen (1984): Population genetic theory of the cost of inbreeding. Amer. Natur. **123**: 642–653.

Feldman, M.W., F.B. Christiansen and L.D. Brooks (1980): Evolution of recombination in a constant environment. Proc. Natl. Acad. Sci. USA **77**: 4838–4841.

Feldman, M.W. and U. Liberman (1986): An evolutionary reduction principle for genetic modifiers. Proc. Natl. Acad. Sci. USA **83**: 4824–4827.

Fisher, R.A. (1930): The Genetical Theory of Natural Selection. Oxford: Clarendon Press.

Fisher, R.A. (1935): The sheltering of lethals. Amer. Natur. **69**: 445–446.

Glesener, R. and D. Tilman (1978): Sexuality and the components of environmental uncertainty: Clues from geographic parthenogenesis in terrestrial animals. Amer. Natur. **112**: 659–673.

Gould, S.J. and R.C. Lewontin (1979): The spandrels of San Marco and the Panglossian paradigm: A critique of the adaptationist programme. Proc. Roy. Soc. Lond. B **205**: 581–598.

Hadeler, K. and U. Liberman (1975): Selection models with fertility differences. J. Math. Biol. **2**: 19–32.

Hamilton, W.D. (1964): The genetical evolution of social behavior. J. theor. Biol. **7**: 1–52.

Hamilton, W.D. (1967): Extraordinary sex ratios. Science **156**: 477–488.

Hamilton, W.D. (1980): Sex versus non–sex versus parasite. Oikos **35**: 282–290.

Hamilton, W.D. (1982): Pathogens as causes of genetic diversity in their host populations. In: R.M. Anderson and R.M. May (Eds.): Population Biology of Infectious Diseases. New York: Springer–Verlag.

Hamilton, W.D., P.A. Henderson and N.A. Moran (1981): Fluctuation of environment and coevolved antagonist polymorphism as factors in the maintenance of sex. In: R.D. Alexander and D.W. Tinkle (Eds.): Natural Selection and Social Behavior. New York: Chiron Press.

Hamilton, W.D. and M. Zuk (1982): Heritable true fitness and bright birds: A role for parasites? Science **218**: 384–387.

Jaenike, J. (1978): An hypothesis to account for the maintenance of sex within populations. Evol. Theor. **3**: 191–194.

Karlin, S. (1968): Equilibrium behavior of population genetic models with nonrandom mating. J. App. Prob. **5**: 487–566.

Karlin, S. (1973): Sex and infinity: A mathematical analysis of the advantages and disadvantages of genetic recombination. In: M. Bartlett and R. Hiorns (Eds.): The Mathematical Theory of the Dynamics of Biological Populations. New York: Academic Press.

Karlin, S. (1975): General two–locus selection models: Some objectives, results and interpretations. Theor. Pop. Biol. 7: 364–398.

Karlin, S. and S. Lessard (1983): On the optimal sex ratio. Proc. Natl. Acad. Sci. USA 80: 5931–5935.

Karlin, S. and S. Lessard (1986): Theoretical Studies on Sex Ratio Evolution. Princeton NJ: Princeton University Press.

Kingman, J.F.C. (1961): A mathematical problem in population genetics. Proc. Camb. Philos. Soc. 57: 574–582.

Lessard, S. (1984): Evolutionary dynamics in frequency dependent two phenotype models. Theor. Pop. Biol. 25: 210–234.

Levin, B.R. (1988): The evolution of sex in bacteria. In: R.E. Michod and B.R. Levin (Eds.): The Evolution of Sex: An Examination of Current Ideas. Sunderland, MS: Sinauer Ass., pp. 194–211.

Levin, D.A. (1975): Pest pressure and recombination systems in plants. Amer. Natur. 109: 437–451.

Lewontin, R.C. (1971): The effect of genetic linkage on the mean fitness of a population. Proc. Natl. Acad. Sci. USA 68: 984–986.

Liberman, U. (1976): Theory of meiotic drive: Is Mendelian segregation stable? Theor. Pop. Biol. 10: 127–132.

Liberman, U. (1988): External stability and ESS: Criteria for initial increase of a new mutant allele. J. Math. Biol. 26: 477–485.

Liberman, U. and M.W. Feldman (1986): A general reduction principle for genetic modifiers of recombination. Theor. Pop. Biol. 30: 341–371.

Liberman, U., M.W. Feldman, I. Eshel and S.P. Otto (1989): Two locus autosomal sex determination. I. On the evolutionary genetic stability of the even sex ratio. Manuscript.

Maynard Smith, J. (1971): The origin and maintenance of sex. In: G.C. Williams (Ed.): Group Selection. Chicago: Aldine Atherton, pp. 165–175.

Maynard Smith, J. (1974): The theory of games and the evolution of animal conflict. J. theor. Biol. 47: 209–221.

Maynard Smith, J. (1975): Evolution of sex. Nature 254.

Maynard Smith, J. (1978): The Evolution of Sex. Cambridge: Cambridge University Press.

Maynard Smith, J. (1982): Evolution and the Theory of Games. Cambridge: Cambridge University Press.

Maynard Smith, J. (1988): Selection for recombination in a polygenic model – The mechanism. Genet. Res. Cam. 51: 59–63.

Maynard Smith, J. and G.A. Parker (1976): The logic of asymmetric contests. Anim. Behav. 24: 159–175.

Maynard Smith, J. and G.R. Price (1973): The logic of animal conflict. Nature 246: 15–18.

Moran, P.A.P. (1964): On the nonexistence of adaptive topographies. Ann. Hum. Genet. 27: 338–343.

Muller, H.J. (1932): Some genetic aspects of sex. Amer. Natur. **66**: 118–138.

Muller, H.J. (1958): Evolution by mutation. Bull. Amer. Math. Soc. **64**: 137–160.

Oster, G., I. Eshel and D. Cohen (1977): Evolution of social insects. Theor. Pop. Biol. **12**: 49–68.

Rice, W.R. (1983): Parent–offspring pathogen transmission: A selective agent promoting sexual reproduction. Amer. Natur. **121**: 187–203.

Seger, J. and W.D. Hamilton (1988): Parasites and sex. In: R.E. Michod and B.R. Levin (Eds.): The Evolution of Sex: An Examination of Current Ideas. Sunderland, MS: Sinauer Ass., pp. 176–193.

Selten, R. (1980): A note on evolutionarily stable strategies in asymmetric animal conflicts. J. theor. Biol. **84**: 93–101.

Selten, R. (1983): Evolutionary stability in extensive two–person games. Math. Soc. Sci. **5**: 269–363.

Sturtevant, A.H. and K. Mather (1938): The interrelations of inversions, heterosis and recombination. Amer. Natur. **72**: 447–452.

Taylor, P.D. (1985): A general mathematical model for sex allocation. J. theor. Biol. **112**: 799–818.

Thompson, V. (1976): Does sex accelerate evolution? Evol. Theory **1**: 131–156.

Tooby, J. (1982): Pathogens, polymorphism, and the evolution of sex. J. theor. Biol. **97**: 557–576.

Trivers, R.L. and H. Hare (1976): Haplodiploidy and the evolution of social insects. Science **191**: 249–263.

Uyenoyama, M.K. (1984): On the evolution of parthenogenesis: A genetic representation of the 'cost of meiosis'. Evolution **38**: 87–102.

Uyenoyama, M.K. and B.O. Bengtsson (1981): Towards a genetic theory for the evolution of the sex ratio. II. Haplodiploid and diploid models with sibling and parent control of the brood sex ratio and brood size. Theor. Pop. Biol. **20**: 57–79.

Uyenoyama, M.K. and B.O. Bengtsson (1982): Towards a genetic theory for the evolution of the sex ratio. III. Parental and sibling control of brood investment ratio under partial sib–mating. Theor. Pop. Biol. **22**: 43–68.

Uyenoyama, M.K. and M.W. Feldman (1981): On relatedness and adaptive topography in kin selection. Theor. Pop. Biol. **19**: 87–123.

Weinshall, D. (1986): Why is a two–environment system not rich enough to explain the evolution of sex? Amer. Natur. **128**: 736–750.

Weinshall, D. and I. Eshel (1987): On the evolution of an optimal rate of sexuality. Amer. Natur. **130**: 578–770.

Williams, G.C. (1966): Adaptation and Natural Selection. Princeton: Princeton Univ. Press.

Williams, G.C. (1975): Sex and Evolution. Princeton: Princeton Univ. Press.

Williams, G.C. (1988): Retrospect on sex and kindred topics. In: R.E. Michod and B.R. Levin (Eds.): The Evolution of Sex: An Examination of Current Ideas. Sunderland, MS: Sinauer Ass., pp. 287–298.

Williams, G.C. and J.B. Mitton (1973): Why reproduce sexually? J. theor. Biol. **39**: 545–554.

EVOLUTIONARY STABILITY AND DYNAMIC STABILITY
IN A CLASS OF EVOLUTIONARY NORMAL FORM GAMES

by

Franz J. Weissing

Abstract:

A complete game theoretical and dynamical analysis is given for a class of evolutionary normal form games which are called 'RSP–games' since they include the well–known 'Rock–Scissors–Paper' game. RSP–games induce a rich selection dynamics, but they are simple enough to allow a global analysis of their evolutionary properties. They provide an ideal illustration for the incongruities in the evolutionary predictions of evolutionary game theory and dynamic selection theory.

Every RSP–game has a unique interior Nash equilibrium strategy which is an ESS if and only if and only if the average binary payoffs of the game are all positive and not too different from one another. Dynamic stability with respect to the continuous replicator dynamics may be characterized by the much weaker requirement that the equilibrium payoff of the game has to be positive. The qualitative difference between evolutionary stability and dynamic stability is illustrated by the fact that every ESS can be transformed into a non–ESS attractor by means of a transformation which leaves the dynamics essentially invariant.

In all evolutionary normal form games, evolutionary stability of a fixed point implies dynamic stability with respect to the continuous replicator dynamics. Due to 'overshooting effects', this is generally not true for the discrete replicator dynamics. In contrast to all game theoretical concepts (including the ESS concept), discrete dynamic stability is not invariant with respect to positive linear transformations of payoffs. In fact, every ESS of an RSP–game can both be stabilized and destabilized by a transformation of payoffs. Quite generally, however, evolutionary stability implies discrete dynamic stability if selection is 'weak enough'.

In the continuous–time case, the interior fixed point of an RSP–game is either a global attractor, or a global repellor, or a global center. In contrast, the discrete replicator dynamics admits a much richer dynamics including stable non–equilibrium behaviour. The occurrence of stable and unstable limit cycles is demonstrated both numerically and analytically.

Some selection experiments in chemostats reveal that competition between different asexual strains of the yeast *Saccharomyces cerevisiae* leads to the same cyclical best reply structure that is characteristic for RSP–games. Possibly, Rock–Scissors–Paper–games are also played in non–human biological populations.

I gratefully acknowledge many stimulating discussions with Josef Hofbauer and Karl Sigmund and the helpful comments from Andreas Dress, Roy Gardner, Peter Hammerstein, John Nachbar, and Reinhard Selten. I also want to thank the Center for Interdisciplinary Research, especially Frau Jegerlehner, for a pleasant time with almost ideal working conditions. Most of all, I have to thank Anke, who often had a hard time while I was struggling around with game equilibrium models.

1. Introduction

There are two main approaches towards the phenotypic analysis of frequency dependent natural selection. First, there is the approach of *evolutionary game theory*, which was introduced in 1973 by John Maynard Smith and George R. Price. In this theory, the dynamical process of natural selection is not modeled explicitly. Instead, the selective forces acting within a population are represented by a fitness function, which is then analysed according to the concept of an evolutionarily stable strategy or ESS. Later on, the static approach of evolutionary game theory has been complemented by a *dynamic stability analysis* of the replicator equations. Introduced by Peter D. Taylor and Leo B. Jonker in 1978, these equations specify a class of dynamical systems, which provide a simple dynamic description of a selection process. Usually, the investigation of the replicator dynamics centers around a stability analysis of their stationary solutions.

Although evolutionary stability and dynamic stability both intend to characterize the long—term outcome of frequency dependent selection, these concepts differ considerably in the 'philosophies' on which they are based. It is therefore not too surprising that they often lead to quite different evolutionary predictions (see, e.g., Weissing 1983). The present paper intends to illustrate the incongruities between the two approaches towards a phenotypic theory of natural selection. A detailed game theoretical and dynamical analysis is given for a generic class of evolutionary normal form games. In spite of its simplicity, this class is rich enough to uncover all kinds of discrepancies between evolutionary stability and dynamic stability. In the course of the analysis some light will be shed on the factors which are responsible for the inconsistencies in the conclusions of the game theoretical and the dynamical approach.

Evolutionary stability and dynamic stability correspond quite well to another if the number of pure strategies is small (Zeeman 1980, Weissing 1983). Discrepancies between these concepts may only be observed at a mixed Nash equilibrium strategy involving at least three pure strategies. Usually some form of cycling takes place around this equilibrium.[1] This suggests to have a closer look at the children's game *Rock–Scissors–Paper*, since this is the prototype example for a game where all these requirements are met.

In its simplest version, the Rock–Scissors–Paper game is modeled as a zero–sum game, which is represented by the payoff matrix

$$A = \begin{bmatrix} 0 & 1 & -1 \\ -1 & 0 & 1 \\ 1 & -1 & 0 \end{bmatrix}. \tag{1.1}$$

It is a characteristic property of this game that strategy i is always the unique best reply to strategy $i+1$ (counted modulo 3): Rock is optimal against Scissors, Scissors is optimal against Paper, and Paper is optimal against Rock. This cyclical pattern is symbolized in the following diagram:

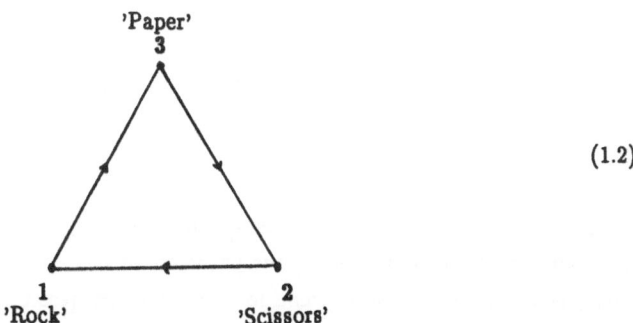

$$(1.2)$$

In fact, (1.2) corresponds to the flow of the replicator dynamics on the border of the strategy simplex (see Figures 1 to 3), and it is plausible that a pattern like this implies the existence of an interior fixed point which is the center of a flow cycling around it.

However, if modeled as a zero–sum game, the Rock–Scissors–Paper game has some undesirable properties, which make it ill–suited as a 'counter–example'. It is easy to show that symmetric constant–sum games do not admit evolutionarily stable strategies. This corresponds well to the fact that interior fixed points of these games are always centers for the continuous and repellors for the discrete replicator dynamics (see Akin & Losert (1984), Hofbauer & Sigmund (1988), as well as Corollary 5.4 and Theorem 6.6 below).

We shall therefore leave the zero–sum context without abandoning the cyclical pattern described by diagram (1.2). Generalizing (1.1), we shall consider payoff matrices of the form

$$A = \begin{bmatrix} a_1 & b_2 & c_3 \\ c_1 & a_2 & b_3 \\ b_1 & c_2 & a_3 \end{bmatrix}, \quad b_i > a_i \geq c_i. \tag{1.3}$$

It will be shown that this is exactly the class of games to which diagram (1.2) does apply (see Theorem 3.4). A symmetric 3x3 normal form game which can be put into form (1.3) will be called a *generalized Rock–Scissors–Paper game* or simply an *RSP–game*. The evolutionary analysis of this class of games will form the subject–matter of this paper.[2]

More specific generalizations of the 'Rock–Scissors–Paper' game have repeatedly entered the literature. In particular, a one–parameter class of RSP–games plays a prominent role among the 'standard examples' of evolutionary game theory. It is the class of ϵ–*perturbed Rock–Scissors–Paper games*, which are characterized by payoff matrices of the form

$$A_\epsilon := \begin{bmatrix} 0 & 1+\epsilon & -1 \\ -1 & 0 & 1+\epsilon \\ 1+\epsilon & -1 & 0 \end{bmatrix}, \quad \epsilon > -1, \tag{1.4}$$

or by rescaled versions thereof.[3] Dating back to a paper of Maynard Smith (1977, quoting Dr. C. Strobeck), these games have been used to illustrate the fact that evolutionary stability need not imply dynamic stability with respect to the discrete replicator dynamics. In addition, they provide a nice illustration for the effects of a change in payoff parameters on the stability of an interior fixed point (see, e.g., Zeeman 1980).

An analysis of ϵ-perturbed Rock–Scissors–Paper games shows, however, that this class of 3x3–games is not rich enough for our purposes. In fact, the notions of evolutionary stability and 'continuous' dynamic stability coincide for these games (see Corollary 6.7). Since we are interested in getting *all* possible types of discrepancies between these concepts, we have to consider RSP–games in their most general form.

Certain aspects of the general class of RSP–games have previously been studied by Zeeman (1980), Weissing (1983), and Hofbauer & Sigmund (1988):
– Zeeman (1980) is mainly concerned with the classification of the phase portraits of evolutionary 3x3–games with respect to the continuous replicator dynamics. As a step towards this, he presents a global analysis of the phase portraits of RSP–games, and he shows that they may be classified according to the sign of a determinant. His results on RSP–games are contained in Section 5 below (Theorem 5.6). The proof given here, however, is constructive and has the advantage of focussing on the differences between evolutionary stability and dynamic stability.
– Complementing Zeeman's results on 'continuous stability', I investigated the conditions for evolutionary stability and discrete local hyperbolic stability (Weissing 1983). These aspects will be analysed in much more detail in Sections 4 and 6 below.
– In their excellent book on the replicator dynamics, Josef Hofbauer and Karl Sigmund (1988) give the most complete survey of stability properties of RSP–games. The results presented – which correspond to Theorems 3.4.2, 4.6.1, 5.6, and 6.8 below – extend to evolutionary stability as well as to dynamic stability with respect to both replicator dynamics. Since Hofbauer and Sigmund are focussing on more general dynamical aspects of natural selection, they do not prove their assertions, although the proofs are often far from being straightforward. Nevertheless, I owe much to their work. My own results and the analysis thereof have been improved considerably by adopting several of the techniques proposed in their book.

In the present paper, RSP–games will be put into a context that is more coherent and systematic than in the publications cited above. To my knowledge, it is the first time that a combined game theoretical and dynamical analysis is given for such a broad class of evolutionary normal form games.

The structure of this paper aims at exemplifying how the general methods developed in Hofbauer & Sigmund (1988) find natural applications in the analysis of concrete examples. By putting them into a broader context, the known features of RSP–games will be presented in a way that makes it easy to generalize them considerably.

In addition to providing new proofs for the known features of RSP–games, the present paper contains several results that do not have counterparts in the literature. In particular, almost all results on the discrete replicator dynamics (Sections 6, 7) seem to be new. For example, the occurence of supercritical Hopf bifurcations and 'closed limit curves' has not been observed before for the discrete replicator dynamics in two dimensions. Also the results on global discrete stability and instability have apparently not been derived before. In a separate paper (Weissing 1990) it will be shown that they apply quite generally to *all* evolutionary normal form games.

The main emphasis of the present paper will be put on elucidating the *qualitative differences* between evolutionary stability and dynamic stability, which are so well exemplified by the class of RSP–games. Among other things, the following results will be derived:

— With respect to the *continous* replicator dynamics, every ESS is a global attractor, but an asymptotically stable fixed point need not be an ESS. However, every asymptotically stable fixed point of an RSP–game — even if it is not an ESS — can be transformed into an ESS of another RSP–game by means of a transformation which leaves the phase portrait 'essentially' invariant. This shows that there is a qualitative difference between evolutionary stability and dynamic stability: the ESS concept is *not* invariant with respect to dynamics–preserving transformations of the state space (see Section 5.).

— With respect to the *discrete* replicator dynamics, evolutionary stability is neither necessary nor sufficient for dynamic stability. In fact, an ESS may be a global repellor, and a global attractor need not be an ESS. In the discrete context, we get an additional qualitative difference between evolutionary stability and dynamic stability: Whereas the ESS concept is invariant with respect to positive linear transformations of the payoff matrix, the stability properties with respect to the discrete replicator dynamics are very strongly affected by them. Indeed, any asymptotically stable fixed point of the discrete replicator equation can be destabilized by means of a positive linear transformation of payoffs (see Section 6).

— For the continuous replicator dynamics, *local* stability properties correspond perfectly to *global* stability properties. Generic Hopf bifurcations and limit cycles do not occur (Theorems 5.6 and 5.8). In the context of the discrete replicator dynamics, the same holds true for the subclass of 'circulant' RSP–games (Theorem 6.8). In general, however, supercritical Hopf bifurcations do occur, a fact which implies the existence of a more complicated attractor encircling the interior fixed point(see Section 7). Because of their simplicity, RSP–games provide one of the rare occasions where this phenonemon can be demonstrated analytically.

Taken together, these features show that RSP–games give an almost ideal illustration of the incongruities between the game theoretical and the dynamical approach towards frequency dependent natural selection. On the one hand, they are simple enough to allow an almost complete analysis of their evolutionary characteristics. In fact, there are no simpler games where all kinds of discrepancies between evolutionary stability and dynamic stability can be observed.[4] On the other hand, this class of games is rich enough to exemplify rather complex dynamical non–equilibrium behaviour like convergence to a closed limit curve.

2. Notation and Basic Definitions

We are interested in a situation where the fitness of individuals is affected by the outcome of a randomly assorted, binary, intra—specific interaction with the structure of an RSP—game. Each participant in an interaction behaves according to one out of three *pure strategies,* which will be called 'Rock', 'Scissors', and 'Paper' and numbered 1, 2, 3, respectively. Calculations with respect to this numbering should always be understood modulo 3. Pure strategies will always be denoted by the letters i and j, and the set of pure strategies will be denoted by I.

Mixed strategies (i.e., frequency distributions over the set of pure strategies) will be identified with the elements of the two—dimensional *strategy simplex*

$$\Delta := \{ \mathbf{p} \in \mathbb{R}^3_+ \mid \textstyle\sum_i p_i = 1 \}. \tag{2.1}$$

Mixed strategies will be denoted by small bold—face letters like $\mathbf{p}$ or $\mathbf{q}$.

The set of pure strategies which is given a positive weight by the mixed strategy $\mathbf{p}$ is called the *support of* $\mathbf{p}$ and denoted by $\mathrm{supp}(\mathbf{p})$:

$$\mathrm{supp}(\mathbf{p}) := \{ i \in I \mid p_i > 0 \}. \tag{2.2}$$

As usual, the pure strategy $i \in I$ will be identified with the mixed strategy $\mathbf{e}^i \in \Delta$, the support of which consists of the single element i. Accordingly, the pure strategies correspond to the three 'corners' of the strategy simplex. A mixed strategy $\mathbf{p} \in \Delta$ will be called a *completeley mixed* or an *interior* strategy, if it has a 'full support', i.e., if $\mathrm{supp}(\mathbf{p}) = I$.

The individuals interacting in an RSP—game form a population, the *state* of which is characterized by the frequency distribution of the pure strategies which are currently used by its members. Accordingly, a population state corresponds to a mixed strategy, the *population strategy.* A population, the state of which is given by population strategy $\mathbf{p} \in \Delta$, will be called a *p—population.*

A 3x3—matrix $A = (a_{ij})$ of the form (1.3) will be interpreted as the *payoff matrix* of an *evolutionary normal form game.* Since a symmetric bimatrix game is completely specified by the payoff matrix of player 1, we shall often identify an evolutionary game with this payoff matrix. The matrix entries are called the *binary payoffs* of the evolutionary game, and a_{ij} should be interpreted as the expected fitness of an individual using pure strategy i due to its interaction with an individual using pure strategy j.

Selection is *frequency dependent* whenever individual fitness is determined by the outcome of an evolutionary game. In fact, the fitness of an individual using pure strategy i depends on $\mathbf{p}$, the population strategy. Assuming that interacting individuals are assorted at random with respect to the strategies used, the individual fitness of an 'i—strategist' in a $\mathbf{p}$—population is given by

$$F_i(\mathbf{p}) := \sum_j a_{ij}p_j = (A\mathbf{p})_i \tag{2.3}$$

($(A\mathbf{p})_i$ denotes the i'th component of the vector $A\mathbf{p}$.). The vector

$$\mathbf{F}(\mathbf{p}) := A\mathbf{p} = \begin{bmatrix} a_1p_1 + b_2p_2 + c_3p_3 \\ c_1p_1 + a_2p_2 + b_3p_3 \\ b_1p_1 + c_2p_2 + a_3p_3 \end{bmatrix}, \tag{2.4}$$

the components of which are given by $F_i(\mathbf{p})$, is called the *fitness vector at p*.

The fitness of an individual behaving according to mixed strategy $\mathbf{q}$ in a p–population is given by

$$\mathcal{F}(\mathbf{q},\mathbf{p}) := \sum_i q_i F_i(\mathbf{p}) = \mathbf{q} \cdot \mathbf{F}(\mathbf{p}) , \tag{2.5}$$

where the dot denotes the Euclidean scalar product. $\mathcal{F}(\mathbf{q},\mathbf{p})$ may also be interpreted as the mean fitness of a sub–population the strategy mix of which is described by $\mathbf{q}$. The function $\mathcal{F}: \Delta \times \Delta \longrightarrow \mathbb{R}$ which is defined by (2.5) is the *fitness function* of the evolutionary game. It corresponds to player 1's *payoff function* of the symmetric bimatrix game (A, A^T), where A^T denotes the transpose of the matrix A.

The *mean fitness* of a p–population is given by

$$\overline{F}(\mathbf{p}) := \sum_i p_i \cdot F_i(\mathbf{p}) = \mathcal{F}(\mathbf{p},\mathbf{p}) , \tag{2.6}$$

and the term $F_i(\mathbf{p}) - \overline{F}(\mathbf{p})$ will be interpreted as the *relative fitness* of strategy i in a p–population.

The *replicator dynamics* intend to give a dynamic model of natural selection. There are two versions of it: a discrete one for the case of discrete and non–overlapping generations, and a continuous one for overlapping generations merging into another. Both versions are based on the assumption that the growth rate of a pure strategy is proportional to its relative fitness. A detailed derivation of the replicator dynamics and a discussion of the underlying assumptions may be found in Weissing (1983) and in Hofbauer and Sigmund (1988).

The continuous version of the replicator dynamics is given by a system of differential equations on the strategy simplex, which is called the *continuous replicator equation*:

$$\dot{p}_i := p_i (F_i(\mathbf{p}) - \overline{F}(\mathbf{p})), \quad i \in I. \tag{2.7}$$

The discrete version of the replicator dynamics may be represented by a system of recursion equations, called the *discrete replicator equation*:

$$p_i' := p_i \frac{F_i(\mathbf{p})}{\overline{F}(\mathbf{p})}, \quad i \in I, \tag{2.8}$$

where $p' \in \Delta$ denotes the population strategy of an offspring population, the parents of which had $p \in \Delta$ as their population strategy.

The term *fitness* has a somewhat different interpretation in the two contexts: In the discrete case, the fitness of a character should be interpreted as the expected number of offspring of the individuals bearing this character. Correspondingly, only non–negative numbers make sense as fitness values. For the continuous replicator equation, there are no such restrictions on the fitness parameters, since the fitness of a character corresponds to its growth rate which may be positive or negative.

In what follows, we are interested in the fixed points of the replicator dynamics given by (2.7) and (2.8). A *fixed point* is a stationary solution of the replicator dynamics, i.e., a population strategy $p \in \Delta$, which satisfies the conditions

$$\dot{p}_i = 0 \quad (\text{or:} \quad p_i' = p_i) \quad \text{for all } i \in I , \tag{2.9}$$

respectively. It is obvious that the discrete and the continuous version of the replicator dynamics have the same set of fixed points, and that a fixed point p^* of (2.7) or (2.8) is characterized by:

$$F_i(p^*) = \overline{F}(p^*) \quad \text{for } i \in \text{supp}(p^*) . \tag{2.10}$$

If p is a pure strategy, (2.10) always holds true. Pure strategies will be called the *trivial fixed points* of the replicator dynamics.

Notice that a completely mixed strategy p^* is an *interior fixed point* if and only if the fitness vector in p^* is a scalar multiple of the vector $1 := (1,1,...,1)$, i.e., if

$$F(p^*) = \lambda 1 \quad (\text{where } \lambda = \overline{F}(p^*)). \tag{2.11}$$

3. Nash Equilibrium Strategies of RSP–Games

In this section it will be shown that RSP–games may be characterized as those evolutionary 3x3 normal form games, for which the border of the strategy simplex does contain neither a Nash equilibrium strategy nor a nontrivial fixed point of the replicator dynamics. From this result it will be easy to derive that every RSP–game has a unique interior Nash equilibrium strategy. Subsequently, the coordinates of the interior fixed point will be characterized in terms of the entries of the payoff matrix.

DEFINITION 3.1: Nash equilibrium strategies.

A mixed strategy p is called a *Nash equilibrium strategy* or simply a *Nash strategy* if the strategy pair (p*,p*) is a symmetric Nash equilibrium point of the symmetric normal form game (A,A^T). Equivalently, p* is a Nash strategy if it is a best reply to itself, i.e. if it satisfies the

Nash condition: $\mathcal{F}(p^*,p^*) \geq \mathcal{F}(q,p^*)$ for all $q \in \Delta$.   (3.1)

For the interpretation of this concept, the reader is referred to van Damme (1987) or to Harsanyi & Selten (1988). The following proposition is a special case of the so–called Fundamental Lemma of non–cooperative game theory:

PROPOSITION 3.2: Characterization of Nash strategies.

p is a Nash equilibrium strategy if and only if the following conditions are satisfied:

$$F_i(p^*) = \overline{F}(p^*) \quad \text{for } i \in \text{supp}(p^*),$$ (3.2a)

$$F_i(p^*) \leq \overline{F}(p^*) \quad \text{for } i \notin \text{supp}(p^*).$$ (3.2b)

A comparison between (3.2) and (2.10) shows that every Nash strategy is a fixed point of the replicator dynamics. For a completely mixed strategy p*, condition (3.2b) is empty. Accordingly, p is an *interior Nash strategy* if and only if it is an interior fixed point of the replicator dynamics, i.e., if and only if (2.11) holds true. On the other hand, fixed points of the replicator dynamics may be characterized as interior Nash strategies of substructures of A (see the proof of Theorem 3.4).

The following existence theorem is of fundamental importance. A proof of part 1. can be found in the classical paper of Nash (1951).

THEOREM 3.3: Existence of Nash equilibrium strategies.

1. For every evolutionary game there exists at least one Nash strategy.
2. If the border of the strategy simplex does not contain a Nash strategy, there exists a *unique* interior Nash strategy.

PROOF of Part 2.:

Since border Nash strategies do not exist, part 1. of the theorem implies the existence of at least one interior Nash strategy. Suppose that there are two interior Nash strategies, p_1^* and p_2^*. By (2.11), this means that the fitness vectors $F(p_1^*)$ and $F(p_2^*)$ are scalar multiples of the vector 1. The linearity of F (see (2.3)) implies that $F(p)$ is also a scalar multiple of 1 for every strategy p of the form: $p = \mu_1 p_1 + \mu_2 p_2^*$.

Correspondingly, all points of the intersection of the line L through p_1^* and p_2^* with the strategy simplex Δ are Nash equilibrium strategies as well. Since the intersection of L with the border of Δ is not empty, the border of Δ contains at least one Nash strategy.

This contradiction shows that the interior Nash strategy is unique.

$* * *$

THEOREM 3.4: Characterization of RSP–games.

1. A symmetric 3x3 normal form game is an RSP–game if and only if it has neither border Nash equilibrium strategies nor nontrivial border fixed points.
2. Every RSP–game has a unique interior Nash equilibrium strategy.[5]

PROOF:

A. By (3.2), a symmetric normal form game $A = (a_{ij})$ does not admit a pure Nash equilibrium strategy if and only if every column of A contains a on–diagonal element which is larger than the corresponding diagonal entry of A, i.e., if and only if the following holds true for every $i \in I$:

$$a_{ki} > a_{ii} \quad \text{for at least one } k \neq i. \tag{3.3}$$

B. A comparison of (2.10) and (3.2) shows that A does not admit a nontrivial border fixed point, if and only if none of its 2x2–restrictions has an interior Nash strategy. Here, a *2x2–restriction* of A is defined to be a 2x2 payoff matrix

$$A_{ij} = \begin{bmatrix} a_{ii} & a_{ij} \\ a_{ji} & a_{jj} \end{bmatrix}, \tag{3.4}$$

which results from A by removing one of its pure strategies from consideration.

It is easy to see that a symmetric 2x2 normal form game A_{ij} has an interior Nash strategy if and only if $(a_{ii}-a_{ji})$ and $(a_{jj}-a_{ij})$ have the same sign. Consequently, A_{ij} does not admit an interior Nash strategy if $(a_{ii}-a_{ji})$ and $(a_{jj}-a_{ij})$ differ in sign. Let us say that the pure strateg i *dominates* the pure strategy j if

$$a_{ii} \geq a_{ji} \quad \text{and} \quad a_{ij} \geq a_{jj}, \tag{3.5}$$

with at least one of these inequalities being strict. Accordingly, A_{ij} does not admit an interior Nash strategy if and only if one of its two pure strategies dominates the other.

C. It is clear from the definition of an RSP–game that none of its three pure strategies is a best reply to itself. Accordingly, RSP–games do not have pure Nash strategies.

On the other hand, all three 2x2–restrictions of an RSP–game have a dominating pure strategy. For example, 'Rock' dominates 'Scissors', if 'Paper' is removed from consideration. Therefore, RSP–games do not admit non–trivial border fixed points.

D. Let now $A = (a_{ij})$ denote a symmetric 3x3 normal form game which has no pure Nash strategies and no non–trivial border fixed points.

Since pure strategy *1* is not a Nash strategy, the first column of A contains an element which is larger than the diagonal entry a_{11}. Without loss in generality, we may assume $a_{31} > a_{11}$. (3.5) applied to the 2x2–restriction A_{13} yields $a_{33} \geq a_{13}$, since otherwise A_{13} would admit an interior Nash strategy.

In view of (3.3), the last inequality implies $a_{23} > a_{33}$, since otherwise strategy *3* would be a pure Nash strategy of A. Applying (3.5) to the 2x2–restriction A_{23}, we get $a_{22} \geq a_{32}$ which in turn implies $a_{12} > a_{22}$. Finally, we get $a_{11} \geq a_{21}$ by applying (3.5) to A_{12}. Summarizing all this, we have derived the following inequalities:

$$a_{31} > a_{11} \geq a_{21}, \quad a_{12} > a_{22} \geq a_{32}, \quad a_{23} > a_{33} \geq a_{13}. \tag{3.6}$$

which imply that A is the payoff matrix of an RSP–game.

E. Since RSP–games do not admit border Nash strategies, Theorem 3.3(2) shows that every RSP–game has a unique interior Nash strategy.[5]

* * *

In order to calculate the Nash strategies of a symmetric normal form game $A = (a_{ij})$, it is often useful to consider a rescaled version of that game. We shall perform a *positive linear transformation* of A, which is given by

$$\tilde{a}_{ij} := \lambda \cdot a_{ij} + \mu_j, \quad \mu = (\mu_1, \mu_2, \mu_3) \in \mathbb{R}^3, \quad \lambda > 0. \tag{3.7}$$

A positive linear transformation is called *homogeneous*, if μ is a constant multiple of the vector 1. Two evolutionary games will be called *pl–equivalent* if one can be transformed into the other by means of a positive linear transformation.

PROPOSITION 3.5: Invariance with respect to positive linear transformations.

Pl–equivalent evolutionary normal form games have the same fixed points and the same Nash equilibrium strategies.

PROOF:

Let $\tilde{A} = (\tilde{a}_{ij})$ be defined by (3.7) and let $\tilde{\mathcal{F}}$ denote the fitness function induced by $\tilde{A}$. A simple calculation shows that $\tilde{\mathcal{F}}$ is related to $\mathcal{F}$ via

$$\tilde{\mathcal{F}}(q,p) = \lambda \mathcal{F}(q,p) + \mu \cdot p, \quad q, p \in \Delta, \tag{3.8}$$

where the dot denotes the Euclidean scalar product. This implies

$$\tilde{\mathcal{F}}(\mathbf{q},\mathbf{p}) - \tilde{\mathcal{F}}(\mathbf{p},\mathbf{p}) = \lambda \cdot (\, \mathcal{F}(\mathbf{q},\mathbf{p}) - \mathcal{F}(\mathbf{p},\mathbf{p}) \,) \,. \qquad (3.9)$$

Setting $\mathbf{q} = \mathbf{e}^i$, and denoting by $\tilde{F}$ and $\overline{\tilde{F}}$ the fitness vector and the mean fitness induced by $\tilde{A}$, we get

$$\tilde{F}_i(\mathbf{p}) - \overline{\tilde{F}}(\mathbf{p}) = \lambda \cdot (\, F_i(\mathbf{p}) - \overline{F}(\mathbf{p}) \,) \,. \qquad (3.10)$$

In view of $\lambda > 0$, (2.10) and (3.2) show that A and $\tilde{A}$ have the same sets of fixed points and Nash equilibrium strategies. * * *

A slightly weaker version of this theorem is essential for 'classical' normative game theory: In normative game theory, payoffs are interpreted in terms of the utility concept of von Neumann and Morgenstern (see Luce & Raiffa 1957). This implies that payoffs are only well–defined up to homogeneous positive linear transformations. Consequently, it is a basic demand of normative game theory that all its solution concepts should be invariant with respect to this class of transformations.

Let us now focus again attention on the class of RSP–games. First, notice that the class of RSP–games is invariant under positive linear transformations of the payoff matrix. According to (1.3), every RSP–game is induced by a triple $(\mathbf{a},\mathbf{b},\mathbf{c})$ of vectors which satisfy

$$\mathbf{a},\mathbf{b},\mathbf{c} \in \mathbb{R}^3, \quad b_i > a_i \geq c_i \text{ for } i \in I. \qquad (3.11)$$

A triple of vectors satisfying (3.11) will be called an *RSP–triple*. Whenever we want to emphasize the dependence of an RSP–game A on its generating RSP–triple $(\mathbf{a},\mathbf{b},\mathbf{c})$ we shall write $A = A(\mathbf{a},\mathbf{b},\mathbf{c})$. Accordingly, $A(\mathbf{a},\mathbf{b},\mathbf{c})$ is given by

$$A(\mathbf{a},\mathbf{b},\mathbf{c}) = \begin{bmatrix} a_1 & b_2 & c_3 \\ c_1 & a_2 & b_3 \\ b_1 & c_2 & a_3 \end{bmatrix} . \qquad (3.12)$$

It will be useful to simplify a given RSP–game $A = A(\mathbf{a},\mathbf{b},\mathbf{c})$ by means of a positive linear transformation, which transforms the diagonal of A into zero. Setting $\lambda = 1$ and $\mu = -\mathbf{a}$ in (3.7), we get a represensation of A in the form

$$A = A_0 + A_1 , \qquad (3.13)$$

where A_0 is pl–equivalent to A and where A_1 denotes the matrix

$$A_1 = A(\mathbf{a},\mathbf{a},\mathbf{a}) = \begin{bmatrix} a_1 & a_2 & a_3 \\ a_1 & a_2 & a_3 \\ a_1 & a_2 & a_3 \end{bmatrix} . \qquad (3.14)$$

Since A_0 is pl–equivalent to A, it is also an RSP–game. It is given by

$$A_0 = A(0,\beta,-\gamma) = \begin{bmatrix} 0 & \beta_2 & -\gamma_3 \\ -\gamma_1 & 0 & \beta_3 \\ \beta_1 & -\gamma_2 & 0 \end{bmatrix}, \tag{3.15}$$

where β and γ denote non–negative vectors which are defined by

$$\beta_i := b_i - a_i > 0, \quad \gamma_i := a_i - c_i \geq 0, \quad i \in I. \tag{3.16}$$

For a given RSP–game, the decomposition (3.13) is unique. A_0 will be called the *interactive component* of A, whereas A_1 will be addressed as the *non–interactive component* of A. If A coincides with A_0, i.e., if its non–interactive component is zero, it will be called an *essential RSP–game* or an RSP–game in *essential form*.

It is obvious from the definition of mean fitness that it splits according to

$$\overline{F}(p) = \overline{F}_0(p) + \overline{F}_1(p), \quad p \in \Delta, \tag{3.17}$$

where $\overline{F}_0$ and $\overline{F}_1$ denote the mean fitness functions induced by A_0 and A_1. $\overline{F}_0(p)$ will be called the *interactive component* of mean fitness at p.

Let $p*$ denote the unique interior Nash strategy of A. Since A is pl–equivalent to A_0, p is also the unique interior Nash strategy of A_0. We shall now characterize $p*$ in terms of the entries of A_0, which are given by (3.16).

For the rest of this paper, the determinant of A_0 will play a crucial role. It is easy to see that it is given by

$$\det(A_0) = \beta_1\beta_2\beta_3 - \gamma_1\gamma_2\gamma_3. \tag{3.18}$$

(a) $\det(A_0) = 0$:

In order to characterize $p*$, let us first consider the case $\det(A_0) = 0$. In view of (3.18), this is equivalent to $\gamma_1\gamma_2\gamma_3 = \beta_1\beta_2\beta_3 > 0$ which implies that the vector γ is strictly positive.

On the other hand, the kernel of A_0, $\ker(A_0)$, is nontrivial, i.e., there exists a real vector $y \in \mathbb{R}^3$, $y \neq 0$, such that $A_0 y = 0$. In coordinates, this is equivalent to

$$\beta_2 y_2 = \gamma_3 y_3, \quad \beta_3 y_3 = \gamma_1 y_1, \quad \beta_1 y_1 = \gamma_2 y_2. \tag{3.19}$$

In view of $\beta_i > 0$ and $\gamma_i > 0$, all components of vectors in $\ker(A_0)$ have the same sign. Accordingly, the linear space $\ker(A_0)$ corresponds to a straight line through 0 which completely belongs to $\mathbb{R}^3_+ \cup \mathbb{R}^3_-$, the union of the positive and the negative orthants of $\mathbb{R}^3$. Each such line has a unique intersection point with the strategy simplex. Let us denote the unique element of

$\Delta \cap \ker(A_0)$ by p^*. Since p^* belongs to the kernel of A_0, the fitness vector at p coincides with the zero vector. In view of (2.11) and Theorem 3.4.2., this implies that p is the unique interior Nash equilibrium strategy of A_0.

We have shown that p^* is characterized by setting $p^* = y$ in (3.19) together with the condition that its components are positive and sum up to 1.

(b) $\det(A_0) \neq 0$:

Let us now assume that $\det(A_0) \neq 0$. This implies the existence of A_0^{-1}, the inverse of the matrix A_0. (2.3), (2.6) and (2.11) applied to A_0 and F_0 lead to

$$p^* = \lambda A_0^{-1} 1, \text{ where } \lambda = \overline{F}_0(p^*) = (1 \cdot A_0^{-1} 1)^{-1} \qquad (3.20)$$

A simple calculation shows that

$$A_0^{-1} 1 = \frac{1}{\det(A_0)} \begin{bmatrix} \beta_2\beta_3 + \gamma_2\gamma_3 + \beta_3\gamma_2 \\ \beta_3\beta_1 + \gamma_3\gamma_1 + \beta_1\gamma_3 \\ \beta_1\beta_2 + \gamma_1\gamma_2 + \beta_2\gamma_1 \end{bmatrix}. \qquad (3.21)$$

Note that for $\det(A_0) = 0$ as well as for $\det(A_0) \neq 0$ the following holds true:

$$\text{sgn}[\overline{F}_0(p^*)] = \text{sgn}[\det(A_0)] = \text{sgn}[\beta_1\beta_2\beta_3 - \gamma_1\gamma_2\gamma_3], \qquad (3.22)$$

i.e., the determinant of A_0 has the same sign as the interactive component of the mean equilibrium payoff.

4. Evolutionary Stability in RSP–Games

In this section, we shall characterize the evolutionarily stable strategies of RSP–games. It will be shown that a Nash equilibrium strategy p^* is an ESS if and only if $\beta_{i+1} > \gamma_i$ holds true for all $i \in I$ and the terms $\beta_{i+1} - \gamma_i$ do not differ too much from one another in a sense to be made precise below. For p to be an ESS it is necessary but not sufficient that $\det(A_0)$, the determinant of the interactive component of the RSP–game A, is strictly positive.

DEFINITION 4.1: Evolutionarily stable strategies.

A mixed strategy $p^* \in \Delta$ is an *evolutionarily stable strategy* or *ESS* if it satisfies the following two conditions:

1. *Nash condition :*
$$F(p^*, p^*) \geq F(q, p^*) \quad \text{for all } q \in \Delta. \qquad (4.1a)$$

2. *ESS condition :*
$$F(p^*, p^*) = F(q, p^*) \text{ for } q \neq p^* \text{ implies } F(p, q) > F(q, q). \qquad (4.1b)$$

For the motivation of this concept see Maynard Smith (1982) or Weissing (1983). Since an ESS is a Nash equilibrium strategy satisfying the additional condition (4.1b), the ESS concept can formally be interpreted as a 'refinement' of the Nash equilibrium concept for symmetric normal form games (e.g., van Damme 1987).

A comparison of (4.1) with (3.9) shows that the concept of an evolutionarily stable strategy is invariant with respect to positive linear transformations of the payoff matrix. This implies that the unique interior Nash strategy of an RSP–game A is an ESS of A if and only if it is an ESS with respect to the interactive component A_0 of A.

The following theorem provides a characterization of evolutionarily stable strategies by means of a single 'local' condition, which plays an important role in the study of dynamic stability with respect to the continuous replicator dynamics. A proof of this theorem may be found in Hofbauer & Sigmund (1988) or in Weissing (1983).

THEOREM 4.2: Characterization of evolutionarily stable strategies.

1. p^* is an ESS if and only if there exists a neighbourhood U of p^* in Δ such that

$$\mathcal{F}(p^*,q) > \mathcal{F}(q,q) \text{ for all } q \in U, \ q \neq p^*. \tag{4.2}$$

2. An interior Nash equilibrium strategy is an ESS if and only if (4.2) holds true globally, i.e., if it holds true for $U = \Delta$.

In Weissing (1990) it is shown that it is useful to introduce some notions of evolutionary instability. Specializing the definitions given there to the case of an *interior* Nash equilibrium strategy we get:

DEFINITION 4.3: Uniform evolutionary instability.

Let p^* denote an interior Nash strategy of the evolutionary normal form game A.
1. p^* is called *uniformly evolutionarily unstable* if the following holds true:

$$\mathcal{F}(p^*,q) \leq \mathcal{F}(q,q) \text{ for all } q \in \Delta. \tag{4.3}$$

2. p^* is called *definitely evolutionarily unstable* if

$$\mathcal{F}(p^*,q) < \mathcal{F}(q,q) \text{ for all } q \in \Delta, \ q \neq p^*. \tag{4.4}$$

3. p^* is called a *definite* Nash strategy, if it is either evolutionarily stable or definitely evolutionarily unstable.
4. p^* is called *neutral Nash strategy* if

$$\mathcal{F}(p,q) = \mathcal{F}(q,q) \text{ for all } q \in \Delta. \tag{4.5}$$

The terminology chosen is motivated by the next proposition where it is shown that the definiteness of an interior Nash strategy corresponds to the definiteness of the quadratic form which is induced by the payoff matrix A. In order to show this, let us interpret the population states $q \neq p$ as disturbances of the equilibrium state p^*. Accordingly, they may be described as displacements of the form $q = p^* + x$, where x is an element of the hyperplane

$$\mathbb{R}_0^n := \{\, x \in \mathbb{R}^n \mid \sum_i x_i = 0 \,\}, \qquad (4.6)$$

which may be interpreted as the 'tangent plane' to the strategy simplex Δ. Representing q in the form $q = p^* + x$ and using the definition of the fitness function, we get:

$$\mathcal{F}(q,q) - \mathcal{F}(p^*,q) = x \cdot A p^* + x \cdot A x. \qquad (4.7)$$

For an interior fixed point, (2.11) yields

$$x \cdot A p^* = \lambda\, x \cdot 1 = 0 \quad \text{for all } x \in \mathbb{R}_0^n. \qquad (4.8)$$

From Theorem 4.2 together with (4.7) and (4.8), we get the following result:

PROPOSITION 4.4: Characterization of interior ESS's.

Let p denote a completely mixed Nash equilibrium strategy of the evolutionary normal form game A. Then the following holds true:

1. p is an interior ESS if and only if the quadratic form induced by A is negative–definite when restricted to $\mathbb{R}_0^n$, i.e., if and only if

$$x \cdot A x < 0 \quad \text{for all } x \in \mathbb{R}_0^n, \ x \neq 0. \qquad (4.9)$$

2. p is uniformly evolutionarily unstable if and only if this quadratic form is positive–semidefinite on $\mathbb{R}_0^n$. It is definitely evolutionarily unstable if and only if the form is positive–definite, and it is a neutral Nash strategy if and only if the form is identical to zero on this subspace.

There is a canonical isomorphism between $\mathbb{R}_0^n$ and $\mathbb{R}^{n-1}$ which may be characterized by the $n \times (n-1)$–matrix

$$P = \begin{bmatrix} 1 & 0 & \cdots & 0 \\ \vdots & \vdots & \ddots & \vdots \\ 0 & 0 & \cdots & 1 \\ -1 & -1 & \cdots & -1 \end{bmatrix}, \quad p_{ij} := \begin{cases} \delta_{ij} & \text{for } i < n \\ -1 & \text{for } i = n \end{cases} \qquad (4.10)$$

(δ_{ij} denotes the 'Kronecker delta'). With the help of P it is possible to transform the quadratic form $\mathbf{x} \cdot A\mathbf{x}$ on $\mathbb{R}_0^n$ to a quadratic form $\mathbf{y} \cdot A \, \mathbf{y}$ on $\mathbb{R}^{n-1}$. According to the transformation rules for quadratic forms, it is given by the $(n-1) \times (n-1)$–matrix

$$A^* = P^T A P . \tag{4.11}$$

It is easy to see that the entries of $A^* = (a_{ij}^*)$ are given by

$$a_{ij}^* := a_{ij} - a_{in} - a_{nj} + a_{nn} . \tag{4.12}$$

Let now A denote an RSP–game, the interactive component of which is given by A_0, and let $\mathbf{p}$ denote the unique interior Nash strategy of these games. Note that

$$P^T A P = P^T A_0 P , \tag{4.13}$$

i.e., the quadratic forms on $\mathbb{R}^{n-1}$ induced by A and A_0 do not differ from another.

Theorem 4.4 shows that the definiteness of the Nash strategy $\mathbf{p}^*$ corresponds to the definiteness of the quadratic form induced by A^* on $\mathbb{R}^{n-1}$. It is well–known that the sign of this form can be derived from the sign of the eigenvalues of the symmetric matrix $S^* := A^* + (A^*)^T$: the quadratic form induced by A^* is positive (negative) definite if and only if all eigenvalues of S^* are positive (negative).

The sign of the eigenvalues of a symmetric 2x2–matrix can easily be derived from the signs of its trace and its determinant. Since the determinant of S^*, $\det(S^*)$, is equal to the product of its eigenvalues while its trace, $\mathrm{tr}(S^*)$, corresponds to their sum, we obtain the following result:

<u>COROLLARY 4.5</u>:

1. The interior fixed point $\mathbf{p}^*$ of an RSP–game is a definite Nash strategy if and only if $\det(S^*) > 0$.
2. $\mathbf{p}^*$ is an ESS if and only if $\det(S^*) > 0 > \mathrm{tr}(S^*)$.
3. $\mathbf{p}^*$ is definitely evolutionarily unstable if and only if $\det(S^*) > 0$ and $\mathrm{tr}(S^*) > 0$.
4. $\mathbf{p}^*$ is a neutral Nash strategy if and only if $\det(S^*) = \mathrm{tr}(S^*) = 0$.

A simple calculation shows that S^* is of the form

$$S^* = \begin{bmatrix} -2\delta_3 & \delta_1 - \delta_2 - \delta_3 \\ \delta_1 - \delta_2 - \delta_3 & -2\delta_2 \end{bmatrix} , \tag{4.14}$$

where the δ_i, $i = 1,2,3$, denote the terms

$$\delta_i := \beta_{i+1} - \gamma_i . \tag{4.15}$$

Trace and determinant of S^* are given by

$$\text{tr}(S^*) = -2(\delta_2+\delta_3) , \qquad (4.16)$$

$$\det(S^*) = 4\delta_2\delta_3 - (\delta_1-\delta_2-\delta_3)^2. \qquad (4.17)$$

Since we are free to rename pure strategies, there is no loss in generality if we assume

$$|\delta_1| \geq |\delta_2| \geq |\delta_3| . \qquad (4.18)$$

In view of the inequality between the arithmetic and the geometric mean it is easy to derive the following implication from (4.17):

$$\det(S^*) > 0 \implies \text{sgn}(\delta_1) = \text{sgn}(\delta_2) = \text{sgn}(\delta_3) \neq 0 . \qquad (4.19)$$

(3.18) together with (4.15) shows that $\det(A_0) > 0$, if all δ_i are positive. On the other hand, $\det(A_0) < 0$ holds true, if all the δ_i are negative. Therefore, (4.19) can be strengthened to

$$\det(S^*) > 0 \implies \text{sgn}(\delta_i) = \text{sgn}(\det(A_0)) \neq 0 \text{ for } i \in I . \qquad (4.20)$$

If all δ_i have the same sign, the determinant of S^* may be written as

$$\det(S^*) = [(\sqrt{|\delta_2|}+\sqrt{|\delta_3|})^2-|\delta_1|] \cdot [|\delta_1|-(\sqrt{|\delta_2|}-\sqrt{|\delta_3|})^2] . \qquad (4.21)$$

It is obvious from (4.18) that the second factor in (4.21) is positive. If all δ_i have the same sign, we therefore get

$$\text{sgn}[\det(S^*)] = \text{sgn}[(\sqrt{|\delta_2|}+\sqrt{|\delta_3|}) - \sqrt{|\delta_1|}] . \qquad (4.22)$$

(4.22) motivates to introduce the term

$$\sigma(A_0) := \frac{\sqrt{|\delta_1|}}{(\sqrt{|\delta_2|} + \sqrt{|\delta_3|})} , \qquad (4.23)$$

which will be called the *skewness* of the RSP–game A_0. In view of (4.18), the skewness of A may be interpreted as a measure for the variance between the three numbers $\sqrt{|\delta_1|}$, $\sqrt{|\delta_2|}$, and $\sqrt{|\delta_3|}$. Note that $\sigma(A_0)$ is smaller than one if and only if

$$(\sqrt{|\delta_2|} + \sqrt{|\delta_3|}) > \sqrt{|\delta_1|} \geq \sqrt{|\delta_2|} \geq \sqrt{|\delta_3|}, \qquad (4.24)$$

i.e., if and only if the three numbers $\sqrt{|\delta_1|}$, $\sqrt{|\delta_2|}$, and $\sqrt{|\delta_3|}$ correspond to the lengths of the sides of a triangle. (Josef Hofbauer made me aware of this fact.)

The combination of (4.20), (4.22), and (4.23) yields

$$\det(S^*) > 0 \iff \sigma(A_0) < 1 \text{ and } \text{sgn}(\delta_i) = \text{sgn}(\det(A_0)) \neq 0. \qquad (4.25)$$

On the other hand, $\det(S^*) = 0$ in combination with $\operatorname{tr}(S^*) = 0$ leads to

$$S^* = 0 \iff \operatorname{sgn}(\delta_i) = \operatorname{sgn}(\det(A_0)) = 0, \ i \in I. \tag{4.26}$$

In view of Corollary 4.5, (4.25) and (4.16) together with (4.26) yield the main result of this section:

<u>THEOREM 4.6</u>: Evolutionary stability in RSP–games.

1. The unique interior Nash strategy $\mathbf{p}^*$ of an RSP–game A is an ESS if and only if

$$\sigma(A_0) < 1 \quad \text{and} \quad \operatorname{sgn}[\beta_{i+1} - \gamma_i] = \operatorname{sgn}[\det(A_0)] > 0 \quad \text{for } i \in I. \tag{4.27}$$

2. $\mathbf{p}^*$ is definitely evolutionarily unstable if and only if

$$\sigma(A_0) < 1 \quad \text{and} \quad \operatorname{sgn}[\beta_{i+1} - \gamma_i] = \operatorname{sgn}[\det(A_0)] < 0 \quad \text{for } i \in I. \tag{4.28}$$

3. $\mathbf{p}^*$ is a neutral Nash strategy if and only if

$$\operatorname{sgn}[\beta_{i+1} - \gamma_i] = \operatorname{sgn}[\det(A_0)] = 0 \quad \text{for } i \in I. \tag{4.29}$$

Note that $\beta_{i+1} - \gamma_i$ corresponds to the *sum* of the binary payoffs which the players get in the RSP–game A_0 whenever an i–strategist is paired with an $(i+1)$–strategist. Accordingly, $\mathbf{p}$ is an ESS if and only if the skewness of A is smaller than one and if the *mean* payoffs to the players in A_0 is always positive whenever different pure strategies meet one another. In view of (3.21), it is a necessary condition for $\mathbf{p}^*$ to be an ESS that the interactive component of mean fitness in $\mathbf{p}^*$ is positive. Notice also that (4.29) is equivalent to

$$A_0 + A_0^T = 0, \tag{4.30}$$

i.e., $\mathbf{p}^*$ is a neutral Nash strategy if and only if A_0 is a *zero–sum game*.

5. Stability with Respect to the Continuous Replicator Dynamics

In this section it will be shown that the stability behaviour of the interior fixed point $\mathbf{p}$ of an RSP–game A depends crucially on the sign of the determinant of its interactive component A_0: When $\det(A_0) > 0$, $\mathbf{p}^*$ is hyperbolically stable and even a global attractor. For $\det(A_0) = 0$, $\mathbf{p}^*$ is a global center, and the interior of the strategy simplex is filled with periodic orbits. When $\det(A_0) < 0$, $\mathbf{p}^*$ is hyperbolically unstable and a global repellor. This characterization will help to elucidate the discrepancies between evolutionary stability and 'continuous' dynamic stability since the sign of $\det(A_0)$ corresponds to the sign of the equilibrium payoff with respect to A_0.

In this section we deal with a continuous dynamical system, which is induced by an autonomous differential equation of the form

$$\dot{\mathbf{p}} = \mathbf{v}^A(\mathbf{p}) \ . \tag{5.1}$$

$\mathbf{v}^A \colon \Delta \to \mathbb{R}_0^n$ denotes a vector field on Δ where for each given point $\mathbf{p} \in \Delta$ the space $\mathbb{R}_0^n$ should be interpreted as the tangent space to Δ at $\mathbf{p}$. The vector field is continuously differentiable and given by the continuous replicator dynamics, i.e., by

$$v_i^A(\mathbf{p}) := p_i \left(F_i(\mathbf{p}) - \overline{F}(\mathbf{p}) \right), \quad i \in I \ . \tag{5.2}$$

A comparison of (5.2) with (3.10) shows that the continous replicator dynamics is essentially invariant with respect to positive linear transformations of the payoff matrix A. In fact, if the payoff matrices A and $\tilde{A}$ are related according to (3.7) we have

$$\mathbf{v}^{\tilde{A}}(\mathbf{p}) = \lambda \, \mathbf{v}^A(\mathbf{p}) \quad \text{for all } \mathbf{p} \in \Delta \ , \tag{5.3}$$

i.e., the tangent vectors given by the two vector fields are collinear for all $\mathbf{p}$. This implies that the dynamical systems induced by these two vector fields have the same phase portrait. The orbits of the two systems coincide – they are only passed with different velocities.

In particular, an RSP–game A is pl–equivalent to its interactive component A_0. In this case, we even have $\lambda = 1$, which implies that A and A_0 induce identical continuous replicator dynamics. Without loss in generality, we may therefore assume that A coincides with A_0, i.e., that it is an RSP–game in essential form.

In the rest of this section, we shall mainly be interested in the stability of fixed points: A fixed point $\mathbf{p}$ of a continuous dynamical system is called (*Lyapunov*) *stable* if orbits starting near $\mathbf{p}^*$ do not go too far away from $\mathbf{p}^*$. A fixed point $\mathbf{p}^*$ is called *attractive* if nearby starting orbits are attracted by $\mathbf{p}^*$. $\mathbf{p}^*$ is called *neutrally stable* if it is stable but not attractive, and it is called *asymptotically stable* if it is both stable and attractive. An interior fixed point $\mathbf{p}^*$ will be called a *global attractor* if it is stable and if it attracts *all* interior orbits. It will be called a *global repellor* if all interior non–equilibrium orbits converge to the boundary of the strategy simplex. Finally, it will be called a *global center* if it is neutrally stable and if the interior of the strategy simplex is filled with periodic orbits. Formal definitions of these concepts of dynamic stability can be found in every textbook on ordinary differential equations (e.g., Hirsch & Smale 1974).

A rather elegant technique for proving stability as well as instability of a fixed point $\mathbf{p}^*$ of a dynamical system is that of constructing a Lyapunov function for it. Formally, a *Lyapunov function* is a smooth scalar function V which has a strict local maximum in $\mathbf{p}^*$ and which has locally a *definite* sign along the orbits of the dynamical system. In order to make the last condition more precise, let $\{\mathbf{p}(t)\}$ denote an orbit of (5.1). At each point $\mathbf{p} = \mathbf{p}(t)$, the derivative of the map $t \mapsto V(\mathbf{p}(t))$ is given by

$$\dot{V}(p) = \sum_i \frac{\partial V}{\partial x_i}(p) \, v_i^A(p) = \text{grad } V(p) \cdot v^A(p). \qquad (5.4)$$

V increases near p^* along the orbits of (5.1), if and only if $\dot{V}(p) > 0$ for all $p \neq p^*$ from a neighbourhood U of p^*. If V has a strict local maximum in p^*, the fixed point p is asymptotically stable in this case. p^* is unstable, if V decreases along nearby orbits, i.e., if $\dot{V}(p) < 0$ for $p \in U$, $p \neq p^*$. It is neutrally stable, if V is a *constant of motion* near p , i.e., if $\dot{V}(p) = 0$ for $p \in U$. All these results and their implications for global stability may be found in many textbooks on dynamical systems (e.g., Bhatia & Szegö 1967).

Let p^* be an interior fixed point of (5.1). The scalar function $V: \Delta \longrightarrow \mathbb{R}$ defined on int(Δ), the interior of Δ, by

$$V(q) := \prod_i (q_i)^{p_i^*} \qquad (5.5)$$

is a promising candidate for a Lyapunov function for the continuous replicator dynamics (5.1). In fact, it has a unique maximum in p^*, and its time derivative at an interior strategy q is given by

$$\dot{V}(q) = V(q) \, (\mathcal{F}(p^*,q) - \mathcal{F}(q,q)). \qquad (5.6)$$

This shows that the function defined by (5.5) is a Lyapunov function for p^*, whenever p is a definite Nash equilibrium strategy. Considering the remarks above, we get the following result, which is an obvious generalization of a theorem first proved by Hofbauer, Schuster & Sigmund (1979) and Zeeman (1980):

THEOREM 5.1: Evolutionary stability and dynamic stability.

For a definite or a neutral Nash equilibrium strategy p^* of an evolutionary normal form game A the notions of evolutionary stability and dynamic stability with respect to (5.1) coincide in the following sense:
1. If p^* is an ESS of A, it is an asymptotically stable fixed point of (5.1).
2. If p^* is definitely evolutionarily unstable, it is a repellor with respect to (5.1).
3. If p^* is a neutral Nash strategy, it is neutrally stable for (5.1).

For an interior definite Nash strategy, the local stability properties derived in Theorem 5.1 translate into global stability properties. In order to see this, let us write q in (5.6) in the form $q = p^* + x$, where $x \in \mathbb{R}_0^n$. Considering (4.7), (4.8), and (4.13), we get

$$\dot{V}(p^* + x) = -V(p^* + x) \, x \cdot Ax = -V(p^* + x) \, x \cdot A_0 x. \qquad (5.7)$$

Proposition 4.4 shows that the scalar function V defined by (5.6) is a *global* Lyapunov function for p , whenever p^* is a definite interior Nash strategy. This yields:

<u>COROLLARY 5.2</u>: Interior ESS's and global dynamic stability.

Let p denote an interior fixed point of the evolutionary normal form game A.
1. If p^* is an ESS of A, it is a global attractor of (5.1).
2. If p^* is definitely evolutionarily unstable, it is a global repellor for (5.1).
3. If p^* is a neutral Nash strategy, it is a global center for (5.1).

Let now A denote an RSP–game, let A_0 be its interactive component, and let δ_i, $i = 1,2,3$, be defined by (4.15), i.e., $\delta_i := \beta_{i+1} - \gamma_i$. Recall that the skewness of A_0, $\sigma(A_0)$, is given by (4.23). Let us set $\sigma(A_0) := 1/2$, if $\delta_1 = \delta_2 = \delta_3 = 0$. In view of Theorem 4.6, A will be called a *balanced RSP–game* if

$$\sigma(A_0) < 1 \text{ and } \operatorname{sgn}[\beta_{i+1} - \gamma_i] = \operatorname{sgn}[\det(A_0)]. \text{ for all } i \in I. \tag{5.8}$$

Together with Theorem 4.6, Corollary 5.2 yields:

<u>PROPOSITION 5.3</u>: Stability in balanced RSP–games.

Let p^* denote the unique interior fixed point of the balanced RSP–game A. Then the following holds true:
1. p is an ESS if and only if $\det(A_0) > 0$. In this case, p^* is a global attractor with respect to the continuous replicator dynamics (5.1).
2. p is definitely evolutionarily unstable if and only if $\det(A_0) < 0$. In this case, it is a global repellor of (5.1).
3. p is a neutral Nash strategy if and only if $\det(A_0) = 0$. In this case, it is neutrally stable. In fact, it is a global center for (5.1).

It will be shown below that the dynamical features described in the preceding theorem are quite typical for RSP–games: the unique interior fixed point is either a global attractor, or a global repellor, or a global center with respect to the continuous replicator dynamics. The perfect correspondence between evolutionary stability and 'continuous' dynamic stability will be lost, however, as soon as we leave the class of balanced RSP–games.

In order to show this, we shall first consider a subclass of the class of balanced RSP–games. An RSP–game A will be called a *perfectly balanced RSP–game*, if the δ_i do not differ from one another, i.e., if

$$\delta_i = \beta_{i+1} - \gamma_i = \text{const} =: \delta. \tag{5.9}$$

Note that we have $\sigma(A_0) = 1/2 < 1$ for a perfectly balanced RSP–game and it is clear that the parameter δ has the same sign as $\det(A_0)$. This justifies the terminology chosen: every perfectly balanced RSP–game is a balanced game.

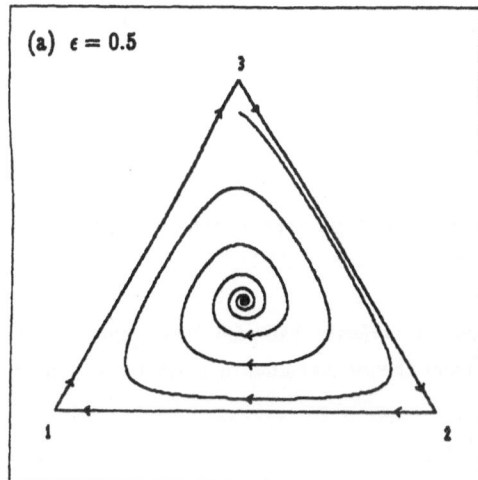 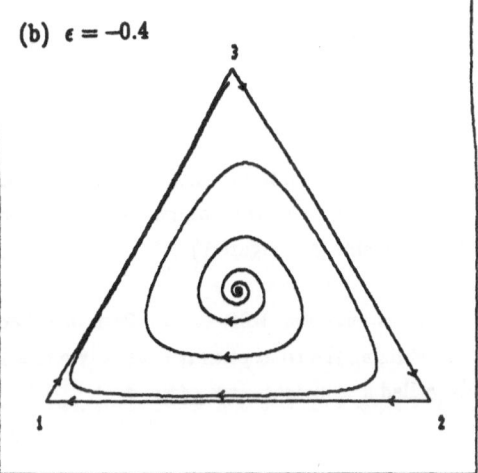

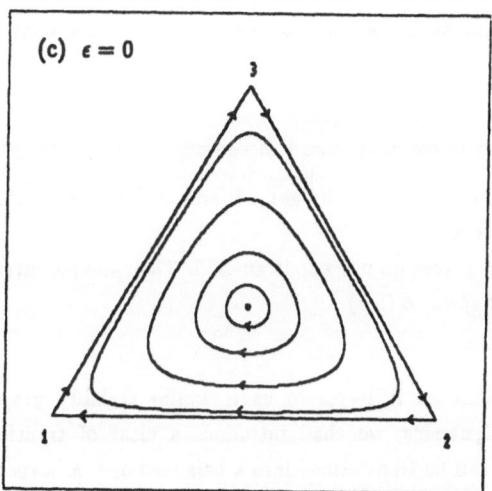

FIGURE 1:   Continuous dynamic stability in ϵ–perturbed Rock–Scissors–Paper games.

(a) $\epsilon > 0$:   p* is an ESS and a global attractor.
(b) $\epsilon < 0$:   p* is definitely evolutionarily unstable and a global repellor.
(c) $\epsilon = 0$:   p* is a neutral Nash strategy and a global center.

The ϵ-perturbed Rock–Scissors–Paper games defined by (1.4) form a typical class of perfectly balanced RSP–games. In that case, δ corresponds to the parameter ϵ, and we have

$$\text{sgn}[\det(A(\epsilon))] = \text{sgn}(\epsilon) .\qquad (5.10)$$

Accordingly, the interior fixed point of an ϵ-perturbed Rock–Scissors–Paper game is an ESS if and only if $\epsilon > 0$, and the notions of evolutionary stability and asymptotic stability with respect to (5.1) coincide (see Figure 1).

Constant–sum RSP–games form another class of perfectly balanced RSP–games. Slightly generalizing the concept as it is usually defined in game theory, a symmetric normal form game will be called a *constant–sum game* if

$$A_0 = -A_0^{\mathrm{T}},\qquad (5.11)$$

i.e., if its interactive component A_0 is a zero–sum game. From what has been shown above, we get:

CoROLLARY 5.4: Stability in constant–sum RSP–games.

1. An RSP–game A is a constant–sum game if and only if it is a balanced RSP–game for which $\det(A_0) = 0$ holds true.
2. Constant–sum RSP–games do not admit an ESS. Their unique interior fixed point is always a global center with respect to (5.1).

In order to show that all RSP–games have similar stability properties with respect to the continuous replicator dynamics, we shall introduce a class of transformations by which every unbalanced RSP–game can be transformed into a balanced one: A *barycentric transformation* of the strategy simplex is a homeomorphism $x \mapsto {}^{\pi}x$ from Δ to itself, which is induced by a positive vector π and which is defined by

$${}^{\pi}x_i := \frac{x_i \pi_i}{x \cdot \pi} \quad (\pi_i > 0 \text{ for all } i \in I) .\qquad (5.12)$$

It is easy to see that — up to a change in velocity — the transformation defined by (5.12) transforms the replicator dynamics (5.2) into another replicator dynamics which is induced by the payoff matrix (see Zeeman 1980)

$$A^{\pi} := A \cdot \text{diag}(\pi^{-1})\qquad (5.13)$$

(π^{-1} denotes the vector $\pi^{-1} := (\pi_1^{-1}, \ldots, \pi_n^{-1})$, and $\text{diag}(y)$ denotes a diagonal matrix, the diagonal of which is given by the vector y).

The elements of $A^\pi = (a^\pi_{ij})$ are characterized by

$$a^\pi_{ij} = a_{ij}\pi_j^{-1}. \tag{5.14}$$

We say that A^π was derived from A *by means of a barycentric transformation.*

We have $\det(A^\pi) = \det(A)\cdot\det[\mathrm{diag}(\pi^{-1})]$. Therefore, the determinants of A and A^π are related via

$$\det(A^\pi) = \frac{1}{\pi_1\pi_2\pi_3}\det(A). \tag{5.15}$$

Since π is a positive vector, it is clear that A^π is an RSP–game in essential form, if and only if A is an RSP–game in essential form.

THEOREM 5.5: Transformation into perfectly balanced games.

Every RSP–game $A = A_0$ in essential form can be transformed into a perfectly balanced RSP–game B_0 by means of a barycentric transformation. B_0 is also a game in essential form, and the determinants of the RSP–games A_0 and B_0 do not differ in sign:

$$\mathrm{sgn}[\det(B_0)] = \mathrm{sgn}[\det(A_0)]. \tag{5.16}$$

PROOF:
Let the RSP–game $A = A_0$ be of the form

$$A_0 = \begin{bmatrix} 0 & \beta_2 & -\gamma_3 \\ -\gamma_1 & 0 & \beta_3 \\ \beta_1 & -\gamma_2 & 0 \end{bmatrix}, \tag{5.17}$$

where $\beta_i > 0$, $\gamma_i \geq 0$ for $i = 1,2,3$. Let us also consider an auxiliary RSP–game in essential form, C_0, which is defined by

$$C_0 = \begin{bmatrix} 0 & \gamma_1 & -\beta_2 \\ -\beta_3 & 0 & \gamma_2 \\ \gamma_3 & -\beta_1 & 0 \end{bmatrix}. \tag{5.18}$$

Let r denote the unique interior fixed point of C_0. By (2.11) this means that $C_0 r^*$ is a scalar multiple of 1:

$$C_0 r^* = \lambda 1, \quad \text{where } \lambda = r^* \cdot C_0 r^*. \tag{5.19}$$

After multiplication with (-1), the coordinate representation of (5.19) is given by

$$\beta_2 r_3^* - \gamma_1 r_2^* = -\lambda,$$
$$\beta_3 r_1^* - \gamma_2 r_3^* = -\lambda, \tag{5.20}$$
$$\beta_1 r_2^* - \gamma_3 r_1^* = -\lambda.$$

(5.20) motivates the following definition of B_0:

$$B_0 := \begin{bmatrix} 0 & \beta_2 r_3^* & -\gamma_3 r_1^* \\ -\gamma_1 r_2^* & 0 & \beta_3 r_1^* \\ \beta_1 r_2^* & -\gamma_2 r_3^* & 0 \end{bmatrix} . \tag{5.21}$$

We have $B_0 = A_0^T$, where the vector r is defined by

$$\pi_1 = (r_2^*)^{-1}; \ \pi_2 = (r_3^*)^{-1}; \ \pi_3 = (r_1^*)^{-1}. \tag{5.22}$$

(5.20) implies that B_0 is a perfectly balanced RSP–game: $\delta = -\lambda$, if $\delta = \delta(B_0)$ is defined by applying (4.15) to B_0. This shows that A_0 can be transformed into a perfectly balanced RSP–game by means of a barycentric transformation.

The only thing that remains to be shown is (5.16). This, however, follows directly from (5.15).

$$* * *$$

Let A_0 and B_0 be given as above. Since the phase portrait of v^{A_0} is the homeomorphic image of the phase portrait of v^{B_0}, we get the following corollary from Theorem 5.3, which – due to its importance – will be stated as a theorem:

THEOREM 5.6: 'Continuous' dynamic stability in RSP–games.

Let p denote the unique interior fixed point of the RSP–game A. Then the following holds true:
1. p is asymptotically stable for (5.1) if and only if $\det(A_0) > 0$. In this case, p^* is even a global attractor for the continuous replicator dynamics.
2. p^* is unstable for (5.1) if and only if $\det(A_0) < 0$. In this case, it is even a global repellor for the continuos replicator dynamics.
3. p is neutrally stable for (5.1) if and only if $\det(A_0) = 0$. In this case, it is a global center for the continuous replicator dynamics.

Theorem 5.6 is illustrated in Figure 2 by a one–parameter family of RSP–games which are given by payoff matrices of the form

$$A_0(\zeta) := \begin{bmatrix} 0 & 1 & \zeta-1 \\ \zeta-1 & 0 & 1 \\ 9 & \zeta-9 & 0 \end{bmatrix}, \ \zeta \le 1 . \tag{5.23}$$

It is easy to see that all these games have the same interior fixed point $p^* = (\frac{1}{3},\frac{1}{3},\frac{1}{3})$, and that the sign of $\det[A_0(\zeta)]$ is equal to the sign of ζ.

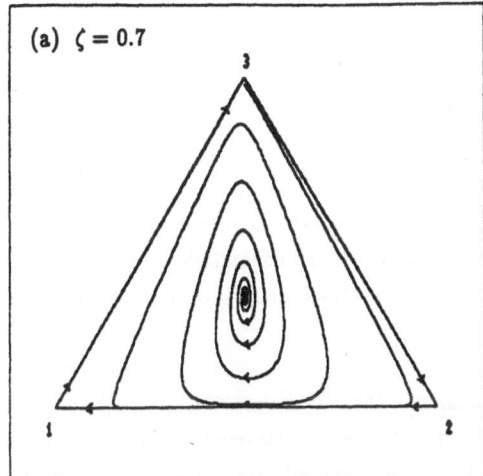

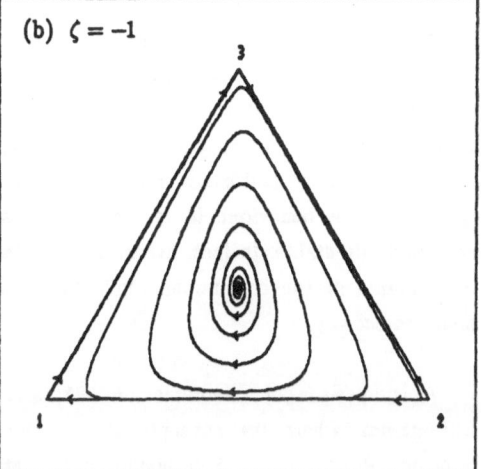

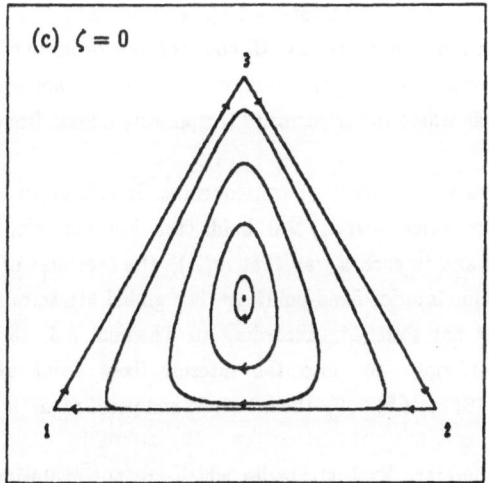

FIGURE 2:   Continuous dynamic stability in the one–parameter family of RSP–games which are given by (5.23).

(a) $\zeta > 0$: p^* is a global attractor but not an ESS.
(b) $\zeta < 0$: p^* is a global repellor but not definitely evolutionarily unstable.
(c) $\zeta = 0$: p^* is a global center but not a neutral Nash strategy.

Note that for $\zeta > 0$, p is a global attractor for the continuous replicator dynamics, but it is not an ESS since $\delta_3 = \zeta - 8$ is a negative number (see Theorem 4.6). A comparison of Figures 1(a) and 2(a) gives an idea of why p* is not an ESS: In Figure 2(a), the interior orbits approach the fixed point p* on spirals with a high 'eccentricity'. All interior trajectories converge to p*, but they do so in a non–monotonical way. Near the 'minor axis' of the spiral, a trajectory comes very close to p . Time and again, however, it departs from p* on its way to the 'major axis' of the spiral. Such a behaviour is typical for attractors of the replicator dynamics which are *not* evolutionarily stable. In fact, evolutionarily stable strategies may be characterized as those stable fixed points of the continuous replicator dynamics which attract nearby orbits (locally) in a *monotonical* way.

A comparison of Theorems 4.6, 5.3, and 5.6 shows that it is just the class of balanced RSP–games, where the concepts of evolutionary stability and 'continuous' dynamic stability coincide. Whenever an RSP–game with $\det(A_0) > 0$ is balanced, the continuous replicator dynamics generates orbits with an 'eccentricity' which is small enough to ensure that the interior fixed point is approached in a monotonical way. If, however, one or more of the δ_i are negative, or if the skewness of A_0 is larger than one, the interior fixed point is not evolutionarily stable since there are always time periods where the trajectories temporarily depart from p*.

These observations have some important consequences: It is easy to construct an RSP–game which is not balanced but for which $\det(A_0) > 0$ holds true. For example, it is sufficient to choose δ_1, δ_2, and δ_3 all positive and in such a way that $\sigma(A)$, the skewness of A, is larger than one. For *any* such game, the unique interior fixed point p* is a global attractor (Theorem 5.6), but it is not an ESS since (4.27) is not satisfied. According to Theorem 5.5, there exists a barycentric transformation which transforms p* into the interior fixed point of a perfectly balanced RSP–game B. Since by (5.16) $\det(B) > 0$, the interior fixed point of B is an ESS (Theorem 5.3).

We have shown that there are Nash strategies which are not evolutionarily stable but which can nevertheless be transformed into an ESS (and vice versa) by means of a barycentric transformation of the state space – although barycentric transformations leave the dynamical features of the continuous replicator dynamics essentially invariant. This result, which is illustrated by Figure 3, will be stated as a Corollary:

COROLLARY 5.7: Evolutionary stability and barycentric transformations.

The concept of evolutionary stability is not invariant with respect to barycentric transformations of the strategy simplex.

We shall close this section by having a look at the linearization of (5.1) at the interior fixed point p . Let

$$D := \mathrm{D}\mathbf{v}^A(\mathbf{p}^*) \qquad (5.24)$$

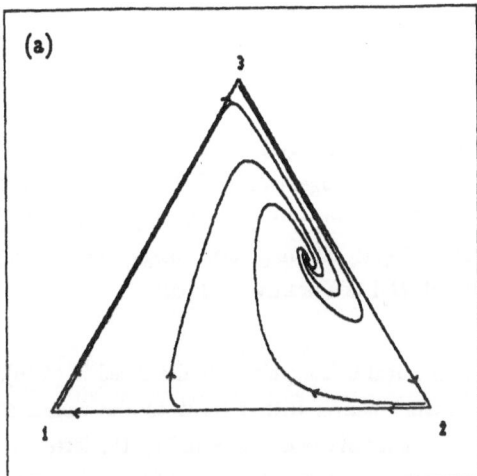 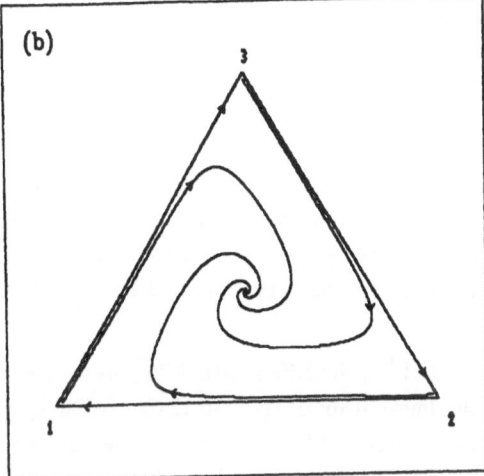

FIGURE 3:   Transformation of an evolutionarily unstable fixed point into an ESS by means of a barycentric transformation of the state space.

(a)   Unbalanced RSP–game with payoff matrix and interior fixed point given by:

$$A = A_0 = \begin{bmatrix} 0 & 1 & 0 \\ 0 & 0 & 1 \\ 5 & 0 & 0 \end{bmatrix}, \quad \mathbf{p}^* = (\tfrac{1}{11}, \tfrac{5}{11}, \tfrac{5}{11}).$$

$\mathbf{p}^*$ is a global attractor, since $\det(A_0) = 5 > 0$.

$\mathbf{p}^*$ is not an ESS, since $\sigma(A_0) = \sqrt{5}/2 > 1$.

(b)   Perfectly balanced RSP–game with payoff matrix and interior fixed point given by:

$$B = B_0 = \begin{bmatrix} 0 & 1 & 0 \\ 0 & 0 & 1 \\ 1 & 0 & 0 \end{bmatrix}, \quad \mathbf{q}^* = (\tfrac{1}{3}, \tfrac{1}{3}, \tfrac{1}{3}).$$

$\mathbf{q}^*$ is a global attractor and an ESS of B.

Notice that B may be obtained from A by means of the barycentric transformation which is induced by the positive vector $\boldsymbol{\tau} = (5,1,1)$.

denote the derivative of the vector field v^A at p^*. It is well–known that the stability behaviour of p with respect to (5.1) is closely related to the sign of the real parts of the eigenvalues of D (see, e.g., Hirsch & Smale 1974): A fixed point is asymptotically stable, if the real parts of all eigenvalues of D are negative, and it is unstable, if at least one of the eigenvalues of D has a positive real part. p^* is called a *hyperbolic fixed point*, if all the eigenvalues of D have a non–zero real part. A hyperbolic fixed point is stable if and only if all the eigenvalues of D are lying in the left half of the complex plane. In such a case, nearby orbits converge to p^* at an exponential rate, and the fixed point will be called *hyperbolically stable*. Slightly abusing terminology, p^* will be called *hyperbolically unstable*, if at least one eigenvalue of D has a positive real part.

If $\mathbb{R}_0^n$ is identified with $\mathbb{R}^{n-1}$ by means of the canonical isomorphism P described in (4.10), the linear map D can be characterized by its *Jacobian matrix* with respect to the canonical coordinates of $\mathbb{R}^{n-1}$. It should cause no confusion, if this matrix is also denoted by the letter D. It is shown in Weissing (1983) that the $(n-1)\times(n-1)$ Jacobian matrix $D = (d_{ij})$ of the vector field (5.2) at an interior fixed point is given by

$$d_{ij} := p_i^* \left(a_{ij}^* - \sum_k p_k^* a_{kj}^* \right),$$ (5.25)

where the summation extends over $k \in I$, $k < n$. $A^* = (a_{ij}^*)$ denotes the $(n-1)\times(n-1)$ matrix given by (4.11). Note that in view of (4.13) it does not matter whether we consider the linearisation with respect to A or with respect to A_0.

It is obvious from (5.25) that the Jacobian D takes an especially simple form, if p^* coincides with the *barycenter* m of the strategy simplex, which is defined by

$$m := \tfrac{1}{3} \cdot 1 = (\tfrac{1}{3},\tfrac{1}{3},\tfrac{1}{3}).$$ (5.26)

Following Zeeman (1980), an RSP–game will be called a *central game*, if its unique interior fixed point coincides with the barycenter m. In view of (2.11), an RSP–game A is a central game if and only if its row sums are equal to one another. The interactive component A_0 of a central RSP–game A is also a central game. More precisely: A is a central game if and only if

$$\eta_i = \beta_i - \gamma_{i+1} = \text{const} =: \eta, \quad \text{for } i \in I.$$ (5.27)

A simple calculation under consideration of (5.25) and (4.13) shows that for a central RSP–game the Jacobian D at the interior fixed point $p^* = m$ is of the form

$$D = \tfrac{1}{9}\begin{bmatrix} -\beta_1 + \beta_3 + \gamma_1 + 2\gamma_3 & 2\beta_2 + \beta_3 + \gamma_2 + 2\gamma_3 \\ -\beta_1 - 2\beta_3 - 2\gamma_1 - \gamma_3 & -\beta_2 - 2\beta_3 + \gamma_2 - \gamma_3 \end{bmatrix}.$$ (5.28)

Using (5.27), it is easy to see that the trace and the determinant of D are given by

$$\text{tr}(D) = -\tfrac{1}{3}\,\eta, \tag{5.29}$$

$$\det(D) = \tfrac{1}{3}(\,\eta^2 + \beta\cdot\gamma\,). \tag{5.30}$$

It is well–known that the trace of a matrix corresponds to the sum of its eigenvalues while the determinant is identical to their product. Note that (5.29) and (5.30) imply

$$\det(D) > [\text{tr}(D)]^2. \tag{5.31}$$

From this, we get immediately that the eigenvalues of D are complex conjugate to another. Correspondingly, their real parts have the same sign as $\text{tr}(D)$. In view of (5.29), the interior fixed point $\mathbf{m}$ is hyperbolically stable if $\eta > 0$, and it is hyperbolically unstable if $\eta < 0$.

For any vector $\mathbf{y} \in \mathbb{R}^3$ let $\bar{y} \in \mathbb{R}$ denote *the arithmetic mean of* $\mathbf{y}$, i.e.,

$$\bar{y} := \tfrac{1}{3}\textstyle\sum_i y_i = \mathbf{m}\cdot\mathbf{y}. \tag{5.32}$$

If $A = A(\mathbf{a},\mathbf{b},\mathbf{c})$ is a central RSP–game, and if β and γ are given by (3.16), we get from (5.27):

$$\eta = \bar{\eta} = \bar{\beta} - \bar{\gamma} = \bar{b} + \bar{c} - 2\bar{a}. \tag{5.33}$$

Combining this with the previous arguments, we have shown that:

PROPOSITION 5.8: Hyperbolic stability in central RSP–games.

Let $A = A(\mathbf{a},\mathbf{b},\mathbf{c})$ be a central RSP–game. Then the following holds true:
1. The interior fixed point $\mathbf{p}^* = \mathbf{m}$ of A is hyperbolically stable with respect to the continuous replicator dynamics if and only if the arithmetic mean of the vectors $\mathbf{b}$ and $\mathbf{c}$ is larger than the arithmetic mean of the vector $\mathbf{a}$, i.e., if

$$\tfrac{1}{2}\cdot(\bar{b} + \bar{c}) > \bar{a}. \tag{5.34}$$

2. $\mathbf{p}$ is hyperbolically unstable if and only if $\bar{a}$ is larger than the mean of $\bar{b}$ and $\bar{c}$.

What has been shown above for central RSP–games can be applied to every RSP–game, since every RSP–game A can be *centralized*, i.e., transformed into a central RSP–game A^τ by means of a barycentric transformation. In fact, A^τ is a central game, if τ is defined by

$$\tau_i := (p_i^*)^{-1}, \quad i \in I, \tag{5.35}$$

where $\mathbf{p}$ denotes the unique interior fixed point of A.

(3.18) and (5.27) imply that η and $\det(A_0)$ have the same sign for central games:

$$\mathrm{sgn}[\det(A_0)] = \mathrm{sgn}(\eta) = \mathrm{sgn}[\frac{\bar{b}+\bar{c}}{2} - \bar{a}] . \qquad (5.36)$$

Correspondingly, the barycenter of a central game is hyperbolically stable if and only if $\det(A_0) > 0$. On the other hand, (5.15) shows that the sign of the determinant of an RSP–game is not changed by a barycentric transformation. Accordingly, the centralization of a given RSP–game and the application of Proposition 5.8 to it yields:

THEOREM 5.9: Hyperbolic stability in RSP–games.

Let A be an RSP–game, and let A_0 denote its interactive component. Then the following holds true:

1. The eigenvalues of the linearization at $\mathbf{p}^*$ are complex conjugate to one another.
2. The unique interior fixed point $\mathbf{p}^*$ of A is hyperbolically stable if and only if $\det(A_0) > 0$.
3. $\mathbf{p}$ is hyperbolically unstable if and only if $\det(A_0) < 0$.

Theorem 5.9 implies that a *Hopf bifurcation* (see, e.g., Marsden & McCracken 1976, Hassard, Kazarinoff & Wan 1981) occurs, whenever a change in one of the parameters of an RSP–game leads to a transition of the determinant of A_0 through zero (see, e.g., Figure 2). Theorem 5.6 shows that RSP–games do not admit *isolated* periodic orbits. This implies that the Hopf bifurcation necessarily is a *degenerate* one, since non–degenerate Hopf bifurcations imply the existence of an attracting or a repelling closed orbit (see, e.g., Marsden & McCracken 1976). The conclusion that RSP–games do not admit non–degenerate Hopf bifurcations for the continuous replicator dynamics does *not*, however, imply that RSP–games form a 'degenerate' class of evolutionary games. In fact, this phenonemon is typical for arbitrary evolutionary 3x3–games: Zeeman (1980) as well as Hofbauer (1981) have shown that *all* Hopf bifurcations are degenerate in the case $n = 3$. Isolated periodic orbits occur only in higher dimensions (i.e., $n \geq 4$).

6. Stability with Respect to the Discrete Replicator Dynamics

In this section, the stability of the interior fixed point $\mathbf{p}^*$ with respect to the discrete replicator dynamics (2.8) will be analysed. It will be shown that – like in the continuous time case – $\mathbf{p}^*$ is a global repellor if $\det(A_0)$, the determinant of the interactive component of A, is smaller than zero. In case that $\det(A_0) > 0$, however, the situation is different. In contrast to the continuous time case, stability of $\mathbf{p}^*$ can be affected by positive linear transformations of payoffs: $\mathbf{p}^*$ can always be stabilized *and* destablized with respect to (2.8) by means of a positive linear transformation of payoffs. In particular, this implies that evolutionary stability of $\mathbf{p}^*$ is neither necessary nor sufficient to ensure stability with respect to the discrete replicator dynamics.

The class of 'circulant' RSP–games will be analysed in some detail. For this class, the situation with respect to the discrete replicator dynamics is very similar to the picture that emerged for the continuous dynamics: local stability of the interior fixed point implies global hyperbolic stability, local instability implies global hyperbolic instability, and the parameter regions for stability and instability are separated by a submanifold of codimension one for which the interior fixed point is a global center.

To begin with, we shall give an equivalent representation of the discrete replicator equation (2.8) in terms of a *difference equation*:

$$\Delta p := p' - p = w^A(p). \tag{6.1}$$

In this form, the discrete replicator dynamics is – like in the continuous case – induced by a vector field $w^A: \Delta \longrightarrow \mathbb{R}_0^n$, the components of which are given by

$$w_i^A(p) := p_i \, \frac{F_i(p) - \overline{F}(p)}{\overline{F}(p)}, \quad i \in I. \tag{6.2}$$

A comparison of (6.2) with (5.2) shows that

$$w^A(p) := \frac{v^A(p)}{\overline{F}(p)}. \tag{6.3}$$

Accordingly, the two vector fields w^A and v^A are collinear, and for each simplex point p the vectors $v^A(p)$ and $w^A(p)$ only differ in their lengths.

A simple calculation shows that (6.2) is 'well–behaved' with respect to barycentric transformations of the state space Δ: a barycentric transformation (5.12) of the strategy simplex transforms (6.2) into another discrete replicator dynamics, which is induced by the payoff matrix (5.13). Since every RSP–game can be centralized, i.e., transformed into a *central* RSP–game by means of a barycentric transformation of Δ, we shall restrict our attention to central games. Recall that an RSP–game is central if and only if its unique interior fixed point p^* coincides with the barycenter m of the simplex.

The *stability concepts* used in this section should be understood analogously to those used in the continuous case. The linearization of (6.1) at $p^* = m$ will be denoted by J_0:

$$J_0 := Dw^A(p^*). \tag{6.4}$$

In view of (6.3) it is easy to see that $J_0 = Dw^A(p^*)$ is related to $D = Dv^A(p^*)$ by means of

$$J_0 = [\, \overline{F}(p) \,]^{-1} \cdot D. \tag{6.5}$$

It is well–known that in the discrete case the stability behaviour of **p** is closely related to the spectral radius of the associated linear map

$$J_1 := J_0 + Id \qquad (6.6)$$

(Id denotes the identity map). In fact, **p*** is asymptotically stable, if all eigenvalues of J_1 have a modulus smaller than one, and it is unstable, if at least one eigenvalue of J_1 exceeds one in modulus. **p** is called *hyperbolically stable*, if all the eigenvalues of J_1 are located strictly within the unit circle, and it is called *hyperbolically unstable*, if at least one of the eigenvalues of J_1 has a modulus larger than one.

As described in Section 2., only non–negative fitness values make sense in the discrete generations context. Accordingly, the RSP–game $A = A(a,b,c)$ will be called *admissible* for the discrete replicator dynamics, if A is a non–negative matrix, i.e., if

$$b_i > a_i \geq c_i \geq 0 \quad \text{for } i \in I. \qquad (6.7)$$

Let $A_0 = A(0,\beta,-\gamma)$ be an essential RSP–game, and let $a \in \mathbb{R}^3$ denote a vector. If a is interpreted as the generating vector for the basic component A_1 of an RSP–game, the pair (A_0,a) induces an RSP–game $A = A(A_0,a)$, which is given by

$$A(A_0,a) := A_0 + A(a,a,a) . \qquad (6.8)$$

The vector $a \in \mathbb{R}^3$ will be called *admissible for* A_0, if $A(A_0,a)$ is an is admissible RSP–game for the discrete replicator dynamics. a is admissible for A_0 if and only if

$$a \geq \gamma \geq 0 , \qquad (6.9)$$

where these vector inequalities should be understood component–wise.

Let now $A = A(a,b,c)$ be a central RSP–game, let J_0 denote the linearization of (6.1) at the interior fixed point $p^* = m$ of A, and let J_1 be given by (6.6). In view of Theorem 5.9(1) and (6.6), the eigenvalues of J_0 (and also those of J_1) are complex conjugate to one another. From this it is easy to derive (see Weissing 1983) that the interior fixed point is hyperbolically stable with respect to (6.1) if and only if the following inequality holds true:

$$\chi(p^*) := \det(D) + \overline{F}(p^*) \cdot tr(D) < 0 . \qquad (6.10)$$

p is hyperbolically unstable, if and only if the inequality–sign in (6.10) is reversed.

It is clear that for any payoff matrix A, $\overline{F}(\mathbf{m})$ corresponds to the arithmetic mean of the entries of A. Together with (5.33) we get

$$\overline{F}(\mathbf{m}) = \tfrac{1}{3}\cdot(\overline{a}+\overline{b}+\overline{c}) = \tfrac{1}{3}\cdot\eta + \overline{a}. \tag{6.11}$$

Remember that $\mathrm{tr}(D)$ and $\det(D)$ are given by (5.29) and (5.30). Together with (6.11) and (5.33), this yields:

$$\chi(\mathbf{p}^*) = \tfrac{1}{3}[\,\tfrac{1}{3}\,\beta\cdot\gamma - \overline{a}\,(\overline{\beta}-\overline{\gamma})\,]. \tag{6.12}$$

Combining all this, we have shown that:

PROPOSITION 6.1: Discrete hyperbolic stability in central RSP–games.

Let A be a central RSP–game and let $\mathbf{p}^* = \mathbf{m}$ denote its unique interior fixed point. Then the following holds true:

1. $\mathbf{p}$ is hyperbolically stable with respect to the discrete replicator equation (6.1) if and only if

$$\overline{a}\,(\overline{\beta}-\overline{\gamma}) > \tfrac{1}{3}\beta\cdot\gamma. \tag{6.13}$$

2. $\mathbf{p}^*$ is hyperbolically unstable with respect to (6.1) if and only if

$$\overline{a}\,(\overline{\beta}-\overline{\gamma}) < \tfrac{1}{3}\beta\cdot\gamma. \tag{6.14}$$

If $\overline{\beta}-\overline{\gamma} \leq 0$, we have $\overline{a} \geq \overline{\gamma} \geq \overline{\beta} > 0$ and $\beta\cdot\gamma > 0$. In view of (6.14), this is sufficient to imply hyperbolical instability of the interior fixed point. If $\overline{\beta}-\overline{\gamma} > 0$, the interior fixed point is hyperbolically stable provided that $\overline{a}$ is 'large enough'. It is clear that one can always achieve this by a positive linear transformation of payoffs. On the other hand, one can also almost always achieve hyperbolic instability by choosing $\overline{a}$ 'small enough'. This will be shown next.

A simple calculation using the equalities $\beta_i = \eta + \gamma_{i+1}$, $i \in I$, yields

$$\tfrac{1}{3}\beta\cdot\gamma = \tfrac{1}{3}\sum_i \gamma_i\gamma_{i+1} + \eta\cdot\overline{\gamma}. \tag{6.15}$$

Together with (6.12) and $\eta = \overline{\beta}-\overline{\gamma}$, we get another formula for $\chi(\mathbf{p}^*)$:

$$\chi(\mathbf{p}^*) = \tfrac{1}{3}[\,\tfrac{1}{3}\sum_i \gamma_i\gamma_{i+1} - \eta\cdot(\overline{a}-\overline{\gamma})\,]. \tag{6.16}$$

Remember that $\mathbf{p}^*$ is hyperbolically stable if and only if $\chi(\mathbf{p}^*) < 0$ and that it is hyperbolically unstable if and only if $\chi(\mathbf{p}^*) > 0$.

If $\eta = \overline{\beta} - \overline{\gamma} \leq 0$, we have $\gamma_{i+1} \geq \beta_i > 0$, $i \in I$. In view of $\overline{a} \geq \overline{\gamma}$, (6.16) implies that p is hyperbolically unstable since

$$\chi(p^*) \geq \tfrac{1}{3} \textstyle\sum_i \beta_i \beta_{i+1} > 0, \quad \text{if } \eta \leq 0 . \tag{6.17}$$

Let us now analyse the case $\eta > 0$. (6.16) shows that p^* is hyperbolically stable if and only if

$$\overline{a} > \overline{\gamma} + \frac{1}{3 \cdot \eta} \cdot [\textstyle\sum_i \gamma_i \gamma_{i+1}] . \tag{6.18}$$

p is hyperbolically unstable if and only if the inequality sign in (6.18) is reversed. (6.18) motivates the definition:

$$\varphi(A_0) := \frac{1}{3 \cdot \eta} \cdot [\textstyle\sum_i \gamma_i \gamma_{i+1}], \quad \text{if } \eta > 0 . \tag{6.19}$$

If $\varphi(A_0) = 0$, p^* is hyperbolically stable if $\overline{a} > \overline{\gamma}$ and it is never hyperbolically unstable. Note that $\varphi(A_0) = 0$ if and only if at least two component of the vector γ are equal to zero. If $\varphi(A_0) > 0$, there are always admissible vectors a such that

$$\overline{\gamma} \leq \overline{a} < \overline{\gamma} + \varphi(A_0) , \tag{6.20}$$

i.e., such that p^* is hyperbolically unstable for the RSP–game $A = A(A_0, a)$. For example, the vector $a = \gamma$ is admissible for A_0, and p^* is hyperbolically unstable for the RSP–game $A(A_0, \gamma)$ if $\varphi(A_0) > 0$. On the other hand, the interior fixed point of $A = A(A_0, a)$ is hyperbolically stable if a is 'large enough' in the sense that

$$\overline{a} > \overline{\gamma} + \varphi(A_0). \tag{6.21}$$

Recall that for central RSP–games we have

$$\text{sgn}(\eta) = \text{sgn}[\det(A_0)] = \text{sgn}[\tfrac{1}{2} \cdot (\overline{b} + \overline{c}) - \overline{a}]. \tag{6.22}$$

Together with the above observations we get:

<u>COROLLARY 6.2</u>: Discrete hyperbolic instability in central RSP–games.

Let $A = A(a,b,c) = A(A_0, a)$ be a central RSP–game and let $p^* = m$.
1. If $\det(A_0) \leq 0$, p^* is hyperbolically unstable with respect to the discrete replicator dynamics (6.1) irrespective of the vector a.
2. If $\det(A_0) > 0$ and $\varphi(A_0) = 0$, p^* is hyperbolically stable with respect to the discrete replicator dynamics if and only if $a \geq \gamma$, $a \neq \gamma$.
3. If $\det(A_0) > 0$ and $\varphi(A_0) > 0$, p^* is hyperbolically unstable with respect to (6.1) if and only if (6.20) holds true. In particular, it is hyperbolically unstable for $a = \gamma$. On the other hand, p is hyperbolically stable if a is large enough in the sense that (6.21) holds true.

All these results can easily be generalized to non–central RSP–games. In fact, every RSP–game $A = A(\mathbf{a,b,c}) = A(A_0,\mathbf{a})$ with interior fixed point $\mathbf{p}$ can be centralized by a barycentric transformation, the generating vector π of which is given by (5.35). It is obvious that A is transformed into the central RSP–game

$$A^\pi = A(\mathbf{a}^\pi,\mathbf{b}^\pi,\mathbf{c}^\pi) = A((A_0)^\pi,\mathbf{a}^\pi), \qquad (6.23)$$

where $\mathbf{a}^\pi$, $\mathbf{b}^\pi$, $\mathbf{c}^\pi$ and $(A_0)^\pi$ are given by

$$a_j^\pi := a_j \cdot p_j^* , \; b_j^\pi := b_j \cdot p_j^* , \; c_j^\pi := c_j \cdot p_j^* , \qquad (6.24)$$

$$\beta_j^\pi := \beta_j \cdot p_j^* , \; \gamma_j^\pi := \gamma_j \cdot p_j^* . \qquad (6.25)$$

Notice that $\text{sgn}[\det(A_0{}^\pi)] = \text{sgn}[\det(A_0)]$, that $\mathbf{a}^\pi$ can be made arbitrarily large by increasing $\mathbf{a}$, and that $\sum_i \gamma_i^\pi \gamma_{i+1}^\pi = 0$ is equivalent to $\sum_i \gamma_i \gamma_{i+1} = 0$. From this we get:

<u>THEOREM 6.3</u>: Discrete hyperbolic stability in RSP–games.

Let A_0 be an RSP–game in essential form, and let $\mathbf{p}^*$ denote its interior fixed point.

1. If $\det(A_0) \leq 0$, $\mathbf{p}^*$ is hyperbolically unstable with respect to the discrete replicator dynamics for all RSP–games that have A_0 as their interactive component.
2. If $\det(A_0) > 0$ and $\sum_i \gamma_i \gamma_{i+1} = 0$, $\mathbf{p}^*$ is hyperbolically stable for all RSP–games $A(A_0,\mathbf{a})$ with $\mathbf{a} \geq \gamma$, $\mathbf{a} \neq \gamma$.
3. If $\det(A_0) > 0$ and $\sum_i \gamma_i \gamma_{i+1} > 0$, there exists a real number $\psi(A_0) > \gamma \cdot \mathbf{p}^* \geq 0$ such that $\mathbf{p}$ is hyperbolically unstable for all admissible RSP–games $A(A_0,\mathbf{a})$ with $\mathbf{a} \cdot \mathbf{p}^* < \psi(A_0)$ and hyperbolically stable for all $A(A_0,\mathbf{a})$ with $\mathbf{a} \cdot \mathbf{p}^* > \psi(A_0)$.

The two main conclusions of Theorem 6.3 will be stated as a corollary:

<u>COROLLARY 6.4</u>: Evolutionary stability and discrete dynamic stability.

1. The concept of *discrete stability*, i.e., stability with respect to (6.1), is *not* invariant with respect to positive linear transformations of the payoff matrix.
2. Evolutionary stability is neither necessary nor sufficient for discrete stability. In fact, an ESS may be hyperbolically unstable with respect to (6.1), and a hyperbolic attractor with respect to (6.1) need not be an ESS.

The potential discrete instability of an evolutionarily stable strategy is illustrated in Figure 4. An orbit starting at $\mathbf{p}_0 = (0.25, 0.25, 0.5)$ is shown there for (a) the discrete replicator dynamics and a small value of $\mathbf{\pi}$, (b) the discrete replicaotr dynamics and a large value of $\mathbf{\pi}$, and (c) the continuous replicator dynamics.

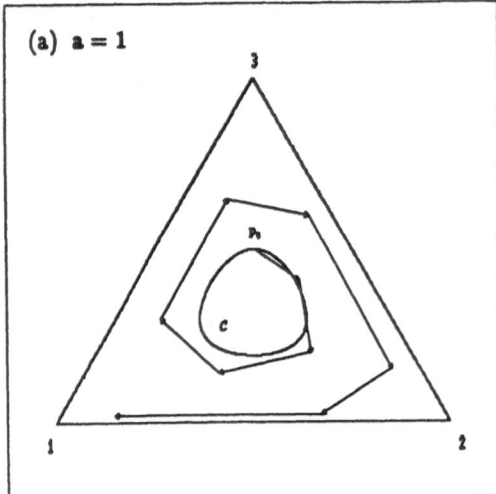

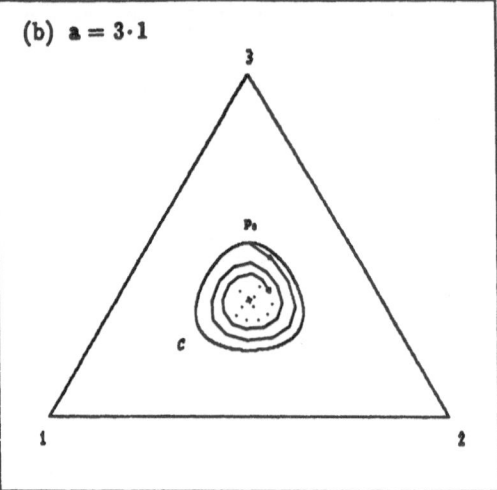

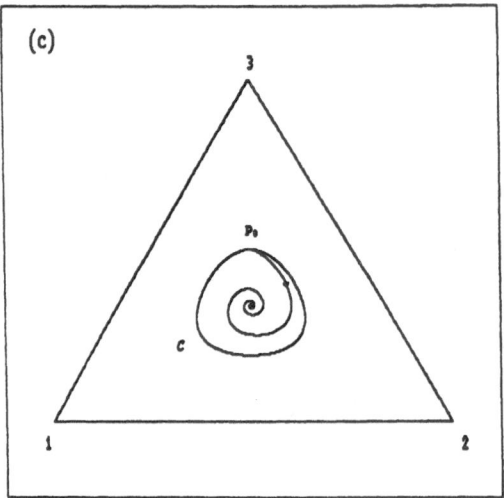

FIGURE 4: Discrete dynamic instability of an evolutionarily stable strategy.
Figure 4 is based on the central RSP–game $A = A(A_0, \mathbf{a})$, where:

$$A_0 = \begin{bmatrix} 0 & 2 & -1 \\ -1 & 0 & 2 \\ 2 & -1 & 0 \end{bmatrix} \quad \text{and} \quad \mathbf{a} = \lambda \cdot \gamma = \lambda \cdot 1.$$

C is a constant–level curve of the scalar function V (see (5.5));
the initial vector $\mathbf{p}_0 := (0.25, 0.25, 0.5)$ belongs to C.

(a) $\mathbf{a} = \gamma$: $\mathbf{p}^*$ is hyperbolically *unstable* with respect to (6.1).
The vector attached to $\mathbf{p}_0$ is so large that it 'overshoots' C.

(b) $\mathbf{a} = 3\gamma$: $\mathbf{p}^*$ is hyperbolically *stable* with respect to (6.1).
Overshooting does not occur.

(c) Orbit of the *continuous* replicator dynamics starting at $\mathbf{p}_0$.

Figure 4 is based on an ϵ–perturbed Rock–Scissors–Paper game with $\epsilon > 0$. As has been shown in Section 5, the interior fixed point $\mathbf{p} = \mathbf{m}$ is an ESS and a global attractor with respect to the continuous replicator dynamics. The initial vector $\mathbf{p}_0$ belongs to the closed curve C, which is a constant–level curve of the scalar function V that was defined in (5.5). Since $\mathbf{p}$ is an ESS, V is a global Lyapunov function for the continuous replicator dynamics.

By definition, this means that the trajectories of (5.1) cross constant–level curves like C from the outside to the inside with respect to C. Equivalently, the vectors of the vector field $\mathbf{v}^A$ point 'inward' for all points on C. Since the vector field $\mathbf{w}^A$ of the discrete replicator dynamics is collinear to $\mathbf{v}^A$, the vectors $\mathbf{w}^A(\mathbf{p})$ also point inward when attached to points $\mathbf{p}$ on C. However, in view of (6.3) and (6.11), the length of a vector $\mathbf{w}^A(\mathbf{p})$ is negatively correlated with $\mathbf{a}$. For small values of $\mathbf{a}$, the length of $\mathbf{w}^A(\mathbf{p})$ may be so large that $\mathbf{p}' = \mathbf{p} + \mathbf{w}^A(\mathbf{p})$ lies outside the closed curve C (see Figure 4(a)). *Overshootings* like this, which are closely connected with the the value of $\mathbf{a}$ (see Figure 4(b)), are the reasons for the potential instability of an ESS with respect to the discrete replicator dynamics.

Theorem 6.3 indicates that for $\det(A_0) > 0$ we can always find an $\mathbf{a}$ that is large enough to prevent overshootings. (In Weissing (1990), this result is generalized to the class of all evolutionary normal form games.) On the other hand – neglecting the border case $\sum_i \gamma_i \gamma_{i+1} = 0$ – overshootings leading to instability do *always* occur if the vector $\mathbf{a}$ is not significantly larger than γ.

In Section 5. we saw that for the continuous replicator dynamics the *local* stability properties of the interior fixed point of an RSP–game correspond perfectly to *global* stability properties. We shall now turn to this question for the case of the discrete replicator dynamics (6.1).

Again, global stability properties will be analysed by constructing suitable Lyapunov functions for (6.1). As in the continuous time case, a *Lyapunov function* is a scalar function $W: \Delta \longrightarrow \mathbb{R}$, which has a strict local maximum in $\mathbf{p}^*$ and which changes in a definite way along the orbits of the dynamical system. The last condition means that the change along orbits, ∇W, should either be positive near $\mathbf{p}^*$, or negative near $\mathbf{p}^*$, or identical to zero near $\mathbf{p}^*$. ∇W is defined by

$$\nabla W(\mathbf{p}) := W(\mathbf{p}') - W(\mathbf{p}) . \tag{6.26}$$

The 'standard' Lyapunov function $V: \Delta \longrightarrow \mathbb{R}$ for the continuous replicator dynamics – which is given by (5.5) – is also very useful in the context of the discrete replicator equation. In fact, V can always be used for demonstrating discrete instability of $\mathbf{p}^*$, if $\mathbf{p}^*$ is uniformly evolutionarily unstable. On the other hand, V is a global discrete Lyapunov function whenever $\mathbf{p}$ is an ESS and the vector $\mathbf{a}$ is 'large enough' to prevent overshootings. These results – which hold true for general evolutionary normal form games – are derived in Weissing (1990). The following theorem combines those results in Weissing (1990) which are relevant for RSP–games:

<u>THEOREM 6.5</u>: Evolutionary stability and global discrete stability.

Let p   denote an interior Nash strategy of an evolutionary normal form game A_0. Then the following holds true:

1. If p   is the unique interior Nash strategy, and if p* is uniformly evolutionarily unstable, it is a global repellor for the discrete replicator dynamics for all games $A = (A_0, a)$ such that the vector a is admissible for A_0.

2. If p   is an ESS, there exists an admissible vector a* such that p* is a global attractor for the discrete replicator dynamics for all games $A = (A_0, a)$ with a ≥ a*.

Theorem 6.5 directly applies to the class of balanced RSP–games, since the interior fixed point of a balanced RSP–game is always either an ESS or uniformly evolutionarily unstable (Proposition 5.3). On the other hand, every RSP–game can be transformed into a balanced one by means of a barycentric transformation of the state space (Theorem 5.5), and barycentric transformations do not change the dynamic features of the discrete replicator dynamics. Consequently, Theorem 6.5 yields:

<u>THEOREM 6.6</u>: Discrete global stability in RSP–games.

Let A_0 be an RSP–game in essential form, and let p* denote its interior fixed point.

1. If $\det(A_0) \leq 0$, p* is a global hyperbolic repellor with respect to the discrete replicator dynamics for any RSP–game $A = A(A_0, a)$ the interactive component ofwhich is given by A_0.

2. If $\det(A_0) > 0$, there exists an admissible vector $a* \in \mathbb{R}^3$ for A_0 such that p* is a global hyperbolic attractor with respect to the discrete replicator dynamics for all RSP–games $A(A_0, a)$ with a ≥ a*.

The example in Figure 4 shows that even for an ESS of a perfectly balanced, central RSP–game the function V is in general not a discrete Lyapunov function. There is, however, a class of RSP–games which is so simple in structure that a global discrete Lyapunov function can be found for all parameter constellations. It is the class of 'circulant' RSP–games, the properties of which will be analysed next.

An RSP–game $A = A(a,b,c)$ will be called a *circulant RSP–game* if the vectors a, b, and c are all scalar multiples of the vector 1. Accordingly, an RSP–game is circulant if and only if its payoff matrix is a 'circulant matrix' (see, e.g., Davis 1979), i.e., if A is of the form

$$A = \begin{bmatrix} a & b & c \\ c & a & b \\ b & c & a \end{bmatrix}, \quad b > a \geq c. \tag{6.27}$$

It is clear that circulant RSP–games are central, perfectly balanced games, and that the class of these games includes the ϵ–perturbed Rock–Scissors–Paper games which are defined by (1.4). Setting

$$\beta := b{-}a, \quad \gamma := a{-}c, \quad \text{and} \quad \delta := \beta{-}\gamma, \tag{6.28}$$

we get from (6.22):

$$\text{sgn}[\, \det(A_0)\,] = \text{sgn}(\delta) = \text{sgn}[\, \tfrac{b+c}{2} - a\,]\,. \tag{6.29}$$

On the other hand, a simple calculation based on (6.12) shows that $\chi(\mathbf{p}^*)$ is given by

$$\chi(\mathbf{p}^*) = \tfrac{1}{3}[\, a^2 - bc\,]\,. \tag{6.30}$$

Now Theorem 5.3, Theorem 6.1, and Theorem 6.6 taken together provide a nice characterization of evolutionary stability and dynamic stability in circulant RSP–games:

<u>COROLLARY 6.7</u>: Stability in circulant RSP–games.

Let A denote a circulant RSP–game given by (6.27). Then the following holds true:
1. The interior Nash strategy $\mathbf{p}^* = \mathbf{m}$ is an ESS if and only if

$$a < (b+c)/2\,, \tag{6.31}$$

   i.e., if and only if a is smaller than the arithmetic mean of b and c.
2. $\mathbf{p}^*$ is a global hyperbolic attractor with respect to the continuous replicator dynamics if and only if (6.31) holds true. It is a global center if $a = (b+c)/2$, and it is a global hyperbolic repellor if $a > (b+c)/2$.
3. $\mathbf{p}$ is hyperbolically stable with respect to the discrete replicator dynamics (6.1) if and only if

$$a^2 < bc, \tag{6.32}$$

   i.e., if a is smaller than the geometric mean of b and c.
   $\mathbf{p}$ is hyperbolically unstable if and only if $a^2 > bc$. If $a \geq (b+c)/2$, it is a *global* hyperbolic repellor for the discrete replicator dynamics.

For circulant RSP–games, it is the inequality between the arithmetic and the geometric mean which leaves room for a discrepancy between 'discrete' and 'continuous' stability to occur. In fact, the interior fixed point is a hyperbolic *attractor* for the continuous replicator dynamics and at the same time a hyperbolic *repellor* for the discrete replicator equation if

$$\sqrt{bc} < a < \tfrac{b+c}{2}\,. \tag{6.33}$$

We shall now prove a theorem which shows that for *circulant* RSP–games there is a perfect correspondence between local discrete stability and global discrete stability:

<u>Theorem 6.8:</u> Global discrete stability in circulant RSP–games.

Let A denote a circulant RSP–game given by (6.27). Let $\mathbf{p}^* = \mathbf{m}$ denote its interior fixed point. Then the following holds true:

1. If $a^2 < bc$, $\mathbf{p}^*$ is a global hyperbolic attractor for the discrete replicator dynamics.
2. If $a^2 > bc$, $\mathbf{p}^*$ is a global hyperbolic repellor for (6.1).
3. If $a^2 = bc$, $\mathbf{p}^*$ is a global center for (6.1), i.e., the interior of the strategy simplex is filled with closed invariant curves encircling $\mathbf{p}^*$.

Figure 5 illustrates how the phase portrait of the discrete replicator dynamics changes near $a = \sqrt{bc}$. In (c), each orbit is iterated for only about 100 generations. Obviously, each of the closed invariant curves surrounding $\mathbf{p}^*$ consists of many orbits of (6.1), since each orbit has only a countable number of elements. In my numerical simulations, I have never observed a finite non–equilibrium orbit corresponding to a periodic trajectory. Instead, the simulations suggest that each orbit is dense on the closed invariant curve to which it belongs.

<u>Proof of Theorem 6.8:</u>
A: In view of Corollary 6.7.3, we may concentrate on the case $a < (b+c)/2$. We shall therefore assume that (6.31) holds true and that the interior fixed point $\mathbf{p}^* = \mathbf{m}$ is an ESS of A.

Theorem 6.8 will be proved by constructing a suitable global Lyapunov function for (6.1). We shall consider the scalar function $W: \text{int}(\Delta) \longrightarrow \mathbb{R}$, which is defined by

$$W(\mathbf{q}) := \frac{\mathbf{q} \cdot A\mathbf{q}}{q_1 q_2 q_3}, \quad \mathbf{q} \in \text{int}(\Delta) . \tag{6.34}$$

(Josef Hofbauer and Karl Sigmund directed my attention to this function.) It will be shown that the interior fixed point $\mathbf{p}^* = \mathbf{m}$ of A is the only critical point of W, and that it is a strict global minimum of W. Moreover, ∇W has a definite sign along the orbits of the discrete replicator dynamics (6.1). In fact, we shall show that the sign of $\nabla W(\mathbf{q})$ is independent of $\mathbf{q}$, and that it is given by

$$\text{sgn}[\nabla W(\mathbf{q})] = \text{sgn}[a^2 - bc] . \tag{6.35}$$

Accordingly, W increases along the orbits of (6.1) if and only if $a^2 > bc$; W decreases along the orbits of (6.1) iff $a^2 < bc$; and it is a constant of motion if and only if $a^2 = bc$. The assertions of Theorem 6.8 follow immediately from these properties.

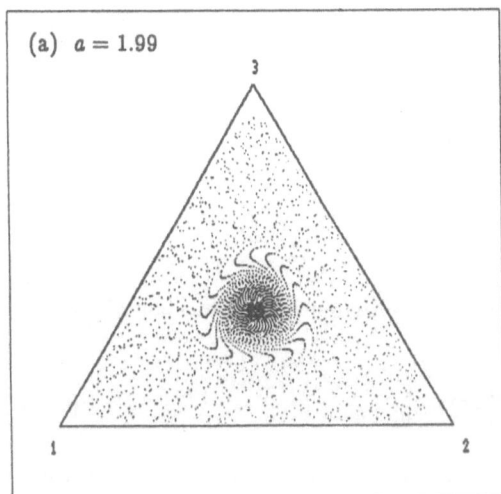

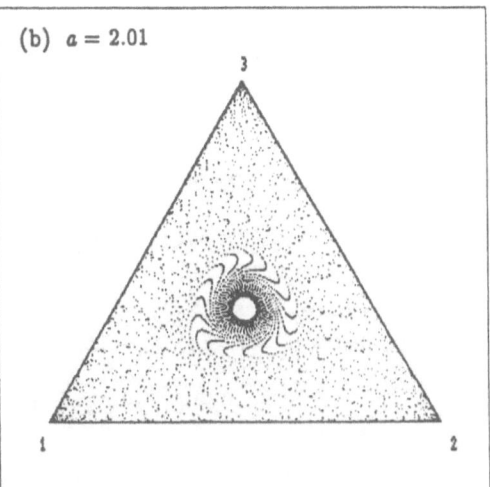

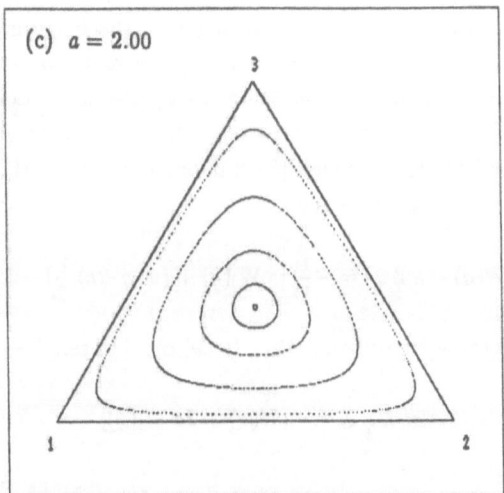

FIGURE 5:   Discrete dynamic stability in a family of circulant RSP–games:

$$A(a) = = \begin{bmatrix} a & 4 & 1 \\ 1 & a & 4 \\ 4 & 1 & a \end{bmatrix}, \quad 4 > a \geq 1 .$$

(a) $a < \sqrt{bc}$: p^* is a global hyperbolic attractor.
      The diagram shows the orbit starting at $p_0 := (0.05, 0.05, 0.90)$.

(b) $a > \sqrt{bc}$: p^* is a global hyperbolic repellor.
      The diagram shows the orbit starting at $p_0 := (0.35, 0.35, 0.30)$.

(c) $a = \sqrt{bc}$: p^* is a global center. Five orbits are shown which
      indicate the closed invariant curves encircling p^*.

B: First, we shall show that W has a global minimum in p^*. A simple calculation yields that $q \cdot Aq$ may be represented in the form

$$q \cdot Aq = a + (b+c-2a)(q_1 q_2 + q_2 q_3 + q_3 q_1). \tag{6.36}$$

Therefore, $W(q)$ may be written as

$$W(q) = a \cdot \frac{1}{q_1 q_2 q_3} + (b+c-2a) \cdot \left(\frac{1}{q_1} + \frac{1}{q_2} + \frac{1}{q_3}\right). \tag{6.37}$$

In view of (6.27) and (6.31) this shows that W is a positive linear combination of the two scalar functions W_1 and W_2, which are given by

$$W_1(q) := \frac{1}{q_1 q_2 q_3} \quad \text{and} \quad W_2(q) := \frac{1}{q_1} + \frac{1}{q_2} + \frac{1}{q_3}. \tag{6.38}$$

Notice that $W_1(q)$ is the inverse of the third power of the geometric mean of q, whereas $W_2(q)$ is the inverse of the harmonic mean of q. It is well–known that the geometric as well as the harmonic mean have a strict global maximum in the barycenter m of the simplex. Correspondingly, W_1 and W_2 both h ve a strict global minimum in m. Being a positive linearcombination of W_1 and W_2, W also has a strict global minimum in $p^* = m$.

C: We shall use the method of Lagrange multipliers in order to show that p^* is the only critical point of W. Therefore, we consider the term

$$\frac{\partial}{\partial q_i}\left[W(q) - \lambda \sum_i q_i\right] = -\frac{1}{q_i}\left[a\, W_1(q) + (b+c-2a)\frac{1}{q_i}\right] - \lambda, \tag{6.39}$$

where λ denotes a Lagrange mulitplier. Setting (6.39) equal to zero for $i \in I$, we get

$$\lambda = \lambda \sum_i q_i = -[W(q) + 2a\, W_1(q)] \tag{6.40}$$

and

$$a \cdot W_1(q) \cdot (1-3q_i) = (b+c-2a) \cdot q_i \cdot [W_2(q) - (q_i)^{-2}]. \tag{6.41}$$

Let us assume that $q \neq m$ and that i minimizes q_i. Then we have $q_i < 1/3$ and $W_2(q) < 3/q_i$ which implies

$$W_2(q) - (q_i)^{-2} < \frac{1}{q_i} \cdot \left(3 - \frac{1}{q_i}\right) < 0. \tag{6.41a}$$

Therefore the right–hand side of (6.41) is negative whereas the left–hand side is negative. This contradiction shows that (6.41) is only compatible with $q = m$: $p^* = m$ is the only critical point of W in $\mathrm{int}(\Delta)$.

D: In order to prove (6.35), we shall first derive an indicator function for the sign of ∇W. Let us define two auxiliary functions $M: \mathbb{R}^3 \longrightarrow \mathbb{R}$ and $P: \mathbb{R}^3 \longrightarrow \mathbb{R}$ by

$$M(z) := z \cdot Az, \quad P(z) := z_1 z_2 z_3, \quad z \in \mathbb{R}^n. \tag{6.42}$$

Using these functions, $W(q)$ may be written as

$$W(q) = \frac{M(q)}{P(q)}, \quad q \in int(\Delta).$$

(6.43)

Let us now choose a fixed strategy $q \in int(\Delta)$, $q \neq m$. Denote the fitness vector at q by $y \in \mathbb{R}^3$, i.e.,

$$y := F(q) = Aq.$$

(6.44)

From the definition (2.8) of q' we get

$$P(q') = P(q) \cdot \frac{P(y)}{(M(q))^3}.$$

(6.45)

This implies

$$W(q') = \frac{M(q')}{P(q')} = W(q) \cdot \frac{M(q')(M(q))^2}{P(y)},$$

(6.46)

and

$$\nabla W(q) := W(q') - W(q) = \frac{W(q)}{P(y)} \cdot [\, M(q')(M(q))^2 - P(y) \,].$$

(6.47)

A few simple calculations yield

$$M(q') \cdot [M(q)]^2 = \Psi(q),$$

(6.48)

where $\Psi(q)$ is given by

$$\Psi(q) := a(\textstyle\sum_i q_i^2 v_i^2) + (b+c)(\textstyle\sum_i q_i q_{i+1} v_i v_{i+1}).$$

(6.49)

Combining (6.47) and (6.48), we get:

$$sgn[\nabla W(q)] = sgn[\Psi(q) - P(y)].$$

(6.50)

E: We shall now derive an explicit representation of the expression $\Psi(q) - P(y)$. The following identities will be used:

$$\textstyle\sum_i q_i^2 q_{i+1}^2 = \sum_i q_i^2 q_{i+2}^2,$$

(6.51)

$$\textstyle\sum_i q_i^3 q_{i+2} = \sum_i q_i q_{i+1}^3,$$

(6.52)

$$q_1 q_2 q_3 = \textstyle\sum_i q_i^2 q_{i+1} q_{i+2} = \sum_i q_i q_{i+1}^2 q_{i+2} = \sum_i q_i q_{i+1} q_{i+2}^2,$$

(6.53)

$$\textstyle\sum_i q_i^3 = \sum_i q_i^4 + \sum_i q_i^3 q_{i+1} + \sum_i q_i q_{i+1}^3,$$

(6.54)

$$\textstyle\sum_i q_i^2 q_{i+1} = \sum_i q_i^2 q_{i+1}^2 + \sum_i q_i^3 q_{i+1} + q_1 q_2 q_3,$$

(6.55)

$$\textstyle\sum_i q_i q_{i+1}^2 = \sum_i q_i^2 q_{i+1}^2 + \sum_i q_i q_{i+1}^3 + q_1 q_2 q_3.$$

(6.56)

(6.51), (6.52), and the last two equalities in (6.53) are consequences of the fact that pure strategies are counted modulo three. The other equalities result by multiplying the left hand sides by the factor $\sum_i q_i$ (which is equal to one). Using (6.51) to (6.56), it is easy to derive the following formulae:

$$
\begin{aligned}
\sum_i q_i^2 v_i^2 \quad &= \quad \sum_i q_i^2 \cdot (aq_i + bq_{i+1} + cq_{i+2})^2 \\
&= \quad a^2 \cdot \left(\sum_i q_i^4 \right) + (b^2 + c^2) \cdot \left(\sum_i q_i^2 q_{i+1}^2 \right) + \\
&\qquad + 2ab \cdot \left(\sum_i q_i^3 q_{i+1} \right) + 2ac \cdot \left(\sum_i q_i q_{i+1}^3 \right) + 2bc \cdot q_1 q_2 q_3 ,
\end{aligned} \tag{6.57}
$$

$$
\begin{aligned}
\sum_i q_i q_{i+1} v_i v_{i+1} \quad &= \quad \sum_i q_i q_{i+1} \cdot (aq_i + bq_{i+1} + cq_{i+2}) \cdot (aq_{i+1} + bq_{i+2} + cq_i) \\
&= \quad (a^2 + bc) \cdot \left(\sum_i q_i^2 q_{i+1}^2 \right) + ac \cdot \left(\sum_i q_i^3 q_{i+1} \right) + \\
&\qquad + ab \cdot \left(\sum_i q_i q_{i+1}^3 \right) + (b^2 + c^2 + ab + ac + bc) \cdot q_1 q_2 q_3 ,
\end{aligned} \tag{6.58}
$$

$$
\begin{aligned}
P(y) \quad &= \quad (aq_i + bq_{i+1} + cq_{i+2}) \cdot (aq_{i+1} + bq_{i+2} + cq_i) \cdot (aq_{i+2} + bq_i + cq_{i+1}) \\
&= \quad abc \cdot \left(\sum_i q_i^3 \right) + (a^2b + b^2c + c^2a) \cdot \left(\sum_i q_i^2 q_{i+1} \right) + \\
&\qquad + (a^2c + b^2a + c^2b) \cdot \left(\sum_i q_i q_{i+1}^2 \right) + (a^3 + b^3 + c^3 + 3abc) \cdot q_1 q_2 q_3 \\
&= \quad abc \cdot \left(\sum_i q_i^4 \right) + \\
&\qquad + (a^2b + b^2c + c^2a + a^2c + b^2a + c^2b) \cdot \left(\sum_i q_i^2 q_{i+1}^2 \right) + \\
&\qquad + (abc + a^2b + b^2c + c^2a) \cdot \left(\sum_i q_i^3 q_{i+1} \right) + \\
&\qquad + (abc + a^2c + b^2a + c^2b) \cdot \left(\sum_i q_i q_{i+1}^3 \right) + \\
&\qquad + (a^3 + b^3 + c^3 + 3abc + a^2b + b^2c + c^2a + a^2c + b^2a + c^2b) \cdot q_1 q_2 q_3
\end{aligned} \tag{6.59}
$$

Combining (6.57), (6.58), and (6.59), we get that $\Psi(q) - P(y)$ is of the form:

$$
\begin{aligned}
\Psi(q) - P(y) \quad = \quad &\sigma_4 \cdot \left(\sum_i q_i^4 \right) + \sigma_3 \cdot \left(\sum_i q_i^3 q_{i+1} \right) + \sigma_2 \cdot \left(\sum_i q_i^2 q_{i+1}^2 \right) + \\
&+ \sigma_1 \cdot \left(\sum_i q_i q_{i+1}^3 \right) + \sigma_0 \cdot q_1 q_2 q_3 ,
\end{aligned} \tag{6.60}
$$

where the coefficients σ_i are given by

$$
\sigma_4 = a^3 - abc = a \cdot (a^2 - bc), \tag{6.61}
$$

$$
\sigma_3 = 2a^2b + (b+c) \cdot ac - (abc + a^2b + b^2c + c^2a) = b \cdot (a^2 - bc), \tag{6.62}
$$

$$
\sigma_2 = a \cdot (b^2 + c^2) + (b+c) \cdot (a^2 + bc) - (a^2b + b^2c + c^2a + a^2c + b^2a + c^2b) = 0, \tag{6.63}
$$

$$
\sigma_1 = 2a^2c + (b+c) \cdot ab - (abc + a^2c + b^2a + c^2b) = c \cdot (a^2 - bc), \tag{6.64}
$$

$$
\begin{aligned}
\sigma_0 &= 2abc + (b+c) \cdot (b^2 + c^2 + ab + bc + ca) - (a^3 + b^3 + c^3 + 3abc + (b+c)(a^2 + bc) + ab^2 + ac^2) \\
&= abc + bc^2 + b^2c - a^3 - a^2b - a^2c = -(a+b+c) \cdot (a^2 - bc).
\end{aligned} \tag{6.65}
$$

F: On the basis of (6.61) to (6.65), it is easy to see that $\Psi(q)-P(y)$ may be put into the form

$$\Psi(q) - P(y) = (a^2-bc)(\ a\cdot\Phi_1(q) + b\cdot\Phi_2(q) + c\cdot\Phi_3(q)\ ),\tag{6.66}$$

where the functions $\Phi_i(q)$ are given by

$$\Phi_1(q) := (\ \textstyle\sum_i q_i^4\ ) - q_1 q_2 q_3,\tag{6.67}$$

$$\Phi_2(q) := (\ \textstyle\sum_i q_i^3 q_{i+1}\ ) - q_1 q_2 q_3,\tag{6.68}$$

$$\Phi_3(q) := (\ \textstyle\sum_i q_i q_{i+1}^3\ ) - q_1 q_2 q_3.\tag{6.69}$$

We want to show that

$$\text{sgn}[\nabla W(q)] = \text{sgn}[\Psi(q)-P(y)] = \text{sgn}[a^2-bc]\tag{6.70}$$

for all $q \neq m$. In view of (6.50) and (6.66), (6.70) is a consequence of

$$\Phi_i(q) > 0 \quad \text{for } i \in I \text{ and } q \neq m.\tag{6.71}$$

The proof of Theorem 6.8 will be completed by showing that (6.71) holds true.

G: Let us first show that $\Phi_1(q) > 0$ holds true for $q \neq m$: Jensen's inequality applied to the convex function $f(y) := y^4$ yields for $n = 3$:

$$\textstyle\sum_i q_i^4 \geq n\cdot(\sum_i \frac{1}{n} q_i)^4 = \frac{1}{27}\cdot(\sum_i q_i)^4 = \frac{1}{27},\tag{6.72}$$

with equality only for $q = m$. On the other hand, we have

$$q_1 q_2 q_3 = P(q) \leq P(m) = \frac{1}{27},\tag{6.73}$$

with equality only for $q = m$. Taken together, (6.72) and (6.73) yield (6.71) for $i = 1$. Considering $i = 2$, a simple calculation using (6.55) shows that

$$\Phi_2(q) = (\textstyle\sum_i q_i^2 q_{i+1}) - (\sum_i q_i q_{i+1})^2.\tag{6.74}$$

Jensen's inequality applied to the convex function $f(y) := y^2$ yields (6.71) for $i = 2$. The corresponding result for $i = 3$ follows immediately in view of the symmetry of the expressions (6.68) and (6.69).

This completes the proof of Theorem 6.8. ***

7. Complex Attractors of the Discrete Replicator Dynamics

In the last section we saw that for *circulant* RSP–games the dynamical features with respect to the discrete replicator equation are 'qualitatively' closely analogous to those obtained for the continuous replicator dynamics. In both cases there are only three types of global dynamic behaviour: the interior fixed point p* is either a global attractor, or a global repellor, or a global center. For the discrete as well as for the continuous dynamics, the parameter region $\mathcal{P}_0$ for which p is a center is a submanifold of codimension one in parameter space which separates the parameter regimes for stability and instability. Whenever $\mathcal{P}_0$ is crossed transversally by a one–parameter family of circulant RSP–games, a Hopf bifurcation does occur which is *degenerate* (see below) since the games in $\mathcal{P}_0$ do not admit *isolated* closed invariant curves encircling the interior fixed point. For circulant RSP–games, the main difference between the discrete and the continuous replicator dynamics is a quantitative one in that the stability of the interior fixed point changes at different parameter constellations for the two dynamics: at $\mathcal{P}_0 = \{ (a,b,c) \mid a = \sqrt{bc} \}$ for the discrete dynamics, and at $\mathcal{P}_0 = \{ (a,b,c) \mid a = (b+c)/2 \}$ for the continuous replicator equation.

In this section, we shall demonstrate that the perfect qualitative correspondence between the continuous and the discrete replicator dynamics does not extend to the class of all RSP–games. This is illustrated by an example in Figure 6. In that example, the interior fixed point is locally hyperbolically unstable with respect to the discrete replicator dynamics, but it is not a *global* repellor. Instead, there exists an invariant simple closed curve $\mathcal{C}$ encircling the interior fixed point which attracts all interior non–equilibrium orbits. $\mathcal{C}$ as well as the region enclosed by it are invariant sets for the discrete replicator equation. This phenonemon will be analysed more closely in the present section.

A simple closed curve which is invariant and attractive (or repulsive) will be called an *attractive (repulsive) closed limit curve* since its properties are similar to those of the limit cycles of a continuous dynamical system. By definition, a *limit cycle* is a simple closed curve which is the α– or the ω–limit set for at least one outside orbit (see e.g. Hirsch & Smale 1976). We shall avoid the term 'cycle' in the discrete time context since a cycle is usually associated with a periodic motion. Obviously, a countable orbit of a discrete dynamical system cannot fill a continuum. Accordingly, each limit curve consists of many orbits and in most cases none of them is a periodic one. Usually, the dynamical behaviour on a closed limit curve consists of a complicated pattern of quasi–periodic motions.

In many cases, closed limit curves encircling a fixed point arise from a *discrete Hopf bifurcation*. Such a bifurcation occurs within a one–parameter family (G_μ) of two–dimensional discrete dynamical systems whenever an associated family (p^*_μ) of fixed points loses its stability at a *critical parameter value* $\mu = \mu_0$ since a pair of complex conjugate eigenvalues cross the unit circle. This means that p^*_μ is a hyperbolic attractor for $\mu < \mu_0$ and a hyperbolic repellor for $\mu > \mu_0$ (a more exact definition of a Hopf bifurcation will be given below).

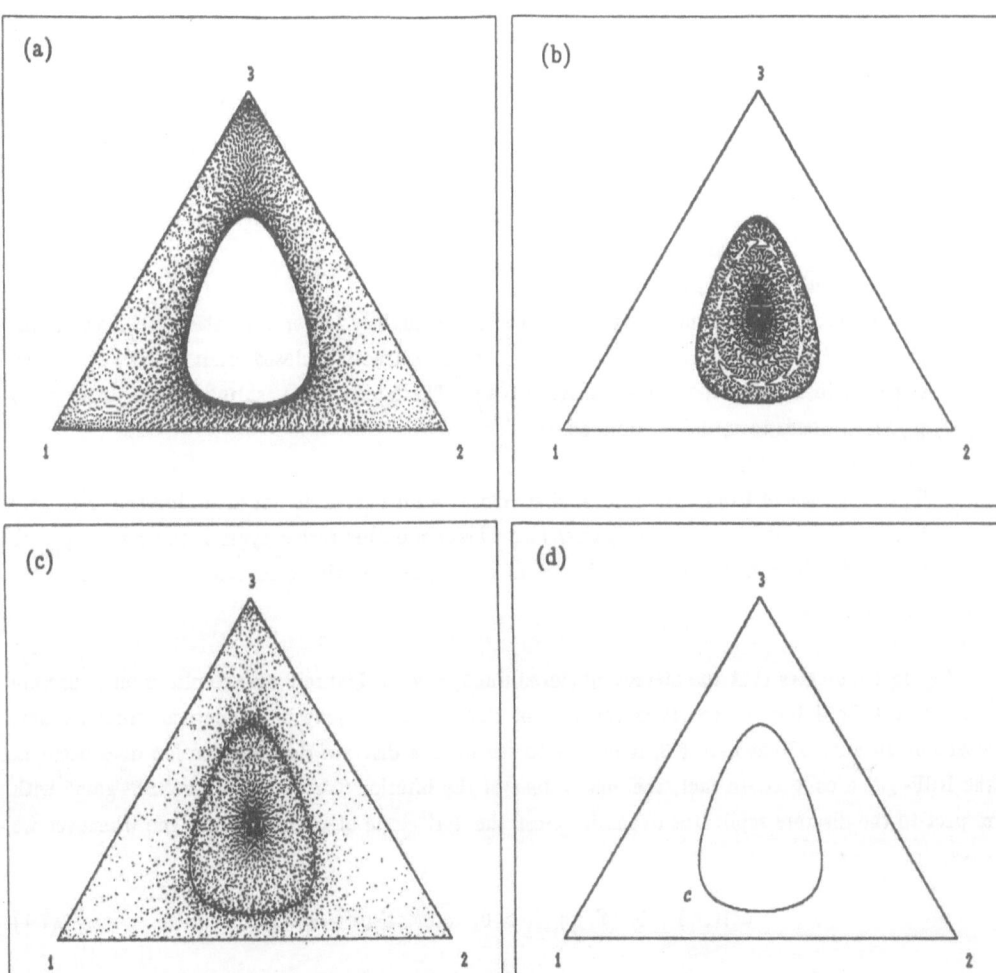

FIGURE 6: Existence of a closed limit curve for the discrete replicator dynamics. This figure shows some orbits of (6.1) for the central RSP–game

$$A = \begin{bmatrix} 10.00 & 11.50 & 9.00 \\ 9.00 & 10.00 & 11.50 \\ 15.85 & 4.65 & 10.00 \end{bmatrix} .$$

(a) Orbit starting at $p = (0.01, 0.01, 0.98)$ (iterated for 20,000 generations).

(b) Orbit starting at $p = (0.33, 0.33, 0.34)$ (36,000 generations; only every third generation is shown).

(c) Superimposition of the orbits in (a) and (b) (only every fifth generation is shown).

(d) Orbit starting at $p = (0.461, 0.461, 0.078)$ which closely approximates the closed limit curve $\mathcal{C}$.

In essence, there are three types of discrete Hopf bifurcations:

(a)   *Subcritical* Hopf bifurcations.

For 'subcritical' parameter values, i.e., for parameter values μ which are slightly smaller than the critical value μ_0, the *stable* fixed point p_μ^* is encircled by a repelling closed limit curve C_μ which circumscribes the domain of attraction of p_μ^*. The family (C_μ) of closed curves shrinks to the fixed point $p_{\mu_0}^*$ (and the domains of attraction get smaller and smaller) if μ tends towards μ_0 from below.

(b)   *Supercritical* Hopf bifurcations.

For 'supercritical' parameter values μ which are slightly larger than the critical value μ_0, the *unstable* fixed point p_μ^* is encircled by an attracting closed limit curve C_μ which circumscribes the 'domain of repulsion' of p_μ^*. The family (C_μ) shrinks to the fixed point $p_{\mu_0}^*$ if μ tends towards μ_0 from above.

(c)   *Degenerate* Hopf bifurcations.

This is a class of Hopf bifurcations where closed limit curves do not arise. Instead, a picture like that obtained for circulant games (see Theorem 6.8) is rather typical: the center $p_{\mu_0}^*$ is surrounded by a family of closed invariant curves, and no closed invariant curves exist near p_μ for $\mu \neq \mu_0$.

Figure 7 indicates that the attracting closed limit curve of Figure 6 also results from a discrete supercritical Hopf bifurcation. It is the aim of this section to give an analytical proof for this assertion. In view of Theorem 6.3, it is easy to see when a discrete Hopf bifurcation does occur in the RSP–game context. In fact, the eigenvalues of the interior fixed point of an RSP–game with respect to the discrete replicator dynamics cross the unit circle of the complex plane whenever we have:

$$\det(A_0) > 0, \quad \sum_i \gamma_i \gamma_{i+1} > 0, \quad \text{and} \quad \mathbf{a} \cdot \mathbf{p}^* = \psi(A_0). \tag{7.1}$$

As in most other applications, it is not difficult to judge whether a discrete Hopf bifurcations occurs. However, it is usually quite intricate to classify a given Hopf bifurcation and to demonstrate that it is of the supercritical type.

The Hopf bifurcation illustrated by Figure 7 occurs in a one–parameter family of discrete replicator dynamics $(\mathbf{?}_\mu)_{\mu>0}$ arising from the family of central RSP–games which are given by

$$A_\mu = \begin{bmatrix} a & b & c \\ c & a & b \\ b+3\mu & c-3\mu & a \end{bmatrix}, \quad b > a \geq c, \ c \geq 3\mu > 0. \tag{7.2}$$

Proposition 6.1 indicates that a discrete Hopf bifurcation occurs at the critical value

$$\mu_0 = \frac{bc-a^2}{b-c}. \tag{7.3}$$

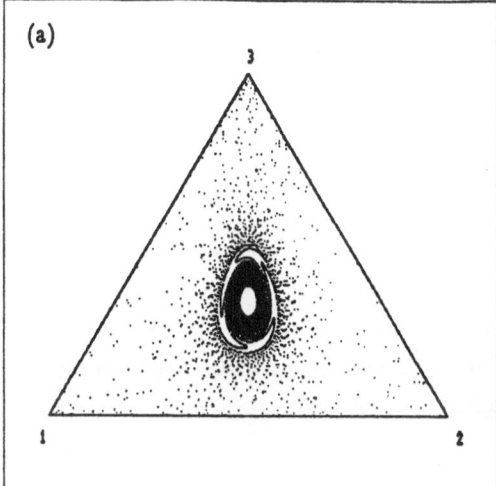

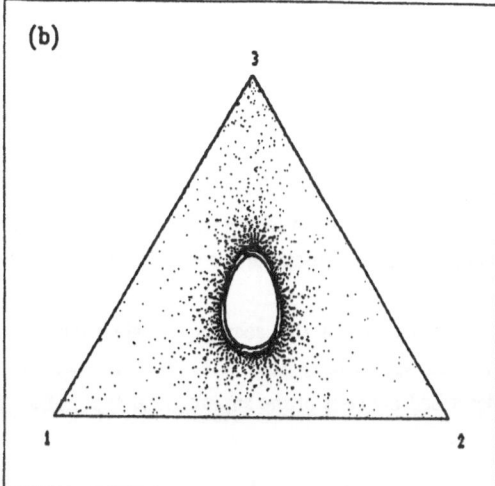

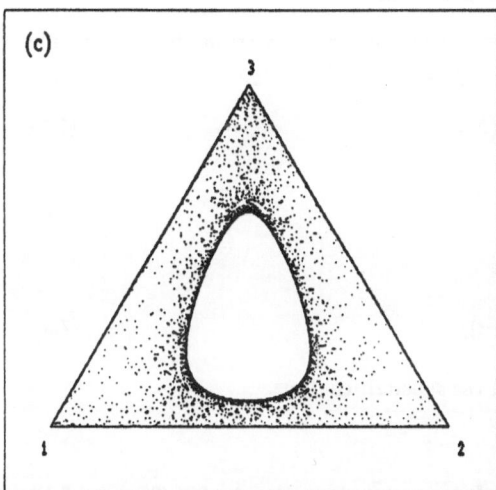

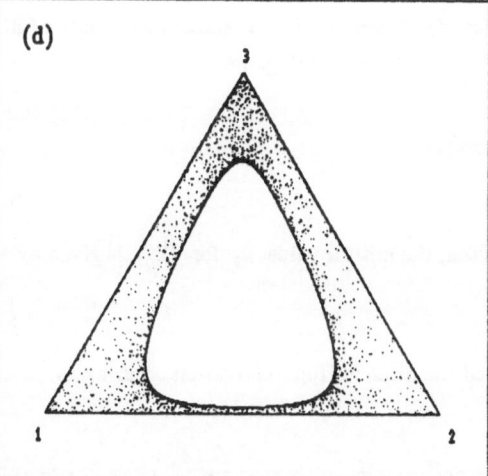

FIGURE 7: A discrete supercritical Hopf bifurcation (occuring at $\mu_0 = 1.40$) which is induced by the one–parameter family of central RSP–games given by:

$$A_\mu = \begin{bmatrix} a & b & c \\ c & a & b \\ b+3\mu & c-3\mu & a \end{bmatrix}, \text{ where } a = 10, b = 11.5, c = 9.$$

Orbit starting at $\mathbf{p} = (0.01, 0.01, 0.98)$ for:

(a) $\mu = 1.401$,
(b) $\mu = 1.41$,
(c) $\mu = 1.45$,
(d) $\mu = 1.50$.

(c) corresponds to the example in Figure 6. See text for details.

In fact, the interior fixed point $p_\mu = m$ of A_μ is a discrete hyperbolic attractor for $\mu < \mu_0$ while it is a hyperbolic repellor for $\mu > \mu_0$. For $\mu = \mu_0$, the eigenvalues of the linearization of $\imath_\mu$ at p_μ^* are one in modulus, i.e., they belong to the unit circle in the complex plane.

The numerical example in Figure 7 suggests that the Hopf bifurcation at μ_0 is a supercritical one: For parameter values μ which are slightly larger than μ_0, an attracting closed limit curve arises, the 'radius' of which increases with increasing distance of μ to μ_0. We have seen that this phenonemon is typical for a supercritical Hopf bifurcation.

In the rest of this section, we shall complement the numerical results of Figures 6 and 7 by proving analytically that — at least for certain parameter constellations — the Hopf bifurcation described above is of the supercritical type. More precisely, we shall show:

THEOREM 7.1: Existence of supercritical discrete Hopf bifurcations.

Let $(\imath_\mu)$ denote the one–parameter family of discrete replicator dynamics arising from the family (A_μ) of central RSP–games which are given by (7.2). Suppose that a is related to b and c via

$$a^2 = bc - \tfrac{1}{2} \cdot (b-c)^2, \tag{7.4}$$

and that

$$b < \tfrac{5}{3} \cdot c. \tag{7.5}$$

Then, the critical value μ_0 for $(\imath_\mu)$ is given by

$$\mu_0 = \tfrac{1}{2} \cdot (b-c), \tag{7.6}$$

and the discrete Hopf bifurcation occuring at μ_0 is of the supercritical type.

The significance of conditions (7.4) and (7.5) will become clear from the proof of Theorem 7.1. (7.4) is not really needed, but the proof is simplified considerably if it holds true. (7.5) ensures that the matrix A_μ remains non–negative for all μ which are slightly larger than μ_0. Notice that the set of parameter constellations (a,b,c) satisfying (7.4) and (7.5) is not empty. For example, the triple $(13,17,11)$ satisfies (7.4) and (7.5) as well as the inequalities in (7.2).

Let us state the main conclusion of Theorem 7.1 as a corollary:

COROLLARY 7.2: Existence of attracting closed limit curves.
There are discrete replicator dynamics which are induced by RSP–games and which admit an attracting closed limit curve.

To my knowledge, this is the first time that closed limit curves are described for the discrete replicator dynamics. There is, however, an 'indirect' proof of their existence in higher dimensions. Hofbauer & Iooss (1984) have shown that a supercritical Hopf bifurcation for the continuous replicator dynamics induces a bifurcation of the same type for the discrete replicator equation if selection is 'weak' enough. This result is not applicable here, since supercritical Hopf bifurcations do not occur for the continuous replicator dynamics if only three pure strategies are involved like in our RSP–games (see Hofbauer 1981).

Before we come to the proof of Theorem 7.1, we shall cite some general results on discrete Hopf bifurcations from dynamical systems theory (see e.g. Guckenheimer & Holmes 1983, Chapter 3.5). Consider a one–parameter family of two–dimensional discrete dynamical systems G_μ that has a smooth family of fixed points p_μ^* at which the eigenvalues of the linearization $DG_\mu(p_\mu^*)$ are complex conjugate to another. Suppose that $\lambda(\mu)$ denotes one of the two eigenvalues of $DG_\mu(p_\mu)$ and that $\lambda(\mu)$ is a smooth function of μ. A *discrete Hopf bifurcation* occurs at the parameter $\mu = \mu_0$, if the family of eigenvalues $\lambda(\mu)$ crosses the unit circle of the complex plane at μ_0 with a positive velocity, i.e., if the following two conditions hold true:

$$|\lambda(\mu_0)| = 1, \tag{7.7}$$

$$\frac{d}{d\mu}(|\lambda(\mu)|)_{\mu=\mu_0} = v > 0. \tag{7.8}$$

(7.7) and (7.8) imply that (for μ near the bifurcation point μ_0) the fixed point p_μ^* is hyperbolically stable for $\mu < \mu_0$ and hyperbolically unstable for $\mu > \mu_0$. Bifurcation structures associated with eigenvalues $\lambda(\mu_0)$ which are third or fourth roots of unity have some special features (see Iooss 1979). Let us neglect these *resonance cases* and assume that

$$\lambda^j(\mu_0) \neq 1 \quad \text{for} \quad j = 1, 2, 3, 4, \tag{7.9}$$

or equivalently

$$\lambda(\mu_0) \notin \mathbb{R}, \quad \lambda(\mu_0) \notin i \cdot \mathbb{R} \quad \text{and} \quad \text{Re}(\lambda(\mu_0)) \neq -\tfrac{1}{2}. \tag{7.10}$$

A famous theorem from the theory of dynamical systems (see Marsden & McCracken 1976, Iooss 1979, Guckenheimer & Holmes 1983) states that under the conditions (7.7), (7.8), and (7.9) there is a smooth change of coordinates H so that the expression of $H \circ G_\mu \circ H^{-1}$ in polar coordinates has the form:

$$H \circ G_\mu \circ H^{-1}(r,\theta) = (r(1+v(\mu-\mu_0)-wr^2), \theta+s+tr^2) + \Omega(r,\theta), \tag{7.11}$$

where

$$s := |\arg(\lambda(\mu_0))| \neq 0 \tag{7.12}$$

denotes the argument of $\lambda(\mu_0)$, whereas $\Omega(r,\theta)$ collects all higher–order terms.

Notice that the first component of (7.11) can be represented in the form

$$r' = r \cdot (1 + v(\mu-\mu_0) - wr^2) + \Omega_1(r,\theta).$$ (7.13)

Consequently, for $w \neq 0$ a third–order approximation may be written as

$$\frac{\Delta r}{wr} = \frac{v}{w} \cdot (\mu-\mu_0) - r^2.$$ (7.14)

Depending on the sign of w, the truncated dynamical system (7.14) has either for all $\mu < \mu_0$ or for all $\mu > \mu_0$ a fixed point $r_\mu^* > 0$ which is given by:

$$r_\mu^* = \sqrt{\frac{v}{w} \cdot (\mu-\mu_0)}, \quad \text{if} \quad \text{sgn}(w) = \text{sgn}(\mu-\mu_0) \neq 0.$$ (7.15)

For those μ for which $\text{sgn}(w) = \text{sgn}(\mu-\mu_0)$, (7.14) can be written in the form

$$\Delta r = rw \cdot ((r_\mu^*)^2 - r^2).$$ (7.14a)

(7.14a) shows that the stability of r_μ^* also depends on the sign of w : r_μ^* is an attractor *if* $w > 0$ and a repellor if $w < 0$. Obviously, r_μ^* is the radius of a circle which is invariant with respect to the third–order approximation of $H \circ G_\mu \circ H^{-1}$, and the circle is an attractive (a repulsive) closed limit curve if and only if r_μ^* is an attractor (a repellor) of (7.14a).

These considerations show that the sign of the *Hopf parameter w* is of crucial importance for the dynamical behaviour of the bifurcating system: The Hopf bifurcation is supercritical if w is positive, and it is subcritical if w is negative. Nothing definite can be said if the Hopf parameter of the bifurcating system is equal to zero. In this case, higher order terms have to be considered. 'Generically', the bifurcation is again either supercritical or subcritical (see Iooss 1979). In applications, however, the condition $w = 0$ often implies that the Hopf bifurcation is a degenerate one (as in the case of circulant RSP–games).

We shall now present a formula (see (7.19)) which allows us to calculate the Hopf parameter of a bifurcating system by having a closer look at the higher order terms of the bifurcating map G_{μ_0} near the fixed point $p_{\mu_0}^*$. This formula may be derived by transforming G_{μ_0} to 'normal form' (see Iooss 1979). A complex coordinate version of it was developed by Wan (1978). In slightly modified form, the real coordinate version given below may be found in Guckenheimer & Holmes (1983).

The calculation of the Hopf parameter w is simplified considerably, if the linear part of the bifurcating system $G_{\mu_0}: \mathbb{R}^2 \longrightarrow \mathbb{R}^2$ at the fixed point $p_{\mu_0}^*$ is in real Jordan form. Let us call the $\mathbb{R}^2$–coordinates *Jordan coordinates* if $DG_{\mu_0}(p_{\mu_0}^*)$ is of the form

$$DG_{\mu_0}(p_{\mu_0}^*) = \begin{bmatrix} \mathrm{Re}(\lambda) & -\mathrm{Im}(\lambda) \\ \mathrm{Im}(\lambda) & \mathrm{Re}(\lambda) \end{bmatrix}, \quad \text{where} \quad \lambda = \lambda(\mu_0).$$ (7.16)

Suppose that a Jordan coordinate representation of G_{μ_0} is given by

$$G_{\mu_0}(x,y) = \left[\begin{matrix} f(x,y) \\ g(x,y) \end{matrix} \right],$$

(7.17)

where f and g are scalar functions. Let us for the rest of this section use the convention that terms like f_i, g_{ij}, or h_{ijj} denote (higher order) partial derivatives of f, g, and h with respect to the ith and jth coordinate which are evaluated at the fixed point $p^*_{\mu_0}$, e.g.:

$$f_{12} := \frac{\partial^2}{\partial x \partial y} f(p^*_{\mu_0}), \text{ or } g_{22} := \frac{\partial^2}{\partial y^2} g(p^*_{\mu_0}).$$

(7.18)

Now the Hopf parameter w can be obtained from the formula (Guckenheimer & Holmes 1983, p.163):

$$w = \frac{1}{16} \cdot \left[\operatorname{Re}(\xi_0 \xi_A \xi_B) + \frac{1}{2} \cdot |\xi_B|^2 + |\xi_C|^2 - \operatorname{Re}(\bar{\lambda} \xi_D) \right],$$

(7.19)

where ξ_0, ξ_A, ξ_B, ξ_C, and ξ_D are given by:

$$\xi_0 = \frac{(1-2\lambda)\bar{\lambda}^2}{1-\lambda},$$

(7.20)

$$\xi_A = \frac{1}{2} \cdot ((f_{11} - f_{22} + 2g_{12}) + i \cdot (g_{11} - g_{22} - 2f_{12})),$$

(7.21)

$$\xi_B = (f_{11} + f_{22}) + i \cdot (g_{11} + g_{22}),$$

(7.22)

$$\xi_C = \frac{1}{2} \cdot ((f_{11} - f_{22} - 2g_{12}) + i \cdot (g_{11} - g_{22} + 2f_{12})),$$

(7.23)

$$\xi_D = (f_{111} + f_{122} + g_{112} + g_{222}) + i \cdot (g_{111} + g_{122} - f_{112} - f_{222}).$$

(7.24)

Notice that $\lambda := \lambda(\mu_0)$ and therefore $|\lambda| = 1$, i.e.,

$$\lambda \cdot \bar{\lambda} = (\operatorname{Re}(\lambda))^2 + (\operatorname{Im}(\lambda))^2 = 1.$$

(7.25)

It is easy to see that this implies

$$|1-\lambda|^2 = 2 \cdot (1 - \operatorname{Re}(\lambda)),$$

(7.26)

and

$$\frac{1-2\lambda}{1-\lambda} = 1 + \frac{1-\lambda}{|1-\lambda|^2} = \frac{3}{2} - \frac{\operatorname{Im}(\lambda)}{2(1-\operatorname{Re}(\lambda))} \cdot i.$$

(7.27)

On the other hand, (7.25) yields

$$\operatorname{Re}(\bar{\lambda}^2) = 2 \cdot (\operatorname{Re}(\lambda))^2 - 1,$$

(7.28)

$$\operatorname{Im}(\bar{\lambda}^2) = 2 \cdot \operatorname{Re}(\lambda) \cdot \operatorname{Im}(\lambda).$$

(7.29)

Applying (7.27), (7.28), and (7.29), it is easy to see that the parameter ξ_0 is of the form:

$$\mathrm{Re}(\xi_0) = \mathrm{Re}(\lambda) \cdot (2 \cdot \mathrm{Re}(\lambda) - 1) - \frac{3}{2}, \tag{7.30}$$

$$\mathrm{Im}(\xi_0) = -\mathrm{Im}(\lambda) \cdot (2 \cdot \mathrm{Re}(\lambda) - 1) - \frac{\mathrm{Im}(\lambda)}{2(1 - \mathrm{Re}(\lambda))}. \tag{7.31}$$

In order to calculate the other parameters ξ_A, ξ_B, ξ_C, and ξ_D, we have to know some of the higher order derivatives of the functions f and g.

If we want to apply (7.19) to a Hopf bifurcation arising in the context of the discrete replicator dynamics, we have to proceed as follows: First, the bifurcating replicator equation has to be transformed to a discrete dynamical system on $\mathbb{R}^2$. Then, we have to calculate the linearization at the interior fixed point. Transforming the linearization into real Jordan form gives us Jordan coordinates and the representation (7.17). On the basis of f and g, we may then calculate the ξ's and the Hopf parameter w.

Let $(\mathbf{z}_\mu)$ denote any one–parameter family of discrete replicator equations $\mathbf{p}' = \mathbf{z}_\mu(\mathbf{p})$ which is induced by a one–parameter family of central evolutionary 3x3 normal form games (A_μ). Then $\mathbf{z}_\mu: \Delta \longrightarrow \Delta$ is a mapping which is given by

$$(\mathbf{z}_\mu(\mathbf{p}))_i = \frac{p_i \cdot (A_\mu \mathbf{p})_i}{\mathbf{p} \cdot A_\mu \mathbf{p}}, \quad \mathbf{p} \in \Delta, \ i \in I. \tag{7.32}$$

Each mapping $\mathbf{z}_\mu$ has an interior fixed point in the barycenter $\mathbf{m}$ of the strategy simplex. Setting $\mathbf{p} = \mathbf{m} + \mathbf{z}$, $\mathbf{z} \in \mathbb{R}_0^3$, we get $\mathbf{z}' = \mathbf{p}' - \mathbf{m} = \mathbf{z}_\mu(\mathbf{m} + \mathbf{z}) - \mathbf{m}$. Accordingly, the discrete dynamical system $\mathbf{z}_\mu$ on Δ can be transformed to an equivalent system S_μ on a subset Δ_S of $\mathbb{R}_0^3$, where Δ_S and $S_\mu: \Delta_S \longrightarrow \Delta_S$ are defined by

$$\Delta_S := \{ \mathbf{z} \in \mathbb{R}_0^3 \mid \mathbf{m} + \mathbf{z} \in \Delta \}, \tag{7.33}$$

$$S_\mu(\mathbf{z}) := \mathbf{z}_\mu(\mathbf{m} + \mathbf{z}) - \mathbf{m}, \quad \mathbf{z} \in \Delta_S. \tag{7.34}$$

Each mapping S_μ has a unique interior fixed point in the zero vector $\mathbf{0}$. If we identify $\mathbb{R}_0^3$ with $\mathbb{R}^2$ by means of the canonical isomorphism $P : \mathbb{R}_0^3 \longrightarrow \mathbb{R}^2$ (see (4.10)), we get an equivalent representation $\mathcal{G}_\mu$ of $\mathbf{z}_\mu$ on a subset $\Delta_\mathcal{G}$ of $\mathbb{R}^2$. Obviously, $\Delta_\mathcal{G}$ and $\mathcal{G}_\mu: \Delta_\mathcal{G} \longrightarrow \Delta_\mathcal{G}$ are given by

$$\Delta_\mathcal{G} := P(\Delta_S), \quad \mathcal{G}_\mu := P^{-1} \circ S_\mu \circ P. \tag{7.35}$$

In essence, $\mathcal{G}_\mu(z_1, z_2)$ results from $S_\mu(z_1, z_2, z_3)$ by replacing z_3 by $-z_1 - z_2$ and by discarding the third component of the vector $S_\mu(\mathbf{z})$.

The linearization of $\mathcal{G}_{\mu_0}$ at its unique interior fixed point 0 (which corresponds to $p = m$) is given by (6.6), (6.5), and (5.25). Let us suppose that the linearization of $\mathcal{G}_{\mu_0}$ is in real Jordan form and that f and g denote the components of $\mathcal{G}_{\mu_0}$ with respect to these Jordan coordinates:

$$\mathcal{G}_{\mu_0}(x,y) = (\ f(x,y),\ g(x,y)\ ). \tag{7.36}$$

It is easy to see that the scalar functions f and g are of the form

$$f(x,y) = \frac{n^f(x,y)}{d(x,y)} - \frac{1}{3}, \tag{7.37}$$

$$g(x,y) = \frac{n^g(x,y)}{d(x,y)} - \frac{1}{3}, \tag{7.38}$$

where the common denominator $d(x,y)$ corresponds to mean fitness. In particular, we have

$$C := d(0) = \overline{F}(m). \tag{7.39}$$

Notice that

$$f(0) = g(0) = 0, \tag{7.40}$$

since 0 is a fixed point of $\mathcal{G}_{\mu_0}$. In order to get the Hopf parameter w, we have to evaluate some higher order partial derivatives of f and g at the interior fixed point 0. Since f and g are rational functions, the expressions for these partial derivatives are quite cumbersome. We shall circumvent this problem by deriving some recursive formulae for the terms f_{ij}, g_{ij}, f_{iij}, and g_{iij} which make it superfluous to calculate the partial derivatives of f and g explicitly.

It is easy to see that the numerators and the denominator of f and g are polynomials of degree two, the derivatives of which are easy to calculate. In particular, all those partial derivatives of n^f, n^g, and d vanish which are of third and higher order. Using this fact together with (7.39) and (7.40), the quotient rule of differentiation yields the following formulae for the partial derivatives of f evaluated at the origin:

$$f_i = \frac{1}{C}\cdot(\ n^f_i - \tfrac{1}{3}\cdot d_i\ ), \tag{7.41}$$

$$f_{ij} = \frac{1}{C}\cdot(\ n^f_{ij} - \tfrac{1}{3}\cdot d_{ij} - f_i d_j - f_j d_i\ ), \tag{7.42}$$

$$f_{iij} = \frac{1}{C}\cdot(\ -f_j d_{ii} - 2f_i d_{ij} - 2f_{ij} d_i - f_{ii} d_j\ ). \tag{7.43}$$

Of course, the corresponding expressions for g_i, g_{ij}, and g_{iij} are completely analogous. Notice that by definition the linarization J_{μ_0} of $\mathcal{G}_{\mu_0}$ at 0 is of the form

$$J_{\mu_0} = \begin{bmatrix} f_1 & f_2 \\ g_1 & g_2 \end{bmatrix}. \tag{7.44}$$

If we are dealing with Jordan coordinates, a comparison of (7.44) and (7.16) yields

$$f_1 = \text{Re}(\lambda), \quad f_2 = -\text{Im}(\lambda), \tag{7.45}$$

$$g_1 = \text{Im}(\lambda), \quad g_2 = \text{Re}(\lambda). \tag{7.46}$$

We shall see later that the calculation of ξ_A, ξ_B, and ξ_C is simplified considerably if these expressions are inserted into (7.42).

All these preparatory remarks hold true for *any* Hopf bifurcation arising in a one–parameter family of discrete replicator equations. We have presented them in some generality in order to indicate how the techniques for calculating the Hopf parameter w can be applied to 3x3 normal form games which go beyond the class of RSP–games. Let us now apply these considerations to the class of RSP–games which is given by (7.2).

PROOF of THEOREM 7.1:

A: Let from now on denote $(\mathbf{z}_\mu)_{\mu>0}$ the one–parameter family of discrete replicator dynamics which is induced by the family (A_μ) of central RSP–games given by (7.2). Let $(\mathcal{G}_\mu)$ be the corresponding family of mappings which is defined by (7.35). With respect to canonical $\mathbb{R}^2$–coordinates, the linearization $D\mathcal{G}_\mu(0)$ of $\mathcal{G}_\mu$ at its unique interior fixed point 0 is characterized by a matrix J_μ which is defined by (6.6), (6.5), and (5.25). Let $\lambda(\mu)$ denote one of the two complex conjugate eigenvalues of J_μ and suppose that $\lambda(\mu)$ is chosen in a way that makes it smoothly dependent on the parameter μ.

From the results of Section 6 we know already that the parameter μ_0 which is defined by (7.3) is a candidate for being the critical value for a discrete Hopfbifurcation, since

$$|\lambda(\mu)| \begin{cases} < 1 & \text{for } \mu < \mu_0 \\ = 1 & \text{for } \mu = \mu_0 \\ > 1 & \text{for } \mu > \mu_0 \end{cases}. \tag{7.47}$$

Of course, it is necessary that A_{μ_0} is an admissible (i.e., a non–negative) payoff matrix for the discrete replicator dynamics. For this to be true, we need $c \geq 3\mu_0$, and we shall even require $c > 3\mu_0$ since we are also interested in μ's which are slightly larger than μ_0. It is easy to see that $c > 3\mu_0$ is equivalent to:

$$bc > a^2 > bc - \tfrac{1}{3} c(b-c). \tag{7.48}$$

Together with (6.6) and (6.5), (5.28) implies that the matrix J_μ is given by

$$J_\mu = \frac{1}{3 \cdot C} \begin{bmatrix} 2a+b-\mu & b-c+\mu \\ -b+c-\mu & 2a+c+\mu \end{bmatrix}, \tag{7.49}$$

where C denotes the mean fitness at the interior fixed point $p^* = m$:

$$C := F(m) = \tfrac{1}{3} \cdot (a+b+c). \tag{7.50}$$

We have already shown that the eigenvalues $\lambda(\mu)$ and $\overline{\lambda}(\mu)$ of J_μ are complex conjugate to another. Consequently, the modulus of $\lambda(\mu)$ corresponds to the square root of the determinant of J_μ, and (7.8) is equivalent to:

$$\tfrac{d}{d\mu} (\det(J_\mu))_{\mu=\mu_0} > 0. \tag{7.51}$$

In view of

$$\tfrac{d}{d\mu} (\det(J_\mu)) = \tfrac{b-c}{C} > 0 , \tag{7.52}$$

all criteria for the occurrence of a discrete Hopf bifurcation at μ_0 are fulfilled.

B: Inspection of (7.49) shows that J_{μ_0} is already in real Jordan form if its diagonal elements are equal to another, i.e., if

$$2a + b - \mu_0 = 2a + c + \mu_0, \tag{7.53}$$

or equivalently

$$\mu_0 = \tfrac{1}{2} \cdot (b-c). \tag{7.54}$$

Accordingly, we are already dealing with Jordan coordinates if (7.3) and (7.53) are satisfied simultaneously, i.e., if a, b, and c are related to another according to:

$$a^2 = bc - \tfrac{1}{4} \cdot (b-c)^2. \tag{7.55}$$

If (7.55) holds, (7.48) is equivalent to

$$c > \tfrac{3}{4} \cdot b. \tag{7.56}$$

Notice that (7.56) corresponds to (7.5), and that (7.55) is identical to (7.4). Accordingly, assumption (7.4) allows us to circumvent the awkward procedure of transforming $\mathcal{G}_{\mu_0}$ to Jordan coordinates, whereas (7.5) guarantees that A_μ is an admissible payoff matrix for the discrete replicator dynamics for all μ which are slightly larger than μ_0. Let us from now on suppose that these two conditions are satisfied.

C: A comparison of (7.49) with (7.16) shows that the eigenvalue $\lambda(\mu_0)$ at the critical value μ_0 is given by

$$\lambda := \lambda(\mu_0) = \tfrac{1}{2 \cdot C} ((a+C) - i \cdot (b-c)). \tag{7.57}$$

(It is important to take that eigenvalue of J_{μ_0} which has a negative imaginary part, since otherwise (7.16) and (7.49) would not correspond to another.)

It is obvious from (7.57) that λ is neither real nor purely imaginary. On the other hand, (7.56) together with $c \leq a$ implies $a+c > b$. It is easy to see that this inequality yields

$$|\text{Im}(\lambda)| < |\text{Re}(\lambda)|. \tag{7.58}$$

(7.58) implies that (7.10) holds true, i.e., that λ is neither a third nor a fourth root of unity: the Hopf bifurcation is non-resonant if (7.4) and (7.5) are satisfied.

D: Let us now consider the coordinate representation (7.36) of $\mathcal{G}_{\mu_0}$. Since the canonical $\mathbb{R}^2$-coordinates are already Jordan coordinates, it is easy to calculate the scalar functions f and g. In view of (7.37) and (7.38), they are determined by:

$$n^f(x,y) := \tfrac{1}{3}\cdot C + \tfrac{1}{3}(2a+b)x + \tfrac{1}{3}(b-c)y + (a-c)x^2 + (b-c)xy, \tag{7.59}$$

$$n^g(x,y) := \tfrac{1}{3}\cdot C - \tfrac{1}{3}(b-c)x + \tfrac{1}{3}(2a+c)y - (b-a)y^2 - (b-c)xy, \tag{7.60}$$

$$d(x,y) := C - (b+c-2a)(x^2+y^2+xy) - \mu_0 \cdot (3x^2 - 3y^2 - x + y). \tag{7.61}$$

Notice that (7.61) yields

$$d_1 = -d_2 = \mu_0 = \tfrac{1}{3}\cdot(b-c). \tag{7.62}$$

E: On the basis of (7.42), we shall now calculate the terms f_{ij} and g_{ij}. In order to do this, let us introduce the notation

$$N^f_{ij} := n^f_{ij} - \tfrac{1}{3}\cdot d_{ij}, \quad N^g_{ij} := n^g_{ij} - \tfrac{1}{3}\cdot d_{ij}. \tag{7.63}$$

Now, (7.42) together with (7.45), (7.46), and (7.62) leads to the expressions

$$f_{11} = \tfrac{1}{C}\cdot(N^f_{11} - 2\cdot\mu_0\cdot\text{Re}(\lambda)), \tag{7.64}$$

$$f_{12} = \tfrac{1}{C}\cdot(N^f_{12} + \mu_0\cdot(\text{Re}(\lambda)+\text{Im}(\lambda))), \tag{7.65}$$

$$f_{22} = \tfrac{1}{C}\cdot(N^f_{22} - 2\cdot\mu_0\cdot\text{Im}(\lambda)), \tag{7.66}$$

$$g_{11} = \tfrac{1}{C}\cdot(N^g_{11} - 2\cdot\mu_0\cdot\text{Im}(\lambda)), \tag{7.67}$$

$$g_{12} = \tfrac{1}{C}\cdot(N^g_{12} - \mu_0\cdot(\text{Re}(\lambda)-\text{Im}(\lambda))), \tag{7.68}$$

$$g_{22} = \tfrac{1}{C}\cdot(N^g_{22} + 2\cdot\mu_0\cdot\text{Re}(\lambda)). \tag{7.69}$$

It is easy to calculate the terms N^f_{ij} and N^g_{ij}. In view of (7.59), (7.60), and (7.61), we get

$$N_{11}^f = \tfrac{1}{3}\cdot(2a + 5b - 7c), \tag{7.70}$$

$$N_{12}^f = -\tfrac{2}{3}\cdot(a - 2b + c), \tag{7.71}$$

$$N_{22}^f = \tfrac{1}{3}\cdot(-4a - b + 5c), \tag{7.72}$$

$$N_{11}^g = \tfrac{1}{3}\cdot(-4a + 5b - c), \tag{7.73}$$

$$N_{12}^g = -\tfrac{2}{3}\cdot(a + b - 2c), \tag{7.74}$$

$$N_{22}^g = \tfrac{1}{3}\cdot(2a - 7b + 5c). \tag{7.75}$$

$\underline{F}$: We shall now calculate the terms ξ_A, ξ_B, and ξ_C by inserting (7.64) to (7.69) into (7.21) to (7.23). The resulting formulae will be simplified by considering

$$\text{Im}(\lambda) = -\mu_0/C \tag{7.76}$$

and the following identities:

$$N_{11}^f + N_{22}^f = N_{12}^f, \tag{7.77}$$

$$N_{11}^g + N_{22}^g = N_{12}^g, \tag{7.78}$$

$$N_{11}^f - N_{22}^f = -3\cdot N_{12}^g, \tag{7.79}$$

$$N_{11}^g - N_{22}^g = 3\cdot N_{12}^f. \tag{7.80}$$

ξ_A, ξ_B, and ξ_C are given by:

$$\text{Re}(\xi_A) = -\frac{1}{2\cdot C}\cdot N_{12}^g + 2\cdot\text{Im}(\lambda)\cdot(\text{Re}(\lambda)-\text{Im}(\lambda)), \tag{7.81}$$

$$\text{Im}(\xi_A) = \frac{1}{2\cdot C}\cdot N_{12}^f + 2\cdot\text{Im}(\lambda)\cdot(\text{Re}(\lambda)+\text{Im}(\lambda)), \tag{7.82}$$

$$\text{Re}(\xi_B) = \frac{1}{C}\cdot N_{12}^f + 2\cdot\text{Im}(\lambda)\cdot(\text{Re}(\lambda)+\text{Im}(\lambda)), \tag{7.83}$$

$$\text{Im}(\xi_B) = \frac{1}{C}\cdot N_{12}^g - 2\cdot\text{Im}(\lambda)\cdot(\text{Re}(\lambda)-\text{Im}(\lambda)), \tag{7.84}$$

$$\text{Re}(\xi_C) = -\frac{5}{2\cdot C}\cdot 2N_{12}^g, \tag{7.85}$$

$$\text{Im}(\xi_C) = \frac{5}{2\cdot C}\cdot 2N_{12}^f. \tag{7.86}$$

$\underline{G}$: Inserting (7.71) and (7.74) into (7.83) to (7.86), we get in view of (7.25):

$$|\xi_C|^2 = \frac{25}{9C^2}\cdot\left[(b-a)^2+(a-c)^2+4(b-c)^2 \right], \tag{7.87}$$

$$\tfrac{1}{2}\cdot|\xi_B|^2 = \frac{2}{25}\cdot|\xi_C|^2 + 4\cdot(\text{Im}(\lambda))^2\cdot\left[\frac{C - 2a}{C} \right]. \tag{7.88}$$

Replacing $\text{Im}(\lambda)$ by (7.76), this implies

$$\tfrac{1}{2}\cdot|\xi_B|^2+|\xi_C|^2 = \frac{1}{C^2}\cdot\left[3(b-a)^2+3(a-c)^2+(\frac{13C-2a}{C})\cdot(b-c)^2 \right]. \tag{7.89}$$

Finally, notice that $\mathrm{Re}(\xi_0\xi_{\mathrm{A}}\xi_{\mathrm{B}})$ is given by

$$\mathrm{Re}(\xi_0\xi_{\mathrm{A}}\xi_{\mathrm{B}}) = \mathrm{Re}(\xi_0)\mathrm{Re}(\xi_{\mathrm{A}})\mathrm{Re}(\xi_{\mathrm{B}}) - \mathrm{Re}(\xi_0)\mathrm{Im}(\xi_{\mathrm{A}})\mathrm{Im}(\xi_{\mathrm{B}}) \tag{7.90}$$
$$- \mathrm{Im}(\xi_0)\mathrm{Re}(\xi_{\mathrm{A}})\mathrm{Im}(\xi_{\mathrm{B}}) - \mathrm{Im}(\xi_0)\mathrm{Im}(\xi_{\mathrm{A}})\mathrm{Re}(\xi_{\mathrm{B}}).$$

In view of (7.19), the only thing that is missing for the calculation of the Hopf parameter w is an expression for the parameter ξ_{D}. This will be derived next.

<u>H:</u> Using (7.43) together with (7.45), (7.46), and (7.62), we get the following formula for ξ_{D}:

$$\mathrm{Re}(\xi_{\mathrm{D}}) = -\frac{1}{C}\cdot(\, \sigma'\cdot\mu_0 + 4\cdot\mathrm{Re}(\lambda)\cdot(d_{11}+d_{22})\,), \tag{7.91}$$

$$\mathrm{Im}(\xi_{\mathrm{D}}) = -\frac{1}{C}\cdot(\, \tau'\cdot\mu_0 + 4\cdot\mathrm{Im}(\lambda)\cdot(d_{11}+d_{22})\,), \tag{7.92}$$

where σ' and τ' are defined by

$$\sigma' := 3(f_{11}-g_{22}) - 2(f_{12}-g_{12}) + (f_{22}-g_{11}), \tag{7.93}$$

$$\tau' := 3(f_{22}+g_{11}) - 2(f_{12}+g_{12}) + (f_{11}+g_{22}). \tag{7.94}$$

Inserting (7.64) to (7.69) into (7.93) and (7.94) and considering (7.76), we get the following simplified expressions for $\mathrm{Re}(\xi_{\mathrm{D}})$ and $\mathrm{Im}(\xi_{\mathrm{D}})$:

$$\mathrm{Re}(\xi_{\mathrm{D}}) = \rho\cdot\mathrm{Re}(\lambda) + \sigma\cdot\mathrm{Im}(\lambda), \tag{7.95}$$

$$\mathrm{Im}(\xi_{\mathrm{D}}) = (\rho+\tau)\cdot\mathrm{Im}(\lambda), \tag{7.96}$$

where ρ, σ, and τ are given by

$$\rho := 16\cdot(\mathrm{Im}(\lambda))^2 - 4\cdot\frac{d_{11}+d_{22}}{C}, \tag{7.97}$$

$$\sigma := \frac{1}{C}\cdot\left[3(\mathrm{N}_{11}^{\mathrm{f}}-\mathrm{N}_{22}^{\mathrm{g}}) - 2(\mathrm{N}_{12}^{\mathrm{f}}-\mathrm{N}_{12}^{\mathrm{g}}) + (\mathrm{N}_{22}^{\mathrm{f}}-\mathrm{N}_{11}^{\mathrm{g}})\right], \tag{7.98}$$

$$\tau := \frac{1}{C}\cdot\left[3(\mathrm{N}_{22}^{\mathrm{f}}+\mathrm{N}_{11}^{\mathrm{g}}) - 2(\mathrm{N}_{12}^{\mathrm{f}}+\mathrm{N}_{12}^{\mathrm{g}}) + (\mathrm{N}_{11}^{\mathrm{f}}+\mathrm{N}_{22}^{\mathrm{g}})\right]. \tag{7.99}$$

From (7.95) and (7.96), we get

$$\mathrm{Re}(\overline{\lambda}\cdot\xi_{\mathrm{D}}) = \rho + \mathrm{Im}(\lambda)\cdot(\sigma\cdot\mathrm{Re}(\lambda) + \tau\cdot\mathrm{Im}(\lambda)). \tag{7.100}$$

<u>I:</u> We are now in the position to calculate the Hopf parameter w by inserting (7.89), (7.90), and (7.100) into (7.19). For example, we get $w = 0.012$ for the parameter constellation $(a,b,c) = (13,17,11)$.

The proof of Theorem 7.1 will be completed by showing that w is positive for *all* parameter constellations satisfying (7.4) and (7.5). Without loss in generality, we may assume $c = 1$, since the discrete replicator dynamics is not affected if the payoff matrix is multiplied by the

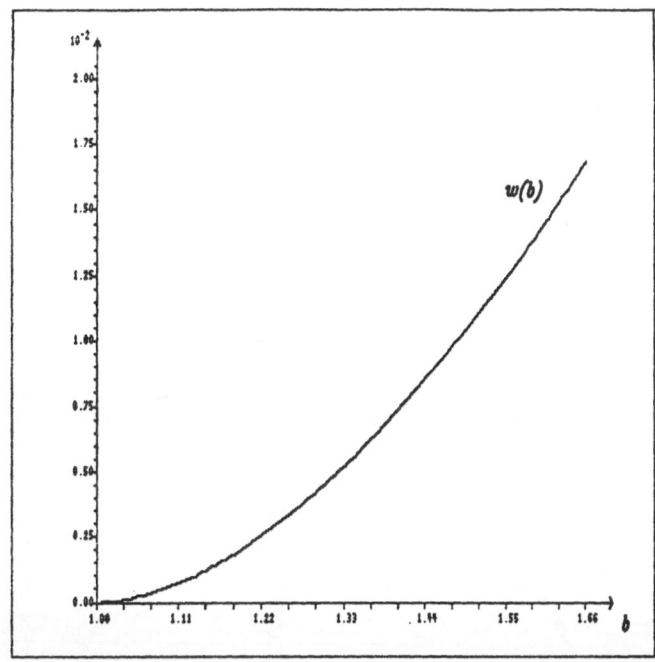

FIGURE 8: With c normalized to one, the Hopf parameter w is depicted as a function of b ($1 < b < 5/3$). For all admissible b, the Hopf parameter is positive, i.e., the Hopf bifurcation is a supercritical one.

positive scalar $1/c$. For any b which is compatible with (7.5) (i.e., $1 < b < 5/3$), the parameter a is fully specified by (7.4). Accordingly, the paramter w may be interpreted as a function of b. In Figure 8, w is depicted as a function of b. Obviously, we have $w(b) > 0$ for all b with $1 < b < 1.667$. This completes the proof.

* * *

From Figures 6 and 7, one might get the impression that the attracting closed limit curve arising from a discrete Hopf bifurcation as described above is a *global* attractor in the sense that it attracts all interior non–equilibrium orbits. Figure 9 shows that this impression is not necessarily correct.

The example in Figure 9 is based on a central RSP–game of the form (7.2) where (a,b,c) is given by (13,17,11). A discrete supercritical Hopf bifurcation occurs at the critical value $\mu_0 = 3$. For admissible parameter values μ which are slightly larger than 3 (Figure 9 shows the case $\mu = 3.25$), an attracting closed orbit C arises which attracts all nearby orbits including all those orbits starting in the region which is enclosed by C. However, C is not a global attractor. Instead, it is encircled by *another* invariant closed curve C' which is a repulsive limit curve for the discrete replicator dynamics. Figure 9 indicates that there are *two* attractors (the limit curve C and the boundary of the strategy simplex) whose domain of attraction is separated by C'.

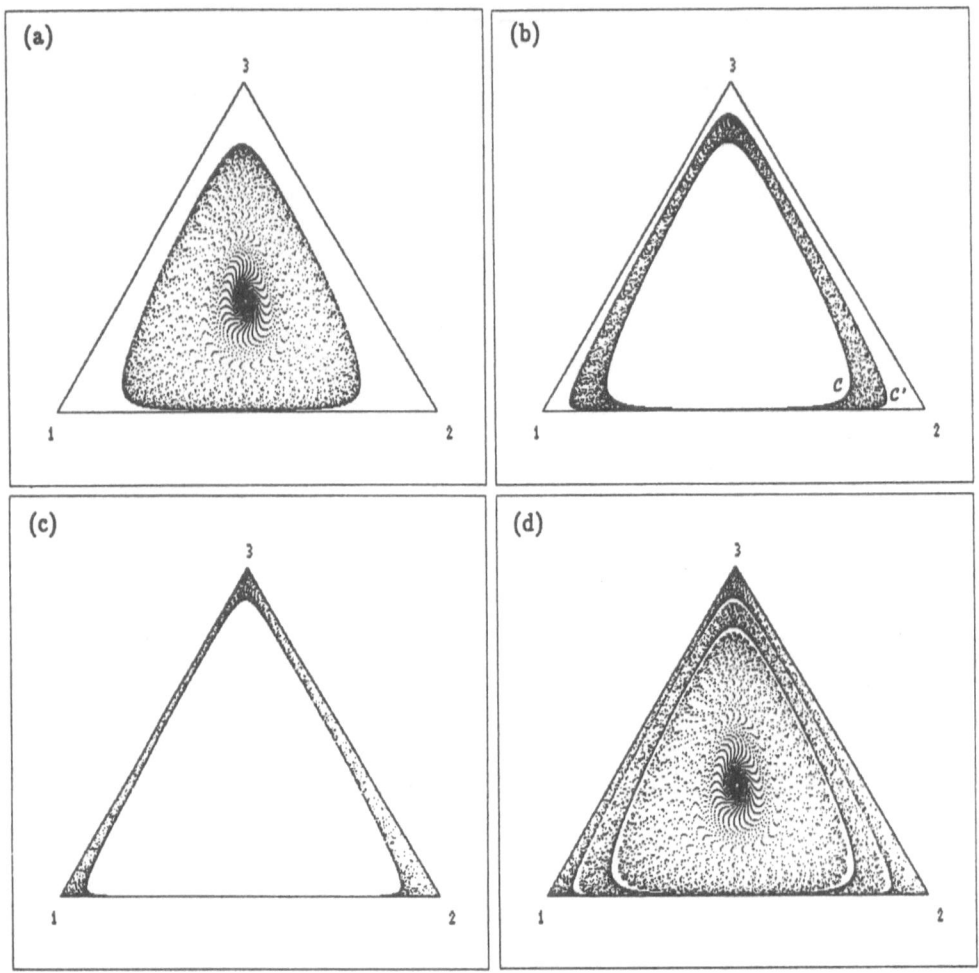

FIGURE 9: A discrete replicator dynamics with two attractors.

This figure shows some orbits of (6.1) for the central RSP–game

$$A = \begin{bmatrix} 13.00 & 17.00 & 11.00 \\ 11.00 & 13.00 & 17.00 \\ 26.75 & 1.25 & 13.00 \end{bmatrix}.$$

(a) Orbit starting at $p = (0.33, 0.33, 0.34)$. It converges to the attracting closed limit curve C.

(b) Orbit starting at $p = (0.05, 0.05, 0.90)$. It is repelled by C' and it converges to the limit curve C.

(c) Orbit starting at $p = (0.049, 0.049, 0.902)$. It is also repelled by C' and it converges to the boundary of the strategy simplex.

(d) Superposition of the three orbits.

8. Are RSP–Games Played in Biological Populations?

In the previous sections, the class of RSP–games was studied for purely theoretical reasons. I hope that it has become clear that these games are ideally suited for exemplifying the various incongruities between the game theoretical and the dynamical approach towards frequency dependent selection. However, RSP–games form a generic class of games and more than 1.5% of all 3x3 normal form games belong to this class.[2] It should, therefore, not be too surprising to find RSP–like game structures in biological populations. It is quite conceivable that, for example, strain A of a bacterial species outcompetes strain B in direct competition, strain B outcompetes strain C and strain C outcompetes strain A. A situation like this has, in fact, been observed during selection experiments in a chemostat.

A 'chemostat' is basically a device that enables a microbial culture to be maintained in permanent exponential growth in a constant and homogeneous environment (see, e.g., Dykhuizen & Hartl 1983). The growth medium is a chemically defined salt solution supplemented with a source of carbon and energy. Concentrations of the components of the fresh medium entering the growth chamber are such that only one (the limiting nutrient) is exhausted by the culture. Competition for the limiting resource leads to selection between different strains of microorganisms. Since it is fairly easy to follow a chemostat for several hundred generations, this device has proven valuable for detecting very slight selective differences between pairs of genotypes. In fact, chemostat experiments are sufficiently reproducible that differences in growth rates as small as 0.5% per generation are readily detected in replicate experiments (Dykhuizen & Dean 1990).

Quite often, long–term selection experiments in a chemostat yield rather unexpected results (see, e.g. Dykhuizen 1990). Charlotte Paquin and Julian Adams (1983), for example, analysed the competition between different asexual strains of the yeast *Saccharomyces cerevisiae* in a glucose–limited chemostat. All strains originated from an ancestral strain and they only differed from another by having accumulated a different number of mutations in the course of the experiment. Let us concentrate on their first experiment with a haploid yeast population and let H_i denote the strain isolated in generation i. (Virtually the same results were obtained for a diploid yeast population.) Paquin and Adams carefully analyzed the pairwise competition between strains H_{30}, H_{133} and H_{203} which, compared to the ancestral strain H_0, presumably had accumulated 0, 2 and 3 adaptive mutations respectively. As expected, each strain had a 'selective advantage' when compared to the previous one: H_{133} outcompeted H_{30} in pairwise competition, and H_{203} in turn outcompeted H_{133}. Rather unexpectedly, however, selection advantage was not transitive since the older strain H_{30} outcompeted the derived strain H_{203} in direct competition. It is tempting to speculate that the yeast populations in Paquin and Adams' experiments evolve under frequency dependent selection with the same intransitivities in pairwise competition that are so characteristic for the class of RSP–games.

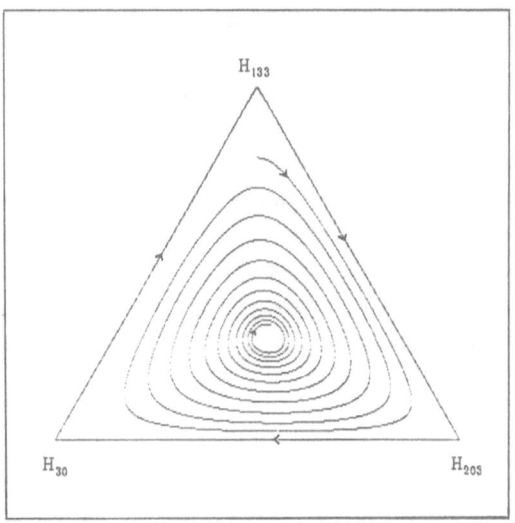

<u>Figure 10:</u>   Orbit of the continuous replicator dynamics induced by the RSP–game (8.1).

The experiments described above were performed with large (about $5 \cdot 10^9$ individuals per population) asexual populations with a very short generation time (about six generations per day). These are precisely the circumstances to which the continuous replicator dynamics does apply. Paquin and Adams (1983, Table 1) determined the relative fitness of the three strains H_{30}, H_{133} and H_{203} in pairwise competition. Their results may be summarized by the payoff matrix

$$A = \begin{bmatrix} 1.00 & 1.18 & 0.88 \\ 0.85 & 1.00 & 1.16 \\ 1.13 & 0.86 & 1.00 \end{bmatrix}, \qquad (8.1)$$

where the pure strategies 1, 2 and 3 correspond to the strains H_{30}, H_{203} and H_{133} respectively. Let us assume that the RSP–matrix A of relative fitness values is pl–equivalent to the matrix of absolute fitness values and that selection is indeed linearly frequency dependent (i.e., that the fitness function is given by (2.3)). Then a stable polymorphism of all three strains is to be expected (see Figure 10). Unfortunately, Paquin and Adams never put all three strains in competition. Thus, it remains unclear whether the theoretical prediction will be experimentally confirmed or not.

Like many other authors in the chemostat literature, Paquin and Adams notice a discrepancy between their results and the 'classical' predictions of (frequency independent) selection theory. However, they do not ascribe their results to frequency dependent selection but to other evolutionary forces like epistatic interactions between different muations. In my view, the possibility of frequency dependent selection is grossly underestimated in the chemostat literature. In a situation where the 'external' (abiotic) environment is held as constant as possible, the selective forces should to a large extent be governed by intraspecific interactions which, almost invariably, lead to frequency dependent selection. Many puzzling results in the chemostat literature (e.g. Helling et al. 1987, Bennett et al. 1990) can be explained much more naturally if frequency dependent selection is taken into consideration.

Notes:

1. It can be shown that a Nash equilibrium in the barycenter m of the strategy simplex is an ESS if and only if the trajectories of the continuous replicator equation starting close to m are attracted *monotonically* by m (with respect to the usual Eucledian metric). One might conjecture that m can only be a non–ESS attractor, if some form of cycling takes place, i.e. if the Jacobian of the continuous replicator dynamics at m is not diagonalizable in the real domain. The following payoff matrix provides a counter–example to that conjecture:

$$A = \begin{bmatrix} 0 & 19 & 1 \\ 19 & 0 & 1 \\ 1 & 19 & 0 \end{bmatrix}.$$

In that example, m is an attractor which is not an ESS. The eigenvalues of the Jacobian at m are real and distinct ($-1/60$ and $-19/60$). Accordingly, the Jacobian is diagonalizable and there is no cycling around m.

2. RSP–games form a generic class of symmetric normal form games, i.e. a class of games of full dimension. If the elements of a 3x3 payoff matrix A are drawn at random from a uniform distribution, an RSP–game will result with probability $1/64$.

3. Adding $\eta := -\epsilon/2$ to the ϵ–perturbed Rock–Scissors–Paper game

$$A_\epsilon = \begin{bmatrix} 0 & 1+\epsilon & -1 \\ -1 & 0 & 1+\epsilon \\ 1+\epsilon & -1 & 0 \end{bmatrix}, \quad \epsilon > -1, \tag{1.4a}$$

yields an RSP–game of the form

$$B_\eta = \begin{bmatrix} \eta & 1-\eta & -1+\eta \\ -1+\eta & \eta & 1-\eta \\ 1-\eta & -1+\eta & \eta \end{bmatrix}, \quad \eta < 1/2. \tag{1.4b}$$

Dividing B_η by $1-\eta$ and setting

$$\delta := \frac{\eta}{1-\eta} = \frac{\epsilon}{2+\epsilon},$$

one arrives at the RSP–game

$$C_\delta = \begin{bmatrix} \delta & 1 & -1 \\ -1 & \delta & 1 \\ 1 & -1 & \delta \end{bmatrix}, \quad -1 < \delta \leq 1, \tag{1.4c}$$

which is payoff equivalent to A_ϵ and B_η.

4. Theorem 3.4. shows that, in evolutionary 3x3 normal form games which are *not* RSP–games, a discrepancy between evolutionary stability and continuous dynamic stability is always associated with the the existence of nontrivial border fixed points of the replicator dynamics. Non–RSP examples are in that sense more complicated than the games analyzed in this paper. However, they may show some additional features that are missing in the context of RSP–games. Zeeman (1980), for instance, presents a nice example which is based on the payoff matrix

$$A = \begin{bmatrix} 0 & 6 & -4 \\ -3 & 0 & 5 \\ -1 & 3 & 0 \end{bmatrix}.$$

Here, the barycenter **m** is an attractor that coexists with the unique ESS which is given by the pure strategy *1*. Although **m** is not an ESS, its domain of attraction is larger than that of the ESS. The fact that **m** is not a global attractor points out another *qualitative* difference between evolutionary stability and dynamic stability: A completely mixed ESS is always a global attractor for the continuous replicator dynamics (see Corollary 5.2) whereas completely mixed non—ESS attractors may coexist with boundary attractors.

5. An alternative proof may be based on the theorem of Poincaré–Hopf (see e.g. Hofbauer & Sigmund 1988, Chapter 19). This theorem can be applied to a slight modification of the flow which is induced by an RSP—game on the border of the strategy simplex (see (1.2)). As a result one gets that the sum of the Poincaré–indices of the interior fixed points of the replicator dynamics is equal to one, the Euler characteristic of the strategy simplex. This implies the existence of at least one interior fixed point which is automatically a Nash equilibrium strategy. In view of the uniqueness result in Theorem 3.3, there exists a unique interior fixed point, and the Poincaré–index of this fixed point is equal to one. If this fixed point is regular, we get some additional information: the fixed point is necessarily either a sink, a source, or a center.

References

Akin, E. (1983): Hopf bifurcation in the two locus genetic model. Providence, R.I.: Memoirs Amer. Math. Soc. No. 284.

Akin, E., and V. Losert (1984): Evolutionary dynamics of zero—sum games. J. Math. Biology 20: 231–258.

Bennett, A.F., K.M Dao and R.E. Lenski (1990): Rapid evolution in response to high—temperature selection. Nature 346: 79–81.

Bhatia, N.P., and G.P. Szegö (1967): Dynamical Systems: Stability Theory and Applications. Berlin: Springer–Verlag.

Damme, E. van (1987): Stability and Perfection of Nash Equilibria. Berlin: Springer–Verlag.

Davis, P.J. (1979): Circulant Matrices. New York: John Wiley & Sons.

Dykhuizen, D.E. (1990): Mountaineering with microbes. Nature 346: 14–15.

Dykhuizen, D.E. and A.M. Dean (1990): Enzyme activity and fitness: Evolution in solution. Trends in Ecology and Evolution 5: 257–262.

Dykhuizen, D.E., and D.L. Hartl (1983): Selection in chemostats. Microbiology Reviews 47: 150–168.

Guckenheimer, J., and P. Holmes (1983): Nonlinear Oscillations, Dynamical Systems, and Bifurcations of Vector Fields. Berlin: Springer–Verlag.

Hassard, B.D., N.D. Kazarinoff, and Y.H. Wan (1981): Theory and Applications of Hopf Bifurcation. Cambridge: Cambridge University Press.

Harsanyi, J.C., and R. Selten (1988): A General Theory of Equilibrium Selection in Games. Cambridge, Mass.: The MIT Press.

Helling, R.B., C.N. Vargas and J. Adams (1987): Evolution of Escherichia coli during growth in a constant environment. Genetics 116: 349–358.

Hirsch, M., and S. Smale (1974): Differential Equations, Dynamical Systems, and Linear Algebra. New York: Academic Press.

Hofbauer, J. (1981): On the occurrence of limit cycles in the Volterra–Lotka equation. Nonlinear Analysis 5: 1003–1007.

Hofbauer, J., and G. Iooss (1984): A Hopf bifurcation theorem for difference equations approximating a differential equation. Monatsh. Mathem. 98: 99–113.

Hofbauer, J., P. Schuster, and K. Sigmund (1981): A note on evolutionarily stable strategies and game dynamics. J. theor. Biol. 81: 609–612.

Hofbauer, J., and K. Sigmund (1988): The Theory of Evolution and Dynamical Systems. Cambridge: Cambridge University Press.

Iooss, G. (1979): Bifurcation of Maps and Applications. Amsterdam: North–Holland.

Iooss, G., and D.D. Joseph (1981): Elementary Stability and Bifurcation Theory. Berlin: Springer–Verlag.

Losert, V., and E. Akin (1983): Dynamics of games and genes: Discrete versus continuous time. J. Math. Biology 17: 241–251.

Luce, R.D., and H. Raiffa (1957): Games and Decisions. New York: John Wiley & Sons.

Marsden, J., and M. McCracken (1976): The Hopf Bifurcation and its Applications. Berlin: Springer–Verlag.

Maynard Smith, J. (1977): Mathematical models in population biology. In: D.E. Matthews (Ed.): Mathematics and the Life Sciences. Lecture Notes in Biomathematics 18. Berlin: Springer–Verlag; pp. 200–221.

Maynard Smith, J. (1982): Evolution and the Theory of Games. Cambridge: Cambridge University Press.

Maynard Smith, J., and G. Price (1973): The logic of animal conflicts. Nature 246: 15–18.

Nash, J. (1951): Non–cooperative games. Annals of Mathematics 54: 286–295.

Paquin, C.E.,and J. Adams (1983): Relative fitness can decrease in evolving asexual populations of S. cerevisiae. Nature 306: 368–371.

Selten, R. (1983): Evolutionary stability in extensive two–person games. Mathem. Social Sciences 5: 269–363.

Taylor, P., and L. Jonker (1978): Evolutionarily stable strategies and game dynamics. Mathem. Biosciences 40: 145–156.

Wan, Y.H. (1978): Computations of the stability condition for the Hopf bifurcation of diffeomorphisms on R^2. SIAM J. Appl. Math. 34: 167–175.

Weissing, F.J. (1983): Populationsgenetische Grundlagen der Evolutionären Spieltheorie. Bielefeld: Materialien zur Mathematisierung der Einzelwissenschaften, Vols 41 & 42.

Weissing, F.J. (1990): On the relation between evolutionary stability and discrete dynamic stability. Manuscript.

Zeeman, E.C. (1980): Population dynamics from game theory. In: Global Theory of Dynamical Systems. Springer Lecture Notes in Mathematics 819: 471–497.

ANTICIPATORY LEARNING IN TWO-PERSON GAMES

by

Reinhard Selten

Summary

A learning process for 2-person games in normal form is introduced. The game is assumed to be played repeatedly by two large populations, one for player 1 and one for player 2. Every individual plays against changing opponents in the other population. Mixed strategies are adapted to experience. The process evolves in discrete time.

All individuals in the same population play the same mixed strategy. The mixed strategies played in one period are publicly known in the next period. The payoff matrices of both players are publicly known.

In a preliminary version of the model, the individuals increase and decrease probabilities of pure strategies directly in response to payoffs against last period's observed opponent strategy. In this model, the stationary points are the equilibrium points, but genuinely mixed equilibrium points fail to be locally stable.

On the basis of the preliminary model an anticipatory learning process is defined, where the individuals first anticipate the opponent strategies according to the preliminary model and then react to these anticipated strategies in the same way as to the observed strategies in the preliminary model. This means that primary learning effects on the other side are anticipated, but not the secondary effects due to anticipations in the opponent population.

Local stability of the anticipatory learning process is investigated for regular games, i.e., for games where all equilibrium points are regular. A stability criterion is derived which is necessary and sufficient for sufficiently small adjustment speeds. This criterion requires that the eigenvalues of a matrix derived from both payoff matrices are negative.

It is shown that the stability criterion is satisfied for 2×2-games without pure strategy equilibrium points, for zero-sum games and for games where one player's payoff matrix is the unit matrix and the other player's payoff matrix is negative definite. Moreover, the addition of constants to rows or columns of payoff matrices does not change stability.

The stability criterion is related to an additive decomposition of payoffs reminiscent of a two way analysis of variance. Payoffs are decomposed into row effects, column effects and interaction effects. Intuitively, the stability criterion requires a preponderance of negative covariance between the interaction effects in both players' payoffs.

The anticipatory learning process assumes that the effects of anticipations on the other side remain unanticipated. At least for completely mixed equilibrium points the stability criterion remains unchanged, if anticipations of anticipation effects are introduced.

1. Introduction

A loose description of the anticipatory learning process investigated here has already been given in the summary. The aim of the paper is a speculative attempt to develop an idealized descriptive theory. The effort remains speculative since the task of experimental validation is left to the future. The theory is idealized since in real laboratory situations populations cannot be very large and individual differences must be expected to matter. Nevertheless, our assumption on the structure of behavior may be basically correct and the stability criterion derived here may have some predictive relevance. At least it may serve to guide the planning of experiments.

1.1 Decision Emergence

The modelling efforts of this paper are based on a picture of limited rationality which involves "decision emergence" rather than "decision making". Mixed strategies are thought of as behavioral dispositions which do not arise from rational calculations. It is assumed that the individual has no direct understanding of the way in which decisions come about. A person without special training does not know how the body performs physiological functions like digestion. Similarly, the decision process has important parts which are *inaccessible* in the sense that introspection is unable to reveal the mechanism. *Decisions are not made, decisions emerge*. The conscious mind supplies inputs to the black box containing the inaccessible parts of the decision mechanism and observes the emergence of decisions as an output.

Anticipation also fits into this picture. An individual can imagine to be in the role of the other player. This amounts to the use of one's own decision mechanism and its inaccessible parts for the prediction of the opponent's behavior. Prediction does not involve conscious calculations, *anticipations emerge* in the same way as decisions do.

A lack of understanding of the internal and the external environment prevents optimization. This explains the possibility of non-optimal behavior. Computational complexity is not the immediate reason for the failure to optimize, even if it may explain the evolutionary adaptiveness of limited rationality.

Of course, an individual who is in the possession of a numerically specified correct model of the learning process should optimize against it. Therefore, the anticipations in our theory should not be misunderstood as representing approximatively correct conscious calculations. In order to emphasize this point, we use the word "anticipation" rather than the customary term "expectation".

1.2 Remarks on the Literature

The notion of an equilibrium point has been introduced as a normative concept (Nash, 1951). Nevertheless, predictions based on equilibrium points (or, more precisely, saddlepoints) are surprisingly successful in some experiments (O'Neill, 1987). Encouraged by the predictive success of a learning theory approach in a different experimental context (Selten and Stoecker, 1986), the author thinks that learning processes are a promising modelling tool for the description of game playing behavior.

The process of fictitious play (Brown, 1951; Robinson, 1951) has not been proposed as a serious model of game learning but rather as an algorithm for the computation of saddlepoints. The usual interpretation as a process of repeated play between the same two individuals does not look plausible. However, a reinterpretation in the population framework employed here is less objectionable.

Shapley has shown instability of fictitious play for a class of 3×3-games; Miyasawa has proved convergence for all 2×2-games (Miyasawa, 1961). Rosenmüller has established conditions for limit cycles (Rosenmüller, 1967). Due to a lack of continuity, it seems to be difficult to obtain general results on stability conditions for fictitious play.

Since Bush and Mosteller published their path-breaking book, many learning models have been presented in the psychological literature (Bush and Mosteller, 1955). Cross considers generalizations of such models which could be applied to game learning (Cross, 1983).

A dynamic process considered in the biological literature as a model of natural selection in symmetric two-person games (Taylor and Jonker, 1978) leads to evolutionary stability (Maynard Smith and Price, 1973; Maynard Smith, 1982) as a sufficient condition for dynamic stability. A reinterpretation as a learning process is possible. This interpretation involves only one population for both players. The symmetry restriction is important. Asymmetric games can be symmetrized by a random assignment of player roles, but evolutionarily stable strategies of the symmetrized game must be pure (Selten, 1980). Further references to the literature can be found in a recent survey article (Bomze, 1986).

Experiments performed by Milinski show that fishes learn to distribute themselves on two food sources in proportion to the amount of nutrition offered per time unit (Milinski, 1979). Game learning processes are of interest for the description of animal behavior. Harley has developed an interesting biological learning model (Harley, 1981). A more careful examination of the convergence properties of this model would be desirable (Selten and Hammerstein, 1984). Crawford has examined a class of reasonable learning models for which he has shown that genuinely mixed equilibrium points are almost always locally unstable (Crawford, 1985).

None of the theories mentioned above involves an element of anticipation. It seems to be possible to construct anticipatory modifications of some of the models in the literature, but no attept in this direction will be made here.

1.3 Linearity Properties

The modelling approach taken in this paper is based on the idea that as much linearity as possible should be imposed on the functional form of dynamic relationships. Non-negativity constraints on probabilities prevent full linearity in the space of mixed strategy combinations. However, it is possible to achieve linearity in every region of the space where a given set of constraints is binding.

Unfortunately, it is not possible to concentratge on the interior of the space of mixed strategies and to ignore border problems unless one is satisfied with the investigation of local stability properties of completely mixed equilibrium points. Border problems matter for other mixed equilibrium points.

As long as we do not have any empirical reason to prefer one functional form over another, it seems to be reasonable to impose linearity properties for the sake of mathematical convenience. Alternatively, one might want to avoid to specify functional forms. It seems to be difficult to obtain explicit stability criteria if one takes this approach. Therefore, a specific functional form is used for the anticipatory learning process proposed here.

2. Regular Equilibrium Points

In the following definitions and notations concerning two person games in normal form will be introduced. The concept of a regular equilibrium point and its properties will be discussed.

2.1 Basic Definitions and Notations

A *two-person game* in normal form can be described as a pair (A,B) of two n×m-matrices. Player 1 has the pure strategies 1,...,n and player 2 has the pure strategies 1,...,m. The matrices

$$A = (a_{ij})_{n \times m} \tag{1}$$

and

$$B = (b_{ij})_{n \times m} \tag{2}$$

are the *payoff matrices* of players 1 and 2, respectively. a_{ij} and b_{ij} are the *payoffs* for players 1 and 2, respectively, if player 1 plays i and player 2 plays j. The set {1,...,n} of player 1's pure strategies is denoted by N and the set {1,...,m} of player 2's pure strategies is denoted by M.

In this paper, all definitions and all statements in lemmata and theorems will be relative to an arbitrary but fixed two-person game (A,B) unless something else is said explicitly. Whenever the word "game" is used without further specification, we mean a two-person game in normal form.

Mixed strategies are described by column vectors with non-negative components summing up to 1. The number of components is n for player 1 and m for player 2. The k-th component is the probability of choosing the pure strategy k. If p denotes a mixed strategy, then p_k denotes the k-th component of p. The same convention is also applied if other letters q, r, or s denote mixed strategies of player 1 or 2. The set of all mixed strategies of player 1 is denoted by P and the set of all mixed strategies of player 2 is denoted by Q. A *mixed strategy pair* (p,q) is defined by p ∈ P and q ∈ Q.

The *payoffs* H(p,q) and K(p,q) of players 1 and 2 for a mixed strategy pair (p,q) are as follows:

$$H(p,q) = p^T Aq \tag{3}$$

$$K(p,q) = p^T Bq. \tag{4}$$

Here and in the remainder of the paper, the upper index T indicates transposition.

Pure strategies are looked upon as special mixed strategies of the same player. The pure strategy k is identified with that mixed strategy, whose k-th component is 1. Accordingly, we use the notation H(i,q) for player 1's payoff obtained, if player 1 plays i and player 2 plays q.

The *support* of a mixed strategy p ∈ P is the set I of all i ∈ N with $p_i > 0$. Analogously, the *support* of q ∈ Q is the set J of all j ∈ M with

$q_j > 0$. If I is the support of p and J is the support of q, then (I,J) is called the *support* of the mixed strategy pair (p,q).

A mixed strategy is called *genuinely mixed* if its support has at least two elements. A mixed strategy pair (p,q) is called *genuinely mixed* if p and q are genuinely mixed. A mixed strategy is called *completely mixed* if its support is the set of all pure strategies of the player concerned. A mixed strategy pair (p,q) is called *completely mixed* if p and q are completely mixed.

2.2 Best Replies

A mixed strategy $r \in P$ of player 1 is a *best reply* of player 1 to $q \in Q$ if we have:

$$H(r,q) = \max_{p \in P} H(p,q). \tag{5}$$

Analogously, a mixed strategy $s \in Q$ is a *best reply* of player 2 to $p \in P$ if we have

$$K(p,s) = \max_{p \in P} K(p,q). \tag{6}$$

It is well known that $r \in P$ is a best reply to $q \in Q$ if and only if every pure strategy i with $r_i > 0$ is a best reply to q. An analogous assertion holds for best replies of player 2.

2.3 Equilibrium Point

An equilibrium point is a mixed strategy pair (r,s) with the property that r is a best reply to s and s is a best reply to r.

Let (r,s) be an equilibrium point and let $E = H(r,s)$ and $F = K(r,s)$ be the corresponding equilibrium payoffs of both players. Let (I,J) be the support of (r,s). With the help of the property of best replies mentioned above, it can be seen that the following equations must be satisfied:

$$\sum_{j \in J} a_{ij} s_j = E \qquad \text{for } i \in I \tag{7}$$

$$\sum_{i \in I} b_{ij} r_i = F \qquad \text{for } j \in J \tag{8}$$

$$\sum_{i \in I} r_i = 1 \tag{9}$$

$$\sum_{j \in J} s_j = 1. \tag{10}$$

We shall use the notation |S| for the number of elements of a finite set S. Eqs.
(7), (8), (9) and (10) can be looked upon as a system of |I| + |J| + 2 variables
r_i with i ∈ I and s_j with j ∈ J and E and F. By assumption, the system has
at least one solution, but it may happen that there are infinitely many solutions
which fill a linear subspace of the (|I|+|J|+2)-dimensional vector space for these
variables.

An equilibrium point (r,s) is calles *isolated* if an open neighborhood V of
(r,s) can be found such that V contains no other equilibrium points. It is clear
that the system (7), (8), (9) and (10) has exactly one solution if (r,s) is isolated.

An equilibrium point (r,s) is called *quasistrong* (Harsanyi, 1973) if in addition
to (7), (8), (9) and (10) the following inequalities are satisfied:

$$\sum_{j \in J} a_{ij} s_j < E \qquad \text{for } i \in N \backslash I \tag{11}$$

$$\sum_{i \in I} b_{ij} r_i < F \qquad \text{for } j \in M \backslash J. \tag{12}$$

These inequalities mean that at the equilibrium point pure strategies with zero
probabilities fail to be best replies.

2.4 Regularity

A game (A,B) is called *regular* if it is isolated and quasistrong. A game (A,B) is
called *regular* if all equilibrium points of (A,B) are regular. It is clear that in
the space of all two-person games the irregular games form a set of lower dimension.
In this sense, one can say that only degenerate cases are excluded by the definition
of regularity.

It is an interesting property of equilibrium points that the system (7), (8),
(9) and (10) is composed of two independent subsystems, namely (7) and (10) on the
one hand and (8) and (9) on the other hand. The payoffs of player 1 determine the
strategy of player 2 and vice versa. We refer to this property as the
heterodependence of equilibrium strategies. Not a player's own payoffs determine his
equilibrium probabilities of pure strategies, but those of the other player.

If (r,s) is regular, then we must have |I| = |J|; otherwise one of both
subsystems would have more unknowns than equations and therefore could not have a
unique solution. It also follows by the definition of regularity that two regular
equilibrium points cannot have the same support. Consequently, a regular game (A,B)
has only a finite number of equilibrium points.

2.5 The Restricted Game

Let I be a non-empty subset of N and let J be a non-empty subset of J. Let $\underline{A}$
and $\underline{B}$ be the submatrices of A and B respectively, resulting from A and B by
the removal of all rows with indices not in I and all columns with indices not in
J. We call the game $(\underline{A},\underline{B})$ the *restriction* of (A,B) to (I,J). In $\underline{A}$ and $\underline{B}$ rows
have new numbers $1,\ldots,|J|$ and columns have new numbers $1,\ldots,|J|$. In the
transition from (A,B) to $(\underline{A},\underline{B})$ pure strategies are renumbered accordingly. the
elements of $\underline{A}$ are denoted by $\underline{a}_{ij}$ and those of $\underline{B}$ by $\underline{b}_{ij}$.

Let (I,J) be the support of a regular equilibrium point (r,s). In this case, the
restriction $(\underline{A},\underline{B})$ to (I,J) is also called the *restricted game* of (r,s). If in (r,s)
the positive components of r and s are renumbered in the same way as the pure
strategies in I and J respectively, and the other components are left out, one
obtains a pair of mixed strategies (r,s) for the restricted game. Obviously, $(\underline{r},\underline{s})$ is
a regular equilibrium point of the restricted game $(\underline{A},\underline{B})$.

A game in which both players have the same number of pure strategies is called
quadratic. The restricted game of a regular equilibrium point is quadratic.

3. The Preliminary Model

As a preparation for the anticipatory learning process to be defined later, a
preliminary model is introduced in this section. The interpretation is based on a
population framework common to both models.

3.1 Population Framework

As has been explained in the summary, learning is supposed to take place in two large
populations, one for player 1 and the other for player 2. The same individual always
plays in the role of the same player. Time is a succession of discrete periods. In
every period, the individuals of both populations are randomly matched into pairs
playing the game. The same game (A,B) is played in every period.

The exposition will be based on the assumption that all individuals in the same
role behave in exactly the same way. In every period all individuals of the same
population play the same mixed strategy, also referred to as the mixed strategy of
the concerned player. Of course, this is a very restrictive assumption. In view of
the linearity properties of our models, a more generous interpretation seems to be
possible which permits some strategic diversity within a population. However, this
idea will not be made precise.

The populations are thought of as infinite. This means that no distinction is
made between the relative frequency of a pure strategy and its probability of being
played. We assume that these frequencies are observed, or, in other words, mixed
strategies are observed before the beginning of the next period.

It is assumed that all individuals know both payoff matrices. Knowledge of own payoffs would be sufficient for the preliminary model, but anticipation requires knowledge of the opponent's payoffs.

The learning process is described by a piecewise linear system of difference equations. The interpretation is based on the decision emergence view explained in the introduction.

3.2 Notation

Time is modelled as a succession of periods $t = 0,1,2,\ldots$. The common mixed strategy in population 1, or, shortly, player 1's mixed strategy at time t is denoted by $p(t)$. Similarly, player 2's mixed strategy at time t is denoted by $q(t)$. The notation $p_i(t)$ is used for the i'th component of $p(t)$. Similarly, $q_j(t)$ is the j-th component of $q(t)$. Often we shall suppress the dependence on t and simply write p and q instead of $p(t)$ and $q(t)$. In such cases, p_+ and q_+ will be used instead of $p(t+1)$ and $q(t+1)$, respectively, while the notation p_{i+} and q_{j+} is used for components. The same conventions will be applied to other time dependent variables.

We shall also use the notation Δp for the first difference vector $p_+ - p$ and Δq for $q_+ - q$. The difference operator Δ is also applied to components; thus Δp_i stands for $p_{i+} - p_i$. The same notation will also be used for other time dependent variables.

3.3 A Constrained Proportionality Principle

In the following, we shall present intuitive arguments for the specific way in which the preliminary model deals with non-negativity constraints. The formulas of the model are hard to interprete directly and may seem to be arbitrary without an intuitive justification. Therefore, we first introduce an intuitively appealing *constrained proportionality principle* which then completely determines the structure of the model.

The game being played is (A,B). We focus on player 1. It will be convenient to use a shorter notation for player 1's payoffs obtained for his pure strategies i against player 2's current strategy q. Define:

$$E_i = H(i,q) \qquad \text{for } i = 1,\ldots,n. \tag{13}$$

The quantities E_i measure the current success of player 1's pure strategies. We want to construct a model where the increase Δp_i of player 1's probability of choosing i is determined by these success measures $E_1,\ldots,E_n$. It is natural to require that ceteris paribus Δp_i should be the greater, the greater E_i is, at least as long as non-negativity constraints do not prevent this. Accordingly, for $p_{i+} > 0$ and $p_{k+} > 0$ we should have:

$$\Delta p_i > \Delta p_k \qquad \text{for } E_i > E_k . \tag{14}$$

Since we want to achieve as much linearity as possible, (14) suggests the following *proportionality principle*:

$$\Delta p_i - \Delta p_k = \alpha(E_i - E_k) \qquad \text{for } p_{i+} > 0 \text{ and } p_{k+} > 0 \tag{15}$$

where α is a positive parameter. (15) can be rewritten as follows:

$$\Delta p_i = \Delta p_k + \alpha(E_i - E_k) \qquad \text{for } p_{i+} > 0 \text{ and } p_{k+} > 0. \tag{16}$$

Obviously, we must have

$$\Delta p_i \geq - p_i. \tag{17}$$

Otherwise, p_{i+} would be negative. If (16) yields a smaller vlaue than $-p_i$ for Δp_i, then (16) cannot apply. In this case, it is natural to require that Δp_i should assume its minimum value $-p_i$. If one wants to achieve as much linearity as possible, deviations from proportionality should not be permitted for any other reason. This leads to the following *constrained proportionality principle*:

$$\Delta p_i = \max[-p_i, \Delta p_k + \alpha(E_i - E_k)] \qquad \text{for } p_{k+} > 0 \text{ and } i = 1, \ldots, n. \tag{18}$$

Assume that the constrained proportionality principle holds. Consider the case $p_{i+} > 0$ and $p_{k+} > 0$ where (18) assumes the form (16). Since the same value of Δp_i is obtained for every k with $p_{k+} > 0$, the expression $\alpha E_k - \Delta p_k$ must assume the same value u for all these k: A number u exists such that the following is true:

$$u = \alpha E_k - \Delta p_k \qquad \text{for all } k \text{ with } p_{k+} > 0. \tag{19}$$

With the help of (19), we obtain an equivalent reformulation of the constrained proportionality principle: A number u exists for which the following is true:

$$\Delta p_i = \max[-p_i, \alpha E_i - u] \qquad \text{for } i = 1, \ldots, n. \tag{20}$$

In the transition from (19) to (20), we have made use of the obvious fact that we must have $p_{k+} > 0$ for at least one $k \in N$. The p_{i+} must sum up to 1. In view of the definition of Δp_i, Eq. (20) can be rewritten as follows:

$$p_{i+} = \max[0, p_i + \alpha E_i - u] \qquad \text{for } i = 1, \ldots, n. \tag{21}$$

This is the reformulation of the constrained proportionality principle to be used in the model. Of course, the same principle is also applied to the Δq_j.

The number u can be thought of as an *auxiliary variable*. It depends on $p_1,\ldots,p_n$ abd $E_1,\ldots,E_n$. As we shall see, u is uniquely determined by the condition that the p_{i+} sum up to 1.

3.4 Definition of the Preliminary Model

In the preliminary model, the mixed strategies p_+ and q_+ for period $t+1$ depend as follows on the mixed strategies p and q for period t:

$$p_{i+} = \max[0, p_i + \alpha H(i,q) - u] \quad \text{for } i = 1,\ldots,n \tag{22}$$

$$q_{j+} = \max[0, q_j + \beta K(p,j) - v] \quad \text{for } j = 1,\ldots,m \tag{23}$$

with

$$\sum_{i=1}^{n} \max[0, p_i + \alpha H(i,q) - u] = 1 \tag{24}$$

$$\sum_{j=1}^{m} \max[0, q_j + \beta K(p,j) - v] = 1 \tag{25}$$

where α and β are positive parameters. The condition that the p_{i+} must sum up to 1 is expressed by (24). The interpretation of (25) is analogous.

Usually, a system of difference equations explicitly describes the dynamic relationships. Eqs. (22) - (25) provide only an implicit description. Nevertheless, the description is complete. This is shown by the following lemma.

Lemma 1: Let α and β be positive constants. Then, for given p and q next period's strategies p_+ and q_+ as well as the auxiliary variables u and v are uniquely determined by (22), (23), (24) and (25). Moreover, p_+, q_+, u and v are continous functions of p and q.

Proof: Let $\underline{c}$ be the minimum of all a_{ij} and let $\bar{c}$ be the maximum of all a_{ij}. Let $\underline{u}$ be $\alpha\underline{c}$ and let $\bar{u}$ be $\alpha\bar{c}$. For all mixed strategy pairs (p,q) the left-hand side of (24) is at least as great as 1 for $u = \underline{u}$ and at most as great as 1 for $u = \bar{u}$. Moreover, the left-hand side of (24) is non-increasing in u. Therefore, for given (p,q) Eq. (24) has a unique solution u which lies in the closed interval $\underline{u} \leq u \leq \bar{u}$. Consider a sequence $(p^1,q^1),(p^2,q^2),\ldots$ of mixed strategy pairs which converges to (p,q). For $k = 1,2,\ldots$ let u^k be the value of u determined by (p^k, q^k). Since the interval $\underline{u} \leq u \leq \overline{u}$ is closed and bounded, the sequence $u^1, u^2_2, \ldots$ has an accumulation point u. It follows by the continuity of H that all accumulation points u of $u^1, u^2, \ldots$ satisfy (24) together with (p,q). Since there is only one such u, the sequence $u^1, u^2, \ldots$ converges to this u. Consequently, u is a continuous function of p and q. An analogous argument shows that v is a continuous function of p and q. It follows by (22) and (23) together with the continuity of H and K that p_+ and q_+ are uniquely determined by p and q and depend continuously on p and q.

3.5 Stationarity

We say that a mixed strategy pair (p,q) is a *stationary point* of the preliminary model if we have $p_+ = p$ and $q_+ = q$ for the strategies p_+ and q_+ determined by (22) to (25). Lemma 2 will show that in the preliminary model equilibrium is necessary and sufficient for stationarity.

Theorem 1: A mixed strategy pair (p,q) is a stationary point of the preliminary model if and only if (p,q) is an equilibrium point. This is true for every parameter pair (α, β) with $\alpha > 0$ and $\beta > 0$ and every game (A,B).

Proof: Let (I,J) be the support of (p,q). (For the definition of "support" see 2.1.) Assume that (p,q) is an equilibrium point. Then, $H(i,q) = H(p,q)$ holds for all $i \in I$ and $K(p,j) = K(p,q)$ holds for all $j \in J$. Therefore,

$$u = \alpha H(p,q) \tag{26}$$

and

$$v = \beta K(p,q) \tag{27}$$

solve (24) and (25). The stationarity of (p,q) follows by (22) and (23).

Now, assume that (p,q) is stationary. Then, it follows by (22) that $\alpha H(i,q) = u$ holds for all $i \in I$. Analogously, $\beta K(p,j) = v$ holds for all $j \in J$. Consequently, we must have (26) and (27). It follows by (22) and (23) together with (26) and (27) that we have:

$$H(i,q) = H(p,q) \qquad \text{for } i \in I \tag{28}$$

$$K(p,j) = K(p,q) \qquad \text{for } j \in J \tag{29}$$

$$H(i,q) \leq H(p,q) \qquad \text{for } i \in N \backslash I \tag{30}$$

$$K(p,j) \leq K(p,q) \qquad \text{for } j \in M \backslash J. \tag{31}$$

This shows that (p,q) is an equilibrium point.

3.6 Local Stability

In this paper, a strong notion of local stability will be applied which includes both attractivity and Liapunov stability. Crawford seems to work with the same kind of local stability even if this is not explicitly spelled out in his paper (Crawford, 1985). It is difficult to exclude the possibility of a snap-back repellor (see Gabisch and Lorenz, 1987:184). A small perturbance may first take the process far away, but eventually it may move to a pure strategy pair from which the stationary point is reached in one step. Clearly, this kind of behavior does not conform to the intuitive notion of local stability.

For every $\varepsilon > 0$, the ε-neighborhood $U_\varepsilon(p,q)$ of (p,q) is defined as the set of all mixed strategy pairs (r,s) whose distance from (p,q) is smaller than ε in the following sense:

$$\sum_{i=1}^{n} (p_i - r_i)^2 + \sum_{j=1}^{m} (q_j - s_j)^2 < \varepsilon^2. \tag{32}$$

Whenever we speak of an *open* neighborhood of (p,q), we mean a neighborhood which is relatively open in the set $P \times Q$ of all mixed strategy pairs. In this sense $U_\varepsilon(p,q)$ is open for every $\varepsilon > 0$.

We say that a stationary point (p,q) is *locally (asymptotically) stable* if for every $\varepsilon > 0$ an open neighborhood V of (p,q) with $V \subseteq U_\varepsilon(p,q)$ can be found such that for

$$(p(0),q(0)) \in V \tag{33}$$

we always have

$$(p(t),q(t)) \in U_\varepsilon(p,q) \quad \text{for } t = 1,2,\dots \tag{34}$$

and

$$\lim_{t \to \infty} (p(t),q(t)) = (p,q). \tag{35}$$

3.7 The Linear System Connected to a Regular Equilibrium Point

Let (r,s) be a regular equilibrium point and let (I,J) be the support of (r,s). As has been pointed out in 2.4, regularity implies $|I| = |J|$. Define $h = |I|$.

If h < n, every ε-neighborhood of (r,s) contains pairs of mixed strategies whose support is different from (I,J). Therefore, the preliminary model is not linear at (r,s). However, if one only considers pairs of mixed strategies (p,q) with the property that (p,q) and (p_+,q_+) have the support (I,J), one obtains a linear relationship between (p_+,q_+) and (p,q). In the following, this linear relationship will be examined in more detail.

Let S be the set of all pairs of mixed strategies (p,q) with support (I,J) for which (22) - (25) yield pairs (p_+,q_+) whose support is (I,J), too. It follows by (22) - (25) that for (p,q) ∈ S we have:

$$p_{i+} = p_i + \alpha H(i,q) - u \quad \text{for } i \in I \tag{36}$$

$$q_{j+} = q_j + \beta K(p,j) - v \quad \text{for } j \in J. \tag{37}$$

Moreover, p and p_+ have the property that the components with indices in I sum up to 1. An analogous statement holds for q and q_+. Summation over i ∈ I in (36) and over j ∈ J in (37) and division by h = |I| = |J| yields:

$$u = \frac{\alpha}{h} \sum_{i \in I} H(i,q) \tag{38}$$

$$v = \frac{\beta}{h} \sum_{j \in J} K(p,j). \tag{39}$$

It is our aim to express the linear relationship between (p_+,q_+) and (p,q) in matrix notation. This is done with the help of the payoff matrices $\underline{A}$ and $\underline{B}$ of the restricted game $(\underline{A},\underline{B})$ of (r,s). For this purpose, we introduce the following *notational conventions:* For every mixed strategy p of player 1 with support I, let $\underline{p}$ be that mixed strategy of player 1 for the restricted game which results from p if first all components with indices not in I are removed and then the remaining components are renumbered from 1,...,h in the order of the previous numbering. For every mixed strategy q of player 2 with support J a mixed strategy $\underline{q}$ of player 2 in $(\underline{A},\underline{B})$ is defined analogously. We call $\underline{p}$ and $\underline{q}$ the *restrictions* of p and q, respectively. The symbols $\underline{p}_+$, $\underline{q}_+$, $\underline{r}$ and $\underline{s}$ stand for the restrictions of p_+, q_+, r and s, respectively. The restriction of a pair (p,q) is the pair $(\underline{p},\underline{q})$ of its restrictions. Thus, $(\underline{r},\underline{s})$ is the restriction of (r,s). As we have seen in 2.5, the pair $(\underline{r},\underline{s})$ is a regular equilibrium point of the restricted game $(\underline{A},\underline{B})$.

Let $\underline{D}$ be the h×h-unit matrix with 1's on the diagonal and zeros everywhere else:

$$\underline{D} = \begin{bmatrix} 1 & \cdot & \cdot & \cdot & 0 \\ \cdot & \cdot & & & \cdot \\ \cdot & & \cdot & & \cdot \\ \cdot & & & \cdot & \cdot \\ 0 & \cdot & \cdot & \cdot & 1 \end{bmatrix} \tag{40}$$

and let $\underline{z}$ be the h-dimensional column vector whose components are all equal to 1:

$$\underline{z} = \begin{bmatrix} 1 \\ \cdot \\ \cdot \\ \cdot \\ 1 \end{bmatrix}. \tag{41}$$

Eqs. (38) and (39) can now be expressed in matrix notation

$$u = \frac{\alpha}{h} \underline{z}^T \underline{A} \underline{q} \tag{42}$$

$$v = \frac{\beta}{h} \underline{z}^T \underline{B}^T \underline{p}. \tag{43}$$

Eqs. (36) and (37) can be rewritten as follows:

$$\underline{p}_+ = \underline{p} + \alpha \underline{A} \underline{q} - \underline{z} u \tag{44}$$

$$\underline{q}_+ = \underline{q} + \beta \underline{B}^T \underline{p} - \underline{z} v. \tag{45}$$

Define

$$Z = \underline{D} - \frac{1}{h} \underline{z} \underline{z}^T \tag{46}$$

and

$$\tilde{A} = Z \underline{A} \tag{47}$$

$$\tilde{B}^T = Z \underline{B}^T. \tag{48}$$

We substitute the right-hand side of (42) and (43) for u and v in (44) and (45) and then make use of (46), (47) and (48). This yields

$$\underline{p}_+ = \underline{p} + \alpha \tilde{A} \underline{q} \tag{49}$$

$$\underline{q}_+ = \underline{q} + \beta \tilde{B}^T \underline{p}. \tag{50}$$

Obviously, $\underline{z}\underline{z}^T$ is the h×h-matrix whose elements are all equal to 1. Therefore, the elements z_{ij} of Z are as follows:

$$z_{ij} = \begin{cases} \dfrac{h-1}{h} & \text{for } i = j \\[2mm] -\dfrac{1}{h} & \text{for } i \neq j. \end{cases} \tag{51}$$

In view of the structure of Z, it is clear that left multiplication by Z amounts to the subtraction of the column average from each element. Therefore, the elements $\tilde{a}_{ij}$ of $\tilde{A}$ depend as follows on those of A:

$$\tilde{a}_{ij} = a_{ij} - \frac{1}{h} \sum_{k=1}^{h} a_{kj} . \tag{52}$$

Similarly, the elements $\tilde{b}_{ij}$ of $\tilde{B}$ are described by (53):

$$\tilde{b}_{ij} = b_{ij} - \frac{1}{h} \sum_{k=1}^{h} b_{ik} . \tag{53}$$

The number $\tilde{a}_{ij}$ can be interpreted as a measure of success of player 1's pure strategy i in (A,B) against player 2's pure strategy j in (A,B). Measuring success in this way involves a comparison with the average payoff for all pure strategies of player 1 in (A,B) against player 2's strategy j in (A,B). The $\tilde{b}_{ij}$ have an analogous interpretation. Accordingly, we call the $\tilde{a}_{ij}$ and $\tilde{b}_{ij}$ *comparative payoffs*. The matrices $\tilde{A}$ and $\tilde{B}$ are the *comparative payoff matrices* of players 1 and 2, respectively.

The game $(\tilde{A}, \tilde{B})$ is called the *comparative payoff game* for (I,J). The definition of the comparative payoff game $(\tilde{A}, \tilde{B})$ by (47) and (48) will also be applied to cases where (I,J) is not necessarily the support of a regular equilibrium point; of course, $|I| = |J| > 0$ must hold. In the case considered here, where (I,J) is the support of a regular equilibrium point (r,s), we also say that $(\tilde{A}, \tilde{B})$ is the *comparative payoff game* of (r,s).

Since (r,s) is an equilibrium point with support (I,J), all components of $A\,s$ are equal. The same is true for $B^T r$. Therefore, it follows by (48) and (49) that we have:

$$\tilde{A}\,s = \begin{bmatrix} 0 \\ \cdot \\ \cdot \\ \cdot \\ 0 \end{bmatrix} \tag{54}$$

$$\tilde{B}^T \underline{r} = \begin{bmatrix} 0 \\ \cdot \\ \cdot \\ \cdot \\ 0 \end{bmatrix}. \tag{55}$$

It follows that $(\underline{r},\underline{s})$ is an equilibrium point of the comparative payoff game of (r,s) with zero equilibrium payoffs for both players. It can also be seen without difficulty that the equation system formed by (54) and (55), together with the conditions that for $\underline{r}$ and $\underline{s}$ components sum up to 1, has a unique solution if and only if the system formed by (7), (8), (9) and (10) has a unique solution. Therefore, $(\underline{s},\underline{r})$ is a regular equilibrium point of the comparative payoff game $(\tilde{A},\tilde{B})$.

With the usual notational conventions for composed vectors and matrices, the system formed by (49) and (50) can now be represented as follows:

$$\begin{bmatrix} \underline{p}_+ \\ \underline{q}_+ \end{bmatrix} = \begin{bmatrix} \underline{p} \\ \underline{q} \end{bmatrix} + \begin{bmatrix} 0 & \alpha\tilde{A} \\ \beta\tilde{B}^T & 0 \end{bmatrix} \begin{bmatrix} \underline{p} \\ \underline{q} \end{bmatrix} \tag{56}$$

where the zeros stand for zero-submatrices. The following lemma summarizes the results obtained above.

Lemma 2: Let (r,s) be a regular equilibrium point and let S be the set of all mixed strategy pairs (p,q) with the property that (p,q) and (p_+,q_+) both have the same support as (r,s). Then, for $(p,q) \in S$ the linear system (56) describes the relationship between $(\underline{p}_+,\underline{q}_+)$ amd $(\underline{p},\underline{q})$. Moreover, $(\underline{r},\underline{s})$ is a stationary point of (56) and a regular equilibrium point of the comparative payoff game $(\tilde{A},\tilde{B})$ of (r,s). At this equilibrium point of $(\tilde{A},\tilde{B})$ both players have zero payoffs.

3.8 The Reduced System

We continue to work with the assumptions of Lemma 2. We shall examine the stability properties of the linear system (56). This is a necessary step in the investigation of local stability in the preliminary model.

The system (56) is subject to the constraints that for both vectors p and q the sum of all h-components must be 1. Let L be the set of all 2h-dimensional vectors whose first h-components sum up to 1 and whose last h-components sum up to 1. We are interested in *stability within* L in the following sense:

$$\lim_{t \to \infty} (\underline{p}(t),\underline{q}(t)) = (\underline{r},\underline{s}) \tag{57}$$

for every initial value

$$(\underline{p}(0),\underline{q}(0)) \in L .\tag{58}$$

Since (56) is linear, stability in this sense is equivalent to *local stability within* L defined as in 3.6, but with the restriction to initial values in L.

In the following, we shall assume $h = |I| > 1$ or, in other words, that (r,s) is genuinely mixed. It will be convenient to replace (56) by an unconstrained system. For this purpose we eliminate p_h and q_h in (56). We substitute the right-hand side of (59) for p_h and the right-hand side of (60) for q_h:

$$p_h = 1 - \sum_{i=1}^{h-1} p_i \tag{59}$$

$$q_h = 1 - \sum_{j=1}^{h-1} q_j. \tag{60}$$

After the elimination of p_h and q_h, we remove the equations for p_{h+} and q_{h+}. We call the system obtained in this way the *reduced system*.

In order to be able to describe the reduced system in more detail, we introduce the following *notational conventions*: $\bar{p}$, $\bar{q}$, $\bar{p}_+$, $\bar{q}_+$, $\bar{r}$ and $\bar{s}$ denote the $(h-1)$-dimensional vectors obtained by the removal of the h-th component from $\underline{p}$, $\underline{q}$, $\underline{p}_+$, $\underline{q}_+$, $\underline{r}$ and $\underline{s}$, respectively. Moreover, let $\bar{A}$ be the $(h-1)\times(h-1)$-matrix whose elements $\bar{a}_{ij}$ are as follows:

$$\bar{a}_{ij} = \tilde{a}_{ij} - \tilde{a}_{ih} \tag{61}$$

for $i = 1,...h-1$ and $j = 1,...h-1$. Similarly, let $\bar{B}$ be the $(h-1)\times(h-1)$-matrix whose elements $\bar{b}_{ij}$ are as follows:

$$\bar{b}_{ij} = \tilde{b}_{ij} - \tilde{b}_{hj} .\tag{62}$$

Let $\bar{a}$ be the column vector formed by the h-1 first components of the last column of $\bar{A}$ and let $\bar{b}$ be the column vector formed by the h-1 first components of the last column of $\bar{B}^T$. It can be seen without difficulty that the reduced system can now be described by the following equation:

$$
\begin{bmatrix} \bar{p}_+ \\ \bar{q}_+ \end{bmatrix} = \begin{bmatrix} \alpha\bar{a} \\ \beta\bar{b} \end{bmatrix} + \begin{bmatrix} \bar{p} \\ \bar{q} \end{bmatrix} + \begin{bmatrix} 0 & \alpha A \\ \beta B^T & 0 \end{bmatrix} \begin{bmatrix} \bar{p} \\ \bar{q} \end{bmatrix}. \tag{63}
$$

Define

$$
R = \begin{bmatrix} 0 & \alpha A \\ \beta B^T & 0 \end{bmatrix} \tag{64}
$$

and

$$
C = \begin{bmatrix} \alpha\bar{a} \\ \beta\bar{b} \end{bmatrix}. \tag{65}
$$

Moreover, let D be the (2h-2)×(2h-2)-unit matrix:

$$
D = \begin{bmatrix} 1 & . & . & . & 0 \\ . & . & & & . \\ . & & . & & . \\ . & & & . & . \\ 0 & . & . & . & 1 \end{bmatrix}. \tag{66}
$$

With this notation, (63) can be expressed as follows:

$$
\begin{bmatrix} \bar{p}_+ \\ \bar{q}_+ \end{bmatrix} = c + (D+R) \begin{bmatrix} \bar{p} \\ \bar{q} \end{bmatrix}. \tag{67}
$$

Obviously, $(\bar{r},\bar{s})$ is a stationary point of the system (67). It is clear that (r,s) is stable within L if and only if $(\bar{r},\bar{s})$ is stable with respect to (67).

Lemma 3:   Let (r,s) be a regular equilibrium point. If (r,s) is a pure strategy combination, then $(\underline{r},\underline{s})$ is stable within L with respect to (56). If (r,s) is genuinely mixed, then $(\underline{r},\underline{s})$ is unstable within L with respect to (56).

Proof: The case of pure strategies   r   and   s   has been discussed. Assume $h = |I| > 1$.   It is sufficient to examine the stability of $(\bar{r},\bar{s})$ with respect to (67).

It is well-known that a system of linear difference equations with constant coefficients is stable if and only if all the eigenvalues of its matrix lie in the interior of the unit circle around the origin in the complex plane. It is also well-known that the sum of all eigenvalues of a quadratic matrix are equal to its *trace*, defined as the sum of all elements on the main diagonal.

The matrix D+R of (66) has 2h-2 eigenvalues. The trace is equal to 2h-2, since all elements on the main diagonal are equal to 1. Therefore, at least one eigenvalue has a real part not smaller than 1. This eigenvalue does not lie in the interior of the unit circle around the origin in the complex plane. $(\bar{r},\bar{s})$ is unstable with respect to (67).

3.9 Movement into the Support of a Regular Equilibrium Point

It will be important for the investigation of local stability properties of the preliminary model that for mixed strategy pairs (p,q) sufficiently near to a regular equilibrium point (r,s) the pair (p_+,q_+) has the same support as (r,s). Loosely speaking, we may say that a solution which converges to (r,s) must move into the support of (r,s).

Lemma 4: Let (r,s) be a regular equilibrium point. Then, for fixed $\alpha > 0$ and $\beta > 0$, a number $\varepsilon > 0$ can be found such that for every pair (p,q) in the ε-neighborhood $U_\varepsilon(r,s)$ of (r,s) the pair (p_+,q_+) determined by (22) to (25) has the same support as (r,s).

Proof: (r,s) is stationary by Theorem 1. Therefore, it follows by (22) and (23) that u and v assume the values H(r,s) and K(r,s) respectively at p = r and q = s. Regularity requires

$$H(i,s) - u < 0 \qquad \text{for } r_i = 0 \tag{68}$$

$$K(r,j) - v < 0 \qquad \text{for } s_j = 0. \tag{69}$$

It follows by the continuity properties of u and v mentioned in Lemma 1 and the continuity of H and K that for sufficiently small ε inequalities (68) and (69) continue to hold in the ε-neighborhood of (r,s). Therefore, for (p,q) in this ε-neighborhood, we have $p_{i+} = 0$ and $q_{j+} = 0$ outside the support of (r,s). By Lemma 1, the strategies p_+ and q_+ depend continuously on p and q. Therefore, for sufficiently small ε, we have $p_{i+} > 0$ and $q_{j+} > 0$ inside the support of (r,s) for (p,q) $\in U_\varepsilon(r,s)$. Consequently, the assertion of the lemma holds.

3.10 Local Stability Properties of the Preliminary Model

With the help of Lemma 3 and Lemma 4 it is now possible to settle the question of local stability or instability in the preliminary model.

Theorem 2: Let (r,s) be a regular equilibrium point. If r and s are pure strategies, then (r,s) is locally stable in the preliminary model described by (22) to (25). If (r,s) is genuinely mixed, then (r,s) is not locally stable in the preliminary model. This is true for all parameter pairs (α, β) with $\alpha > 0$ and $\beta > 0$.

Proof: Let ε be a number with the property spelled out by Lemma 4. Consider first the case where r and s are pure strategies. Then, a solution (p(t),q(t)) with (p(0),q(0)) in $U_\varepsilon(r,s)$ immediately moves to (r,s) and stays there forever. Therefore, in this case (r,s) is locally stable.

Now assume that (r,s) is genuinely mixed. Suppose that (r,s) is locally stable. Then, an open neighborhood V of (r,s) with $V \subseteq U_\varepsilon(r,s)$ can be found such that every solution starting there converges to (r,s) and stays in $U_\varepsilon(r,s)$. In particular this is true for solutions which start in $V \cap S$, where S is the set of all mixed strategy pairs with the same support as (r,s). These solutions are described by the linear system (57) which is unstable by Lemma 3. It follows that (r,s) fails to be locally stable with respect to the preliminary model.

Remark: The usual local stability criteria are based on differentiability at the stationary point and therefore cannot be applied to the preliminary model. Lemma 4 overcomes this difficulty.

4. The Anticipatory Learning Process

The preliminary model will now be modified by the introduction of anticipation. The resulting modified model is the *anticipatory learning process*. In this model, the individuals do not react directly to the observed strategies p and q. They first form anticipated strategies $p_>$ and $q_>$. Anticipated strategies depend on p and q in the same way as p_+ and q_+ depend on p and q in the preliminary model. We may say that anticipations follow the preliminary model.

The reactions follow the preliminary model, too. p_+ is computed in the same way as in the preliminary model, but on the basis of p and $q_>$ instead of p and q. Analogously, q_+ is computed on the basis of $p_>$ and q.

Anticipations on the other side are not anticipated. Nevertheless, anticipations are approximately correct if α and β are small.

4.1 Definition of the Anticipatory Learning Process

We continue to use the notation introduced in 3.2. In addition to p, q, p_+ and q_+, the anticipated strategies $p_>$ and $q_>$ appear in the definition of the anticipatory learning process. We also use the notation $p_>(t)$ and $q_>(t)$ where time dependence needs to be emphasized. As in the case of "+", lower indices placed before ">" indicate components. The following equations describe the *anticipatory learning process*:

$$p_{i>} = \max \left[0, p_i + \alpha H(i,q) - u_>\right] \qquad \text{for } i = 1,\ldots,n \qquad (70)$$

$$q_{j>} = \max \left[0, q_j + \beta K(p,j) - v_>\right] \qquad \text{for } j = 1,\ldots,m \qquad (71)$$

$$\sum_{i=1}^{n} \max \left[0, p_i + \alpha H(i,q) - u_>\right] = 1 \qquad (72)$$

$$\sum_{j=1}^{m} \max \left[0, q_j + \beta K(p,j) - v_>\right] = 1 \qquad (73)$$

$$p_{i+} = \max \left[0, p_i + \alpha H(i,q_>) - u\right] \qquad \text{for } i = 1,\ldots,n \qquad (74)$$

$$q_{j+} = \max \left[0, q_j + \beta K(p_>,j) - v\right] \qquad \text{for } j = 1,\ldots,m \qquad (75)$$

$$\sum_{i=1}^{n} \max \left[0, p_i + \alpha H(i,q_>) - u\right] = 1 \qquad (76)$$

$$\sum_{j=1}^{m} \max \left[0, q_j + \beta K(p_>,j) - v\right] = 1. \qquad (77)$$

As before, α and β are positive constants. The column vector $p_>$ of the $p_{i>}$ is player 1's *anticipated strategy* and the column vector of the $q_{j>}$ is player 2's *anticipated strategy*.

<u>Lemma 5</u>: Let α and β be positive constants. Then, for given p and q the anticipated strategies $p_>$ and $q_>$ and next period's strategies p_+ and q_+ as well as the auxiliary variables u, v, $u_>$ and $v_>$ are uniquely determined by (70) to (77). Moreover, $p_>$, $q_>$, p_+, q_+, u, v, $u_>$ and $v_>$ depend continuously on p and q.

<u>Proof</u>: The assertion is an immediate consequence of Lemma 1.

4.2 A Best Reply Property of Stationary Points

In the following, we shall obtain a first result on stationary points of the anticipatory learning process. The investigation of stationarity in the anticipatory learning process is more difficult than in the preliminary model. Finally, we shall show that for sufficiently small α and β equilibrium is necessary and sufficient for stationarity in regular games. However, several auxiliary results must be obtained before this can be done. A numerical example will show that the restriction to sufficiently small α and β cannot be avoided.

<u>Lemma 6</u>: Let (p,q) be a stationary point of the anticipatory learning process. Then, p is a best reply to $q_>$ and q is a best reply to $p_>$. This is true for all $\alpha > 0$ and $\beta > 0$ and for all games (A,B).

<u>Proof</u>: The proof makes use of (74) to (77) only. The argument is essentially the same as in the proof of Lemma 2 and will not be repeated here.

4.3 <u>Equilibrium as a Sufficient Condition of Stationarity</u>

It is easy to answer this question.

<u>Lemma 7</u>: Let (p,q) be an equilibrium point. Then, (p,q) is a stationary point of the anticipatory learning process. Moreover, we have $p_> = p$ and $q_> = q$. This is true for all $\alpha > 0$ and $\beta > 0$ and all games (A,B).

<u>Proof</u>: It follows by Lemma 2 that (p,q) is stationary in the preliminary model. Therefore, we can conclude by (70) to (73) that we have $p_> = p$ and $q_> = q$. It follows by Lemma 6 and the definition of an equilibrium point that (p,q) is stationary.

4.4 <u>A Counterexample</u>

In the following, we shall exhibit an example of stationary points which are not equilibrium points. Figure 1 shows a 2×2-game. Assume $\alpha = \beta = .5$. Let (p,q) be a mixed strategy pair with

$$p_1 = q_1 = x \tag{78}$$

where x is a number with

$$0 \leq x \leq 1. \tag{79}$$

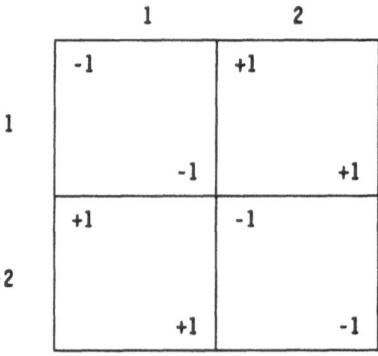

<u>Figure 1</u>: <u>A counterexample</u>: Player 1's strategies correspond to rows and player 2's strategies correspond to columns. Payoffs of player 1 are shown in the upper left corner and those of player 2 in the lower right corner of a field corresponding to a pure strategy combination.

We have

$$H(1,q) = K(p,1) = 1-2x \tag{80}$$

$$H(2,q) = K(p,2) = 2x-1. \tag{81}$$

Eqs. (38) and (39) yield $u_> = v_> = 0$. We obtain:

$$p_{1>} = q_{1>} = x + \frac{1}{2} - x = \frac{1}{2} \tag{82}$$

$$p_{2>} = q_{2>} = 1 - x + x - \frac{1}{2} = \frac{1}{2} . \tag{83}$$

Both pure strategies of player 1 yield the same payoff against $q_>$. Therefore, p is a best reply to $q_>$. Similarly, q is a best reply to $p_>$. It follows by Lemma 6 that (p,q) is stationary with respect to the anticipatory learning process with $\alpha = \beta = .5$. This is true for all x with $0 \leq x \leq 1$, but only for $x = 1/2$ the mixed strategy pair (p,q) is an equilibrium point.

The game of Figure 1 has three equilibrium points, two in pure strategies and one in mixed strategies. Each of these equilibrium points is regular. The game of Figure 1 is regular.

Of course, the example depends on the special value of α and β. However, the pure strategy pairs $(1,1)$ and $(2,2)$, which are connected to the extreme points of the interval $0 \leq x \leq 1$, are stationary for every pair (α,β) with $\alpha \geq .5$ and $\beta \geq .5$. We shall show this for the example of $(1,1)$. One obtains:

$$p_{1>} = \max \left[0, 1 - \alpha - u_>\right] \tag{84}$$

$$p_{2>} = \min \left[1, \max (0, \alpha - u_>)\right] . \tag{85}$$

This yields:

$$p_{1>} = \begin{cases} 1-\alpha & \text{for } \alpha \leq 1 \\ 0 & \text{for } \alpha \geq 1 \end{cases} \tag{86}$$

$$p_{2>} = \begin{cases} \alpha & \text{for } \alpha \leq 1 \\ 1 & \text{for } \alpha \geq 1 . \end{cases} \tag{87}$$

In both cases we have $p_{2>} \geq p_{1>}$. Similarly, $\beta \geq .5$ yields $q_{2>} \geq q_{1>}$. Therefore, player 1's pure strategy 1 is a best reply to $q_>$ and player 2's pure strategy 1 is a best reply to $p_>$. It follows by Lemma 7 that the pure strategy pair $(1,1)$ is stationary.

The example shows that assertions on equilibrium as a necessary condition for stationarity must be restricted to sufficiently small α and β.

4.5 A Property of Regular Equilibrium Points

In the following, we shall derive an auxiliary result on regular equilibrium points. In order to do this, it is necessary to classify stationary points (p,q) according to the support of (p,q) and $(p_>,q_>)$.

Let (p,q) be a mixed strategy pair and let $(p_>,q_>)$ be the pair of anticipated strategies associated to (p,q). Moreover, let (I,J) be the support of (p,q) and let $(I_>,J_>)$ be the support of $(p_>,q_>)$. We call $(I,J,I_>,J_>)$ the *type* of (p,q). The symbol T is used for the *type function* which assigns the type of (p,q) to every (p,q):

$$(I,J,I_>,J_>) = T(p,q). \tag{88}$$

The following lemma shows that for sufficiently small α and β a regular equilibrium point is the only stationary point of its type.

Lemma 8: Let (r,s) be a regular equilibrium point. Then, numbers $\alpha_0 > 0$ and $\beta_0 > 0$ can be found such that for $0 < \alpha < \alpha_0$ and $0 < \beta < \beta_0$ the following is true: (r,s) is the only stationary point of the anticipatory learning process, whose type is $T(r,s)$.

Proof: Let $(I,J,I_>,J_>)$ be the type of (r,s). Since (r,s) is regular, we have $|I| = |J|$. In view of Lemma 7, the anticipated strategies associated with (r,s) are $r_> = r$ and $s_> = s$. Therefore, we have $I_> = I$ and $J_> = J$. Without loss of generality, we assume $I = \{1,...,h\}$ and $J = \{1,...,h\}$.

Assume that (p,q) is a stationary point of the anticipatory learning process with $T(p,q) = T(r,s)$. We use the notational conventions of 3.7. In view of (49) and (50) together with (70) to (73) we have

$$p_> = p + \alpha\tilde{A}q \tag{89}$$

$$q_> = q + \beta\tilde{B}^T p . \tag{90}$$

Since (p,q) is stationary, (74) to (77) yield:

$$\tilde{A}q + \beta\tilde{A}\tilde{B}^T p = 0 \tag{91}$$

$$\tilde{B}^T p + \alpha\tilde{B}^T\tilde{A}q = 0 . \tag{92}$$

In addition to (70) and (71) we have

$$\sum_{i=1}^{h} p_i = 1 \tag{93}$$

$$\sum_{j=1}^{n} q_j = 1. \tag{94}$$

Consider the equation system formed by (70), (71), (72) and (73). In the limiting case $\alpha = \beta = 0$ this system has a unique solution, namely $\underline{p} = \underline{r}$ and $\underline{q} = \underline{s}$. This is a consequence of the fact that $(\underline{r},\underline{s})$ is a regular equilibrium point of $(\tilde{A},\tilde{B})$.

The condition for unique solvability of the system (91) to (94) requires that certain determinants do not vanish. Since all these determinants depend continuously on α and β, we can find numbers α_0 and β_0 such that the system continues to have a unique solution for $0 < \alpha \leq \alpha_0$ and $0 < \beta \leq \beta_0$. Therefore, for α and β in these intervals $\underline{p} = \underline{r}$ and $\underline{q} = \underline{s}$ is the only solution of the system. We can conclude that the assertion of the lemma is true.

4.6 Equilibrium as a Necessary Condition

It will be shown that in regular games for sufficiently small α and β a stationary point of the anticipatory learning process must be an equilibrium point. In order to do this, we have to exclude the possibility that for arbitrarily small α and β stationary points can be found which fail to be equilibrium points. Since the number of possible types is finite, it is sufficient to investigate this problem for a fixed type. Lemma 8 already answers this question for types of equilibrium points. We have to show that for other types no stationary point exists for sufficiently small α and β.

Lemma 9: Let (A,B) be a regular game. Let $(I,J,I_>,J_>)$ be a possible type of a mixed strategy pair such that no equilibrium point is of this type. Then, numbers $\alpha_0 > 0$ and $\beta_0 > 0$ can be found such that for $0 < \alpha \leq \alpha_0$ and $0 < \beta \leq \beta_0$ the anticipatory learning process has no stationary point of the type $(I,J,I_>,J_>)$.

Proof: Assume that the assertion does not hold. Let $(I,J,I_>,J_>)$ be a type for which this is the case. Then, we can find monotonically decreasing sequences $\alpha_1,\alpha_2,\ldots$ and $\beta_1,\beta_2,\ldots$ converging to zero such that for $k = 1,2,\ldots$ the anticipatory learning process with $\alpha = \alpha_k$ and $\beta = \beta_k$ has a stationary point (p^k,q^k) of the type $(I,J,I_>,J_>)$. In view of the compactness of $P \times Q$ and the possibility of selecting a convergent subsequence of the (p^k,q^k), it can be assumed that the sequence $(p^1,q^1),(p^2,q^2),\ldots$ converges to a pair (p,q) of mixed strategies:

$$\lim_{k \to \infty} (p^k, q^k) = (p, q). \tag{95}$$

Assume that the sequences of the α_k, β_k and (p^k, q^k) have the properties described above.

Let $(I', J', I'_>, J'_>)$ be the type of (p, q). Since a sequence of positive numbers may converge to zero, the type of (p, q) may be different from $(I, J, I_>, J_>)$, but the sets $I', J', I'_>, J'_>$ must be subsets of $I, J, I_>, J_>$, respectively.

For $k = 1, 2, \ldots$ let $p_{k>}$ and $q_{k>}$ be the anticipated strategies and let $u^k, v^k, u^k_>, v^k_>$ be the values of the auxiliary variables connected to (p^k, q^k) by (70) to (77) with $\alpha = \alpha_k$ and $\beta = \beta_k$. It follows by (72) and (73) that we have

$$u^k_> = \frac{\alpha}{|I_>|} \sum_{i \in I_>} H(i, q^k) \tag{96}$$

$$v^k_> = \frac{\beta}{|J_>|} \sum_{j \in J_>} K(p^k, j). \tag{97}$$

In view of (95) and the continuity of H and K, the sequences $u^1_>, u^2_>, \ldots$ and $v^1_>, v^2_>, \ldots$ converge. We have:

$$u_> = \lim_{k \to \infty} u^k_> = \frac{\alpha}{|I_>|} \sum_{i \in I_>} H(i, q) \tag{98}$$

$$v_> = \lim_{k \to \infty} v^k_> = \frac{\beta}{|J_>|} \sum_{j \in J_>} K(p, j). \tag{99}$$

Since the sequences $\alpha_1, \alpha_2, \ldots$ and $\beta_1, \beta_2, \ldots$ converge to zero, it follows by (70) and (71) that the sequence of the $(p^k_>, q^k_>)$ converges to (p, q):

$$\lim_{k \to \infty} (p^k_>, q^k_>) = (p, q). \tag{100}$$

In view of Lemma 6, the stationarity of (p, q) has the consequence that p^k is a best reply to $q^k_>$ and q^k is a best reply to $p^k_>$. Since the property of being a best reply can be expressed by weak inequalities, it is preserved in the limit. It follows by (100) that (p, q) is an equilibrium point. In view of the regularity of (A, B), it is clear that (p, q) is a regular equilibrium point.

The regularity of (p,q) implies the following inequalities:

$$H(i,q) < H(p,q) \quad \text{for} \quad i \in N \backslash I' \tag{101}$$

$$K(p,j) < K(p,q) \quad \text{for} \quad j \in M \backslash J' . \tag{102}$$

In view of the continuity of H and K, inequalities (101) and (102) permit the conclusion that a k_0 exists such that similar inequalities hold for $k > k_0$:

$$H(i,q_>^k) < H(p^k,q_>^k) \quad \text{for} \quad i \in N \backslash I' \tag{103}$$

$$K(p_>^k,j) < K(p_>^k,q) \quad \text{for} \quad j \in N \backslash J' . \tag{104}$$

Since p^k is a best reply to $q_>^k$, inequality (103) cannot hold for $i \in I$. In view of $I' \subseteq I$, this has the consequence $I' = I$. Similarly, we can conclude $J' = J$.

It follows by (100) that for sufficiently great k components $p_{i>}^k$ of $p_>^k$ with $i \in I$ must be positive. This yields $I_> \supseteq I$. Similarly, we obtain $J_> \supseteq J$.

We shall show $I = I_>$ and $J = J_>$. Consider a pure strategy $i \in I_>$. In view of $p_i = o$ and $p_{i>} > 0$, it follows by (70) that we must have:

$$H(i,q^k) - u_>^k > 0 \quad \text{for} \quad i \in I_> \tag{105}$$

for $k = 1,2,\ldots$. Since (p,q) is an equilibrium point, it follows by (98) that the following is true:

$$u_> \leq H(p,q). \tag{106}$$

In view of the continuity of H, inequalities (105) and (106) yield:

$$H(i,q) \geq H(p,q) \quad \text{for} \quad i \in I_> . \tag{107}$$

It follows by the regularity of (p,q) that every $i \in I_>$ belongs to I. In view of $I_> \supseteq I$, we have $I = I_>$. Similarly, one obtains $J = J_>$.

We have shown that all (p^k,q^k) are of the same type $(I,J,I_>,J_>)$ as (p,q). This contradicts the assumption that no equilibrium point is of the same type as the (p^k,q^k). Consequently, the assertion of the lemma holds.

Theorem 3: Let (A,B) be a regular game. Then, numbers $\alpha_0 > 0$ and $\beta_0 > 0$ can be found such that for $0 < \alpha \leq \alpha_0$ and $0 < \beta \leq \beta_0$ the following is true: A mixed strategy pair is stationary with respect to the anticipatory learning process described by (70) to (77) if and only if (p,q) is an equilibrium point.

Proof: The stationarity of equilibrium points follows by Lemma 7. By Lemma 9, a stationary point must have the same type as some equilibrium point. Since the number of possible types is finite, the theorem follows by Lemma 8.

4.7 The Linear System Connected to a Regular Equilibrium Point

Let (r,s) be a regular equilibrium point. We shall proceed in a similar fashion as in 3.7. Let (I,J) be the support of (r,s) and define $h = |I|$. Since (r,s) is regular, we also have $|J| = h$.

The set S will from now on be defined as the set of all mixed strategy pairs (p,q) such that (p,q) as well as $(p_>,q_>)$ and (p_+,q_+) determined by (70) to (77) have the support (I,J) of (r,s). We continue to use the notational conventions of 3.7 and extend them to $p_>$ and $q_>$. In the anticipatory learning process, the connection between $(p_>,q_>)$ and (p,q) is the same one as between (p_+,p_+) and (p,q) in the preliminary model. Moreover, in the anticipatory learning process p_+ depends on $(p,q_>)$ in the same way as on (p,q) in the preliminary model. An analagous statement holds for q_+. Therefore, for $(p,q) \in S$, Eq. (56) permits the following conclusions:

$$\begin{bmatrix} p_> \\ q_> \end{bmatrix} = \begin{bmatrix} p \\ q \end{bmatrix} + \begin{bmatrix} 0 & \alpha\tilde{A} \\ \beta\tilde{B}^T & 0 \end{bmatrix} \begin{bmatrix} p \\ q \end{bmatrix} \tag{108}$$

where the zeros stand for $h \times h$-zero-matrices.

$$\begin{bmatrix} p_+ \\ q_+ \end{bmatrix} = \begin{bmatrix} p \\ q \end{bmatrix} + \begin{bmatrix} 0 & \alpha\tilde{A} \\ \beta\tilde{B}^T & 0 \end{bmatrix} \begin{bmatrix} p_> \\ q_> \end{bmatrix}. \tag{109}$$

Eqs. (108) and (109) together yield:

$$\begin{bmatrix} p_+ \\ q_+ \end{bmatrix} = \begin{bmatrix} p \\ q \end{bmatrix} + \begin{bmatrix} 0 & \alpha\tilde{A} \\ \beta\tilde{B}^T & 0 \end{bmatrix} \begin{bmatrix} p \\ q \end{bmatrix} + \begin{bmatrix} 0 & \alpha\tilde{A} \\ \beta\tilde{B}^T & 0 \end{bmatrix}^2 \begin{bmatrix} p \\ q \end{bmatrix}. \tag{110}$$

This system of difference equations has the same significance for the anticipatory learning process as (56) has for the preliminary model. Also with respect to (110) we are interested in stability within L, defined in the same way as in 3.7.

4.8 The Reduced System

We proceed in the same way as in 3.8. If (r,s) is a pair of pure strategies, then $(\underline{r},\underline{s})$ is trivially stable within L, since L consists of a single point. In the following, we assume $h = |I| = |J| > 1$.

In the same way as in 3.8 we eliminate p_h and q_h and remove the equations for $p_{h>}$ and $q_{h>}$ in (108). The $(h-1)$-dimensional vectors resulting from $\underline{p}_>$ and $\underline{q}_>$ by taking away the h-th component are denoted by $\bar{p}_>$ and $\bar{q}_>$, respectively. One obtains:

$$\begin{bmatrix} \bar{p}_> \\ \bar{q}_> \end{bmatrix} = c + (D+R) \begin{bmatrix} \bar{p} \\ \bar{q} \end{bmatrix} \tag{111}$$

with D and R defined as in 3.8. Similarly, we eliminate $p_{h>}$ and $q_{h>}$ in (109) and remove the equations for p_{h+} and q_{h+}. This yields

$$\begin{bmatrix} \bar{p}_+ \\ \bar{q}_+ \end{bmatrix} = c + \begin{bmatrix} \bar{p} \\ \bar{q} \end{bmatrix} + R \begin{bmatrix} \bar{p}_> \\ \bar{q}_> \end{bmatrix}. \tag{112}$$

Define

$$g = (D+R)c. \tag{113}$$

In view of (111) and (112), the relationship between $(\bar{p}_+,\bar{q}_+)$ and $(\bar{p},\bar{q})$ can be expressed as follows:

$$\begin{bmatrix} \bar{p}_+ \\ \bar{q}_+ \end{bmatrix} = g + (D+R+R^2) \begin{bmatrix} \bar{p} \\ \bar{q} \end{bmatrix}. \tag{114}$$

We call this system of difference equations the *reduced system* of (r,s) in the anticipatory learning process. The system (110) is stable within L, if and only if the reduced system (114) is stable.

4.9 The Interaction Covariance Matrix

The stability of the reduced system depends on the eigenvalue of the matrix R. As we shall see, these eigenvalues are closely connected to those of another matrix, which is interesting in view of its interpretation. In order to explain the meaning of this "interaction covariance matrix", we first introduce a decomposition of a payoff matrix into a "biadditive" part and an "interaction" part. The same kind of decomposition underlies statistical procedures of two-way analysis of variance (see, e.g., Darlington, 1975). There the word "interaction" refers to effects which are not attributed to rows or columns, but to the interaction of both. If payoff matrices are decomposed in this way, we can also speak of a separation of non-interactive and interactive effects of pure strategy choices.

Let (r,s) be a genuinely mixed regular equilibrium point and let $(\underline{A},\underline{B})$ be the restricted game of (r,s). In view of the regularity of (r,s), both players have the same number of pure strategies. Let h be this number. $\underline{A}$ and $\underline{B}$ are $h \times h$-matrices.

An $h \times h$-matrix X is called *biadditive* if constants $u_1,\ldots,u_h$ and $v_1,\ldots,v_h$ can be found such that the elements x_{ij} of X satisfy the condition

$$x_{ij} = u_i + v_j \tag{115}$$

for $i = 1,\ldots,h$ and $j = 1,\ldots,h$. The u_i are called *row effects* and the v_i are *column effects*. The payoff matrices $\underline{A}$ and $\underline{B}$ are not biadditive, but we can still try to attribute as much payoff variance as possible to additively composed row and column effects; the remaining payoff parts are then defined as "interaction effects". We decompose $\underline{A}$ accordingly:

$$\underline{a}_{ij} = u_i + v_j + \hat{a}_{ij} \tag{116}$$

for $i = 1,\ldots,h$ and $j = 1,\ldots,h$. We minimize interaction effects in the sense of the least square criterion. Define:

$$Q = \sum_{i=1}^{h} \sum_{j=1}^{h} (\underline{a}_{ij} - u_i - v_j)^2 . \tag{117}$$

The row and column effects u_i and v_i are determined in such a way that Q is minimized. At the minimum, the partial derivatives of Q with respect to the u_i and v_j must be zero. This yields:

$$u_i = \frac{1}{h} \sum_{j=1}^{h} (\underline{a}_{ij} - v_j) \qquad \text{for } i = 1,\ldots,h \tag{118}$$

$$v_j = \frac{1}{h} \sum_{i=1}^{h} (\underline{a}_{ij} - u_i) \qquad \text{for } j = 1,\ldots,h. \tag{119}$$

These equations do not uniquely determine the u_i and the v_j. However, as we shall see, they uniquely determine the *interaction effects*:

$$\hat{a}_{ij} = a_{ij} - u_i - v_j \ . \tag{120}$$

Define

$$a_i = \frac{1}{h} \sum_{j=1}^{h} a_{ij} \tag{121}$$

$$a'_j = \frac{1}{h} \sum_{i=1}^{h} a_{ij} \tag{122}$$

$$a = \frac{1}{h^2} \sum_{i=1}^{h} \sum_{j=1}^{h} a_{ij}. \tag{123}$$

Obviously, a_i is the i-th *row average*, a'_j is the j-th *column average* and a is the *overall average* of A. It follows by (118) and (119) that we have:

$$u_i = a_i - \frac{1}{h} \sum_{j=1}^{h} v_j \qquad \text{for } i = 1,\ldots,h \tag{124}$$

$$v_j = a'_j - \frac{1}{h} \sum_{i=1}^{h} u_i \qquad \text{for } j = 1,\ldots,h. \tag{125}$$

Summation over i in (118) or over j in (119) yields:

$$\sum_{i=1}^{h} u_i + \sum_{j=1}^{h} v_j = ha. \tag{126}$$

Therefore, (124) and (125) permit the following conclusion:

$$a_i + v_j = a_i + a'_j - a. \tag{127}$$

Eq. (127) describes the elements $u_i + v_j$ of the *biadditive part* of A. We are more interested in the *interaction part* $\hat{A}$ with the elements $\hat{a}_{ij}$. In view of (120) and (127) we have

$$\hat{a}_{ij} = a_{ij} - a_i - a'_j + a \tag{128}$$

for $i = 1,\ldots,h$ and $j = 1,\ldots,h$. Player 2's payoff matrix $\underline{B}$ can be decomposed analogously. Let $\underline{b}_i$ the i-th row average, $\underline{b}'_j$ the j-th column average and $\underline{b}$ the overall average of $\underline{B}$. The minimization of interaction effects in the sense of the least square criterion yields:

$$\hat{b}_{ij} = \underline{b}_{ij} - \underline{b}_i - \underline{b}'_j + \underline{b} \tag{129}$$

for $i = 1,\ldots,h$, and $j = 1,\ldots,h$. The h×h-matrix of the $\hat{b}_{ij}$ is denoted by $\hat{B}$. We call $\hat{A}$ and $\hat{B}$ the *payoff interaction matrices* of (r,s) for players 1 and 2, respectively.

It will be analytically advantageous to describe the matrices $\hat{A}$ and $\hat{B}$ as transformations of the payoff matrices $\underline{A}$ and $\underline{B}$ of the restricted game. Consider the matrix Z defined by (46) in 3.7. As we shall see, we have:

$$\hat{A} = Z\underline{A}Z \tag{130}$$

$$\hat{B} = Z\underline{B}Z. \tag{131}$$

Left multiplication by Z amounts to the subtraction of the column average from each element. Right multiplication by Z amounts to the subtraction of the row average from each element. It can be seen immediately that the consecutive performance of both operations yields the results described by (128) and (129).

It is useful to point out two properties of Z. It is clear by (46) that Z is symmetric:

$$Z^T = Z . \tag{132}$$

We have:

$$Z^2 = \underline{D} - \frac{2}{h} zz^T + \frac{1}{h^2} zz^Tzz^T . \tag{133}$$

Since the scalar product z^Tz is equal to h, the last term on the right-hand side is equal to zz^T/h. This yields:

$$Z^2 = Z . \tag{134}$$

Eq. (134) is expressed by saying that the matrix Z is *idempotent*.

Consider the matrix

$$C = \hat{A}\hat{B}^T . \tag{135}$$

The element c_{ij} of C is the covariance of player 1's payoff interaction effects

$\hat{a}_{ik}$ in row i with player 2's payoff interaction effects $\hat{b}_{jk}$ in row j. Accordingly, we call C the *interaction covariance matrix* of (r,s).

In view of (130), (131) and (134), we have:

$$C = Z\underline{A}Z\underline{B}^T Z . \tag{136}$$

The definition of the comparative payoff matrices $\tilde{A}$ and $\tilde{B}$ by (47) and (48) shows that instead of (136) we can write:

$$C = \tilde{A}\tilde{B}^T Z . \tag{137}$$

It can be seen with the help of (128) and (129) that column sums and row sums in $\tilde{A}$ and $\tilde{B}$ are zero. This conclusion can also be reached by (130) and (131) in view of the effects of left and right multiplication by Z. In the same way, it follows by (136) that the row sums and column sums of C are equal to zero.

The interaction covariance matrix C can also be connected to the matrices $\bar{A}$ and $\bar{B}$ defined by (61) and (62) in 3.8. In order to reveal this connection, we first derive relationships of $\tilde{A}$ and $\tilde{B}$ with $\hat{A}$ and $\hat{B}$, respectively. Define:

$$V = \begin{bmatrix} 1 & \cdots & 0 & 0 \\ \vdots & & \vdots & \vdots \\ \vdots & & \vdots & \vdots \\ 0 & \cdots & 1 & 0 \end{bmatrix} . \tag{138}$$

This $(h-1)\times h$-matrix V is obtained from $(h-1)\times(h-1)$-unit matrix $\underline{D}$ by the addition of an h-th column, whose elements are all zero. Define:

$$W = \begin{bmatrix} 1 & \cdots & 0 \\ \vdots & & \vdots \\ \vdots & & \vdots \\ 0 & \cdots & 1 \\ -1 & \cdots & -1 \end{bmatrix} \tag{139}$$

This $(h-1)\times h$-matrix W results from $\underline{D}'$ by the addition of an h-th row, whose elements are all equal to -1. With the help of (61) and (62), it can be seen without difficulty that we have:

$$\bar{A} = V\tilde{A}W \tag{140}$$

$$\bar{B}^T = V\tilde{B}^T W. \tag{141}$$

Since left multiplication by Z amounts to the subtraction of column averages, we have:

$$ZW = W. \tag{142}$$

In view of the definitions of $\tilde{A}$ and $\tilde{B}$ by (47) and (48), it follows by (130) and (131) that $\tilde{A}W$ and $\tilde{B}^TW$ can be replaced by $\hat{A}W$ and $\hat{B}^TW$, respectively. We obtain:

$$\tilde{A} = V\hat{A}W \tag{143}$$

$$\tilde{B}^T = V\hat{B}^TW. \tag{144}$$

The reduced system (115) depends on R only, and the definition of R by (64) shows that R is determined by $\tilde{A}$, $\tilde{B}$, α and β. Therefore, (143) and (144) reveal an interesting property of the reduced system. The reduced system is fully determined by the payoff interaction matrices $\hat{A}$ and $\hat{B}$. Row effects and column effects in the payoff matrices $\underline{A}$ and $\underline{B}$ of the restricted game do not influence the dynamics of the reduced system. Only interaction effects matter.

Eqs. (143) and (144) permit the derivation of a relationship between $\tilde{A}\tilde{B}^T$ and the interaction covariance matrix C. We have:

$$\tilde{A}\tilde{B}^T = V\hat{A}WV\hat{B}^TW . \tag{145}$$

It can be seen easily that WV has the following structure:

$$WV = \begin{bmatrix} 1 & \cdot & \cdot & \cdot & 0 & 0 \\ \cdot & & & & \cdot & \cdot \\ \cdot & & & & \cdot & \cdot \\ \cdot & & & & \cdot & \cdot \\ 0 & \cdot & \cdot & \cdot & 1 & 0 \\ -1 & \cdot & \cdot & \cdot & -1 & 0 \end{bmatrix} . \tag{146}$$

The h×h-matrix WV can be described as obtained from W by adding an h-th column of zeros. In view of the fact that right multiplication by Z amounts to the subtraction of row averages, it can be seen easily that the following is true:

$$WVZ = Z . \tag{147}$$

It follows by (131) that we have:

$$WV\hat{B}^T = WVZ\underline{B}^TZ = \hat{B}^T . \tag{148}$$

This together with the definition of C by (135) permits the following conclusion from (145):

$$\tilde{A}\tilde{B}^T = VCW \tag{149}$$

With the help of this relationship, the stability properties of the reduced system will be connected to the eigenvalues of the interaction covariance matrix C.

4.10 Eigenvalue Properties

Lemma 10 reveals the connection between the eigenvalues of the matrix R which appears in the reduced system and those of the interaction covariance matrix C. Later it will be shown that the stability of the reduced system requires that all eigenvalues of C are negative.

Lemma 10: Let (r,s) be a genuinely mixed regular equilibrium point. Let $\tilde{A}, \tilde{B}, \bar{A}, \bar{B}$ and R be defined as in 3.7 and 3.8 and let C be the interaction covariance matrix of (r,s). The following statements (i) to (v) hold:

(i)   If λ is an eigenvalue of R, then $-\lambda$ is an eigenvalue of R.

(ii) No eigenvalue of R is equal to zero.

(iii) λ is an eigenvalue of R if and only if for some eigenvalue η of $\bar{A}\bar{B}^T$ the equation

$$\lambda^2 = \alpha\beta\eta \tag{150}$$

   holds.

(iv) η is an eigenvalue of $\bar{A}\bar{B}^T$ if and only if η is a non-vanishing eigenvalue of C.

(v) If $\underline{v}$ is an eigenvector of C, connected to a non-vanishing eigenvalue η of C, then the sum of all components of $\underline{v}$ is zero.

Proof: Assume that λ is an eigenvalue of R. Consider an eigenvector of R associated with λ, let v be the vector of the h-1 first components and let w be the vector of the h-1 last components of this eigenvector. In view of the definition of R we have:

$$\begin{bmatrix} 0 & \alpha\bar{A} \\ \beta\bar{B}^T & 0 \end{bmatrix} \begin{bmatrix} v \\ w \end{bmatrix} = \begin{bmatrix} \alpha\bar{A}w \\ \beta\bar{B}^T v \end{bmatrix} = \lambda \begin{bmatrix} v \\ w \end{bmatrix} \tag{151}$$

with

$$v \neq 0 \quad \text{or} \quad w \neq 0 \tag{152}$$

since at least one component of an eigenvector must be different from zero. (151) yields

$$\lambda v = \alpha \bar{A} w \tag{153}$$

$$\lambda w = \beta \bar{B}^T v \ . \tag{154}$$

Eqs. (153) and (154) permit the following conclusion

$$\begin{bmatrix} 0 & \alpha \bar{A} \\ \\ \beta \bar{B}^T & 0 \end{bmatrix} \begin{bmatrix} v \\ \\ -w \end{bmatrix} = \begin{bmatrix} -\alpha \bar{A} w \\ \\ \beta \bar{B}^T v \end{bmatrix} = -\lambda \begin{bmatrix} v \\ \\ -w \end{bmatrix}. \tag{155}$$

This shows that $-\lambda$ is an eigenvalue of R. We have proved (i).

Assume $\lambda = 0$. In view of (153) and (154) we have:

$$\bar{A} w = 0 \tag{156}$$

$$\bar{B}^T v = 0. \tag{157}$$

Define

$$\underline{v} = W v \tag{158}$$

$$\underline{w} = W w \tag{159}$$

where W is the matrix defined by (139). Obviously, $\underline{v}$ and $\underline{w}$ are h-dimensional vectors, whose first h-1 components agree with those of v and w, respectively. For both $\underline{v}$ and $\underline{w}$ the sum of all components is zero. In view of (140) and (141), Eqs. (156) and (157) yield:

$$V \tilde{A} \underline{w} = 0 \tag{160}$$

$$V \tilde{B}^T \underline{v} = 0. \tag{161}$$

Since column sums in $\tilde{A}$ and $\tilde{B}$ are zero, the vectors $\tilde{A}\underline{w}$ and $\tilde{B}^T\underline{v}$ have the property that the sum of all components is zero. It follows by the definition of (138) of V that (160) and (161) cannot hold, unless we have:

$$\tilde{A} \underline{w} = 0 \tag{162}$$

$$\tilde{B}^T \underline{v} = 0. \tag{163}$$

It follows by (162) and (163) that for sufficiently small ε the mixed strategy pairs $(\underline{p}, \underline{q})$ with

$$\underline{p} = \underline{r} + \varepsilon \underline{v} \tag{164}$$

$$\underline{q} = \underline{s} + \varepsilon \underline{w} \tag{165}$$

is an equilibrium point of the comparative payoff game $(\tilde{A}, \tilde{B})$. (152) permits the conclusion that, contrary to Lemma 2, the equilibrium point $(\underline{r}, \underline{s})$ fails to be isolated. Consequently, zero cannot be an eigenvalue of R. We have proved (ii).

We continue to assume that λ is an eigenvalue of R. We want to show that (150) holds for some eigenvalue η of $\bar{A}\bar{B}^T$, or, in other words, that λ^2 is an eigenvalue of $\alpha\beta\bar{A}\bar{B}^T$. It follows by (153) and (154) that we have

$$\lambda^2 v = \alpha\bar{A}\lambda w = \alpha\beta\bar{A}\bar{B}^T v \tag{166}$$

$$\lambda^2 w = \beta\bar{B}^T\lambda v = \alpha\beta\bar{B}^T\bar{A}w. \tag{167}$$

In view of (152), it follows by (166) and (167) that λ^2 is an eigenvalue of $\alpha\beta\bar{A}\bar{B}^T$ or of $\alpha\beta\bar{B}^T\bar{A}$. It is well-known that a change of the order of factors in a product of two matrices does not change the eigenvalue of the matrix product. We can conclude that λ^2 is an eigenvalue of $\alpha\beta\bar{A}\bar{B}^T$ and that (150) is satisfied for some eigenvalue η of $\bar{A}\bar{B}^T$.

Before we turn our attention to the reverse direction of (iii), we show that $\eta = 0$ cannot be an eigenvalue of $\bar{A}\bar{B}^T$. Assume that $\bar{A}\bar{B}^T$ has an eigenvalue zero. Then, either $\bar{A}$ or $\bar{B}^T$ must have a vanishing determinant. This has the consequence that either (156) or (157) has a non-trivial solution. As we have seen above, the existence of such a non-trivial solution permits that $(\underline{r}, \underline{s})$ fails to be a regular equilibrium point of the comparative payoff game $(\tilde{A}, \tilde{B})$. It follows that zero cannot be an eigenvalue of $\bar{A}\bar{B}^T$.

We now assume that λ is not necessarily an eigenvalue of R and that (150) holds for some eigenvalue η of $\bar{A}\bar{B}^T$. Let v be an eigenvalue of $\bar{A}\bar{B}^T$ connected to η. In view of (150), we have

$$\lambda^2 v = \alpha\beta\bar{A}\bar{B}^T v. \tag{168}$$

In view of (133) and (135), we have

$$w = \frac{\beta}{\lambda} B^T v . \tag{169}$$

Eqs. (168) and (169) yield (153) and (154). It follows that (151) holds with $v \neq 0$. We can conclude that λ is an eigenvalue of R.

We now turn our attention to (v). Let $\underline{y}$ be an eigenvector of C, connected to a non-vanishing eigenvalue η of C. In view of (136) and (132), we have

$$\eta\underline{y} = ZC\underline{y}. \tag{170}$$

In view of the effect of left multiplication by Z, the sum of all components of $\underline{y}$ must be zero. We have proved (v).

It remains to show (iv). We continue to assume that η is a non-vanishing eigenvalue of C and that $\underline{y}$ is an associated eigenvector of C. Define

$$v = V\underline{y}. \tag{171}$$

In view of (138), the (h-1)-dimensional vector v is the vector of the h-1 first components of $\underline{y}$. Since by (v) the component sum of $\underline{y}$ is zero, (139) has the following consequence:

$$\underline{y} = Wv. \tag{172}$$

In view of (172), we have:

$$\eta\underline{y} = C\underline{y} = CWv. \tag{173}$$

With the help of (171), we obtain

$$\eta v = \eta V\underline{y} = VCWv. \tag{174}$$

In view of (149) we can conclude

$$\eta v = \bar{A}\bar{B}^{\mathsf{T}}v . \tag{175}$$

Since the component sum of $\underline{y}$ is zero, $v \neq 0$ follows by $\underline{y} \neq 0$. Therefore, (175) shows that η is an eigenvalue of $\bar{A}\bar{B}^{\mathsf{T}}$.

Now, assume that η is an eigenvalue of $\bar{A}\bar{B}^{\mathsf{T}}$ and that v is an associated eigenvector of $\bar{A}\bar{B}^{\mathsf{T}}$. Let $\underline{y}$ be defined by (172). With the help of (171) and (172), we can derive (174) from (175) and (173) from (174). This shows that η is an eigenvalue of C. We have proved the lemma.

4.11 A Stability Criterion for the Reduced System

In the following, we shall derive a result on the stability properties of the reduced system (115). It will be shown that for sufficiently small α and β the reduced system is stable if and only if all non-vanishing eigenvalues of the interaction covariance matrix are negative.

A condition which requires negative eigenvalues is much stronger than a codition which requires negative real parts. However, it is not true that the negativity condition on the non-vanishing eigenvalues of the interaction covariance matrix is satisfied only in degenerate cases. This question will be discussed in Section 5.

<u>Lemma 11</u>: Let (r,s) be a genuinely mixed regular equilibrium point. Then, numbers $\alpha_0 > 0$ and $\beta_0 > 0$ can be found such that for $0 < \alpha < \alpha_0$ and $0 < \beta < \beta_0$ the reduced system (115) of (r,s) in the anticipatory learning process is stable if and only if all eigenvalues of the interaction covariance matrix   C   of (r,s) are negative.

<u>Proof</u>: It follows by a well-known theorem on matrix polynomials that μ is an eigenvalue of $D+R+R^2$ if and only if for some eigenvalue λ of R we have:

$$\mu = 1 + \lambda + \lambda^2 . \tag{176}$$

Therefore, the system (115) is stable if and only if all eigenvalues λ of R satisfy the following condition

$$|1 + \lambda + \lambda^2| < 1 . \tag{177}$$

In view of (iii) in Lemma 10, we can look at the eigenvalues λ of R as functions of α and β. Let λ be an eigenvalue of R and let η be the eigenvalue of $\bar{A}\bar{B}^T$ connected to λ by (150). Let $\rho > 0$ and ϕ with $0 \leq \phi < 2\pi$ be determined by the following equation

$$\lambda = \rho\sqrt{\alpha\beta} \, (\cos \phi + i \sin \phi). \tag{178}$$

In view of (150), Moivre's Theorem yields

$$\eta = \rho^2(\cos 2\phi + i \sin 2\phi). \tag{179}$$

This shows that ρ and ϕ depend only on η and not on α and β. For every η, we obtain two functions of the form (178) of α and β. We use the symbol "Re" in order to indicate the real part of the expression in brackets following after "Re". (178) yields:

$$\text{Re}(\lambda + \lambda^2) = \rho\sqrt{\alpha\beta} \, (\cos \phi + \rho\sqrt{\alpha\beta} \cos 2\phi). \tag{180}$$

This shows that for sufficiently small $\alpha\beta$ the real part of $\lambda+\lambda^2$ has the same sign as the real part of λ. Consequently, for sufficiently small $\alpha\beta$ the real part of $1+\lambda+\lambda^2$ is greater than 1 if λ has a positive real part. It follows that for sufficiently small $\alpha\beta$ inequality (177) cannot be satisfied if at least one eigenvalue λ of R has a positive real part. In view of (i) in Lemma 10, this is the case if at least one eigenvalue of R has a non-vanishing real part. It follows by (iii) and (iv) in Lemma 10 that for sufficiently small α and β the negativity of all non-vanishing eigenvalues η of the interaction covariance matrix C of (r,s) is a necessary condition for the stability of the reduced system (114).

Now, assue that all non-vanishing eigenvalues η of C are negative. In view of Lemma 10, this has the consequence that (178) assumes the following form:

$$\lambda = \gamma \sqrt{\alpha\beta} \ i \tag{181}$$

where γ is a positive or negative constant. Equation (181) yields:

$$|1 + \lambda + \lambda^2| = \sqrt{(1-\gamma^2\alpha\beta)^2+\gamma^2\alpha\beta} \tag{182}$$

$$|1 + \lambda + \lambda^2| = \sqrt{1-\gamma^2\alpha\beta(1-\gamma^2\alpha\beta)} \ . \tag{183}$$

This shows that (177) is satisfied for $\gamma^2 < 1/\alpha\beta$. It follows that for sufficiently small α and β the negativity of all non-vanishing eigenvalues η of the interaction covariance matrix C of (r,s) is a sufficient condition for the stability of the reduced system.

Remark: From what has been said at the end of 4.8, it follows under the assumptions of Lemma 10 that for sufficiently small α and β the linear system (110) of (r,s) in the anticipatory learning process is stable within L if and only if all eigenvalues of the interaction covariance matrix C of (r,s) are negative.

4.12 Local Stability Under Sufficiently Slow Anticipatory Learning

Local stability conditions for the anticipatory learning process can be obtained only for sufficiently small α and β. In the case of small values of α and β, the learning process is slow in the sense that the changes from period to period are small. This interpretation underlies a convenient way of speaking which will facilitate the statement of stability results: We say that an equilibrium point (r,s) is *locally stable under sufficiently slow anticipatory learning* if numbers $\alpha_0 > 0$ and $\beta_0 > 0$ can be found such that (r,s) is locally stable with respect to the anticipatory learning process for all parameter pairs (α,β) with $0 < \alpha < \alpha_0$ and $0 < \beta < \beta_0$.

Theorem 4: Let (r,s) be a regular equilibrium point of (A,B). If (r,s) is an equilibrium point in pure strategies, then (r,s) is always locally stable under sufficiently slow anticipatory learning. If (r,s) is a genuinely mixed equilibrium point, then (r,s) is locally stable under sufficiently slow anticipatory learning if and only if all eigenvalues of the interaction covariance matrix C of (r,s) are negative.

Proof: As a first step in the proof of the theorem, we show that for fixed $\alpha > 0$ and $\beta > 0$ a number $\varepsilon > 0$ can be found such that for every pair (p,q) in the ε-neighborhood $U_\varepsilon(r,s)$ of (r,s) the pairs $(p_>,q_>)$ and (p_+,q_+) determined by (70) to (77) have the same support as (r,s).

By Lemma 4 we can find a number $\delta > 0$ such that for $(p,q) \in U_\delta(r,s)$ the pair of next period's strategies in the preliminary model has the support of (r,s). Let δ be a number of this kind. In view of the continuity statements of Lemma 5, we can find an $\varepsilon > 0$ with $\varepsilon \leq \delta$ such that for (p,q) in $U_\varepsilon(r,s)$ the pairs $(p,q_>)$ and $(p_>,q)$ are in $U_\delta(r,s)$. Let ε be a number of this kind. Since p_+ depends in the same way on $(p,q_>)$ as it depends on (p,q) in the preliminary model and since an analogous statement holds for q_+, it follows that for $(p,q) \in U_\varepsilon(r,s)$ the pair (p_+,q_+) has the same support as (r,s). In view of the fact that $(p_>,q_>)$ depends in the same way on (p,q) as (p_+,q_+) does in the preliminary model, it follows by $\varepsilon \leq \delta$ that ε meets the requirements of the statement to be proved.

Consider the case $h = 1$. Here, for $(p,q) \in U_\varepsilon(r,s)$ the pair (p_+,q_+) is nothing else than (r,s) since no other pair has the same support as (r,s). This proves the assertion of the theorem for the case $h = 1$.

From now on assume $h > 1$. It is clear that (r,s) cananot be locally stable unless $(\underline{r},\underline{s})$ is stable within L with respect to the linear system (110) described in 4.7. In view of Lemma 10 and the remark after Lemma 11, it follows that (r,s) cannot be locally stable unless all non-vanishing eigenvalues of C are negative. It remains to show that this condition is sufficient for local stability if α and β are sufficiently small.

Assume that non-vanishing eigenvalues of C are negative. It follows by Lemma 11 that for sufficiently small $\alpha > 0$ and $\beta > 0$ the pair $(\underline{r},\underline{s})$ is stable within L with respect to (110). Let α and β be sufficiently small in this sense. Since $(\underline{r},\underline{s})$ is stable within L, we can find a number $\underline{\delta} > 0$ with $\underline{\delta} \leq \varepsilon$ such that the following is true: Let $(p(0),q(0))$ be a pair in $U_{\underline{\delta}}(r,s)$ with the same support as (r,s); then every $(p(t),q(t))$ in the solution starting with $(p(0),q(0))$ is in $U_\varepsilon(r,s)$ and has the same support as (r,s). Here, ε is the number which has been chosen above. The choice of δ guarantees that all $(p(t),q(t))$ remain in $U_\varepsilon(r,s)$ as long as they have the same support as (r,s). The choice of δ guarantees that by induction all $(p(t),q(t))$ have indeed the same support as (r,s). Let $\underline{\delta}$ be a number which meets the requirement.

In view of the continuity statements of Lemma 5, it is now clear that for initial values $(p(0),q(0))$ in a sufficiently small neighborhood of (r,s) the solution starting there will remain in $U_\varepsilon(r,s)$ and converge to (r,s). Moreover, ε can be chosen arbitrarily small. This shows that (r,s) is locally stable. We have proved the theorem.

5. Special Stability Conditions

In games with numerically specified payoffs, the eigenvalue of the interaction covariance matrix of a regular equilibrium point can be computed numerically in order to decide the question of local stability under sufficiently slow anticipatory learning. However, the criterion of Theorem 4 cannot be easily applied to game models whose payoffs depend on unknown parameters. Therefore, we are interested in special stability conditions which permit conclusions based on the structure of numerically unspecified payoff functions.

5.1 Biadditive Equivalence

Let A and F be two matrices with the same number of rows and the same number of columns. We say that A and F are *biadditively equivalent* if we have

$$A = \gamma F + X \tag{184}$$

with $\gamma > 0$ where X is a biadditive matrix. It is clear that the relationship of "biadditive equivalence" is symmetric and transitive.

Let (A,B) and (F,G) be two games such that each of both players has the same number of pure strategies in both games. We say that (A,B) is *biadditively equivalent* to (F,G) if A is biadditively equivalent to F, and B is biadditively equivalent to G.

Let ($\underline{A},\underline{B}$) be the restricted game of a regular equilibrium point (r,s) of a game (A,B). Assume that ($\underline{A},\underline{B}$) is biadditively equivalent to (F,G):

$$\underline{A} = \gamma_1 F + X \tag{185}$$

$$\underline{B} = \gamma_2 G + Y \tag{186}$$

with $\gamma_1 > 0$ and $\gamma_2 > 0$ where X and Y are biadditive matrices. Since right multiplication by Z amounts to the subtraction of row averages and left multiplication to the subtraction of column averages, it follows that ZXZ and ZYZ vanish. Therefore, we have:

$$\hat{A} = \gamma_1 ZFZ \tag{187}$$

$$\hat{B} = \gamma_2 ZGZ \tag{188}$$

$$C = \gamma_1 \gamma_2 ZFG^T Z. \tag{189}$$

This shows that up to a positive factor the interaction covariance matrix of (F,G) agrees with that of (A,B). Consequently, the criterion of Theorem 4 can be applied to (F,G) instead of (A,B). We have obtained the following result:

Theorem 5: Let (r,s) be a genuinely mixed regular equilibrium point and let (<u>A</u>,<u>B</u>) be the restricted game of (r,s). Moreover, let (F,G) be biadditively equivalent to (<u>A</u>,<u>B</u>). Then, (r,s) is locally stable under sufficiently slow anticipatory learning if and only if all non-vanishing eigenvalues of the interaction covariance matrix

$$C = Z F G^T Z \tag{190}$$

of (F,g) are negative.

5.2 Special Cases of Local Stability

We shall prove a theorem which lists several conditions on the restricted game which yield local stability under sufficiently slow anticipatory learning. One of these conditions requires that the restricted game is biadditively equivalent to a zero-sum game. A *zero-sum game* is a game (A,B) with

$$A = -B. \tag{191}$$

We shall now introduce two other classes of games which also give rise to special stability conditions. Let (A,B) be a quadratic game or, in other words, a game where both players have the same number h of pure strategies. In a game of this kind, a player is called a *pursuer* if his payoff matrix is <u>D</u> and he is called an *evader* if his payoff matrix is -<u>D</u>. A *pursuit game* is a quadratic game with a pursuer and an *evasion game* is a quadratic game with an evader.

The names "pursuer" and "evader" are suggested by the interpretation of pure strategies as locations to be chosen by the players. A pursuer is exclusively motivated to catch the other player and an evader has no other goal than to avoid his opponent.

A pursuit game or an evasion game (A,B) is called *negative definite* if AB is symmetric and negative definite. In the case of a pursuit game (A,B) with A = <u>D</u>, this means that B is negative definite, whereas in an evasion game (A,B) with A = -<u>D</u> negative definiteness requires that B is positive definite.

A *2×2-game* is a game in which each of both players has two pure strategies. One of the special stability conditions requires that the restricted game is a 2×2-game without pure equilibrium points.

Theorem 6: Let (r,s) be a regular equilibrium point and let (<u>A</u>,<u>B</u>) be the restricted game of (r,s). The equilibrium point is locally stable under sufficiently slow anticipatory learning if one of the following five conditions (i) to (v) is satisfied:

(i)   (r,s) is an equilibrium point in pure strategies.

(ii)   (<u>A</u>,<u>B</u>) is a 2×2-game without equilibrium points in pure strategies.

142

(iii) $(\underline{A},\underline{B})$ is biadditively equivalent to a zero-sum game.

(iv) $(\underline{A},\underline{B})$ is biadditively equivalent to a negative definite pursuit game.

(v) $(\underline{A},\underline{B})$ is biadditively equivalent to a negative definite evasion game.

Proof: In view of Theorem 4, the assertion holds in the case of (i). Assume (ii). We explore the consequences of the absence of pure equilibrium points. No pure strategy can be dominating in $(\underline{A},\underline{B})$ since otherwise $(\underline{A},\underline{B})$ would have an equilibrium point in pure strategies. Therefore,

$$\underline{a}_{11} \geq \underline{a}_{21} \tag{192}$$

implies

$$\underline{a}_{22} > \underline{a}_{12} \tag{193}$$

and consequently

$$\underline{a}_{11} + \underline{a}_{22} - \underline{a}_{12} - \underline{a}_{21} > 0. \tag{194}$$

Moreover, (192) implies

$$\underline{b}_{11} < \underline{b}_{12} . \tag{195}$$

Otherwise, $(1,1)$ would be an equilibrium point. In view of the absence of dominance, (195) yields:

$$\underline{b}_{21} > \underline{b}_{22} . \tag{196}$$

It follows by (195) that we have:

$$\underline{b}_{11} + \underline{b}_{22} - \underline{b}_{12} - \underline{b}_{21} < 0 . \tag{197}$$

Eqs. (194) and (197) yield

$$(\underline{a}_{11} + \underline{a}_{22} - \underline{a}_{12} - \underline{a}_{21})(\underline{b}_{11} + \underline{b}_{22} - \underline{b}_{12} - \underline{b}_{21}) < 0. \tag{198}$$

In the case $\underline{a}_{11} < \underline{a}_{21}$, the same inequality can be derived in a similar way. As we shall see, (198) has the consequence that the non-vanishing eigenvalues of C is negative. (C has only one non-vanishing eigenvalue.) With the help of (128) in Section 4.9 we obtain

$$\hat{a}_{11} = \underline{a}_{11} - \frac{\underline{a}_{11} + \underline{a}_{12}}{2} - \frac{\underline{a}_{11} + \underline{a}_{21}}{2} + \frac{\underline{a}_{11} + \underline{a}_{12} + \underline{a}_{21} + \underline{a}_{22}}{4} \tag{199}$$

$$\hat{a}_{11} = \frac{\underline{a}_{11} + \underline{a}_{22} - \underline{a}_{12} - \underline{a}_{21}}{4} \tag{200}$$

$$\hat{a}_{22} = \hat{a}_{11} \tag{201}$$

$$\hat{a}_{12} = \hat{a}_{21} = -\hat{a}_{11} \; . \tag{202}$$

Analogous formulas hold for the elements of $\hat{B}$. It can be seen easily that C has the following structure

$$C = \begin{bmatrix} c & -c \\ -c & c \end{bmatrix} \tag{203}$$

with

$$c = 2\hat{a}_{11}\hat{b}_{11} \; . \tag{204}$$

It follows by (198) that c is negative. It can be seen immediately that C has two eigenvalues, namely 0 and $2c$. This shows that the non-vanishing eigenvalue of C is negative.

We now turn our attention to (iii). In view of Theorem 5, it is sufficient to look at the eigenvalues of C in the zero-sum case. Assume $\underline{A} = -\underline{B}$. we have:

$$\lambda v^T v = -v^T \underline{A} Z \underline{A}^T Z v. \tag{205}$$

Obviously, C is symmetric. Therefore, all eigenvalues of C are real. Let λ be a non-vanishing eigenvalue of C and let v be an associated eigenvector. We have:

$$w = \underline{A}^T Z v \; . \tag{206}$$

Left multiplication by v^T yields:

$$\lambda v v^T = -v^T \underline{A} Z \underline{A}^T Z v. \tag{207}$$

Define

$$w = \underline{A}^T Z v. \tag{208}$$

Since Z is symmetric and idempotent, we obtain

$$\lambda v^T v = -w^T w. \tag{209}$$

In view of the fact that the components of v are real, it follows by $v \neq 0$ that vv^T is positive. Moreover, w^Tw is non-negative. It follows by $\lambda \neq 0$ that λ is negative.

We now turn our attention to (iv). It is sufficient to consider the case $\underline{A} = \underline{D}$. The other case in which player 2 is the pursuer is analogous. since Z is idempotent, we have

$$C = Z\underline{B}^T Z. \tag{210}$$

Let λ be a non-vanishing eigenvalue of C and let v be an eigenvector associated with λ. Since C is a matrix product beginning with Z, the sum of all components of v is zero and we have:

$$v = Zv. \tag{211}$$

Since $\underline{B}$ is symmetric, C is symmetric, too. Therefore, all eigenvalues of C are real and v has real components. With the help of

$$\lambda v = Z\underline{B}v \tag{212}$$

we obtain:

$$\lambda v^T v = v^T \underline{B} v. \tag{213}$$

In view of $v \neq 0$ and the negative definiteness of $\underline{B}$, it is clear that $v^T v$ is positive and the right hand side of (213) is negative. It follows by $\lambda \neq 0$ that λ is negative.

5.3 Bailiff and Poacher

In the following, Theorem 6 will be applied to a simple example. Imagine a situation where a bailiff (player 1) is responsible for three fish ponds 1, 2 and 3. A poacher (player 2) wants to steal fish in one of the ponds. The bailiff can watch only one pond and the poacher can steal at only one pond. This means that each of both players has three pure strategies, namely 1, 2 and 3. the bailiff has no other interest than to catch the poacher. If both go to the same pond, the poacher is caught by the bailiff, the bailiff gets a payoff of +1 and the poacher gets a payoff of -1. If both choose different ponds, the bailiff gets a payoff of zero and the poacher gets a payoff of u_1, u_2, u_3, depending on the pond; the ponds may be different with respect to the value of a successful act of poaching. The game is shown in Figure 2.

poacher

	1	2	3
bailiff 1	1 $\quad$ -1	0 $\quad$ u_2	0 $\quad$ u_3
2	0 $\quad$ u_1	1 $\quad$ -1	0 $\quad$ u_3
3	0 $\quad$ u_1	0 $\quad$ u_2	1 $\quad$ -1

$u_i > 0$
for $i = 1, 2, 3$

Figure 2: Bailiff and poacher. In every field, player 1's payoff is indicated in the upper left corner and player 2's payoff is shown in the lower right corner.

The parameters u_1, u_2 and u_3 are assumed to be positive. Under appropriate additional conditions, the game has a regular completely mixed equilibrium point (p*,q*) whose probabilities for pure strategies are as follows:

$$p_i^* = 1 - \frac{\frac{2}{1+u_i}}{\frac{1}{1+u_1} + \frac{1}{1+u_2} + \frac{1}{1+u_3}} \tag{214}$$

$$q_i^* = \frac{1}{3} \tag{215}$$

for $i = 1, 2, 3$. It can be seen easily that (p*,q*) is a regular equilibrium point, if p_1^*, p_2^* and p_3^* are positive. Assume that this is the case. We are going to show that (p*,q*) is locally stable under sufficiently slow anticipatory learning.

If in each column i of Figure 2, the constant u_i is subtracted from player 2's payoff, one receives the biadditively equivalent game shown in Figure 3. Obviously, this game is a negative definite pursuit game. It follows by Theorem 6 that the completely mixed equilibrium point (p*,q*) of the game of Figure 2 is locally stable under sufficiently slow anticipatory learning.

Of course, (p*,q*) is not necessarily an equilibrium point of the game of Figure 3. It is not important for the application of Theorem 6 whether this is the case or not.

"Bailiff and poacher" is not a zero-sum game. However, the interests of both players are opposed in the sense that the bailiff wants to catch the poacher and the poacher wants to avoid being caught. Game situations with a similar flavor seem to offer a good chance for local stability under sufficiently small anticipatory learning.

	1	2	3
1	1 $\qquad$ $-1-u_1$	0 $\qquad$ 0	0 $\qquad$ 0
2	0 $\qquad$ 0	1 $\qquad$ $-1-u_2$	0 $\qquad$ 0
3	0 $\qquad$ 0	0 $\qquad$ 0	1 $\qquad$ $-1 -u_3$

Figure 3: A game which is biadditively equivalent to the game of Figure 2.

5.4 The 3×3-case

It is interesting to look at the special case of a restricted game with three pure strategies for each of both players. Consider a 3×3-interaction covariance matrix C. Since all row sums and column sums of C are zero, it is possible to characterize a 3×3-interaction matrix by four parameters, namely the three diagonal elements, in the following denoted by c_1, c_2 and c_3, and by a "skewness parameter" η:

$$C = \begin{bmatrix} c_1 & c_3 - \frac{s}{2} + \eta & c_2 - \frac{s}{2} - \eta \\ c_3 - \frac{s}{2} - \eta & c_2 & c_1 - \frac{s}{2} + \eta \\ c_2 - \frac{s}{2} + \eta & c_1 - \frac{s}{2} - \eta & c_3 \end{bmatrix} \qquad (216)$$

where

$$s = c_1 + c_2 + c_3 \qquad (217)$$

is the trace of C. It can be seen easily that every 3×3-interaction covariance matrix can be described in this way. Let C_{ii} be the adjunct of the diagonal element of c_i. We have:

$$C_{11} = c_2 c_3 - \left[c_1 - \frac{s}{2} \right]^2 + \eta^2 \tag{218}$$

$$C_{22} = c_1 c_3 - \left[c_2 - \frac{s}{2} \right]^2 + \eta^2 \tag{219}$$

$$C_{33} = c_1 c_2 - \left[c_3 - \frac{s}{2} \right]^2 + \eta^2 . \tag{220}$$

The non-vanishing eigenvalues λ of C are roots of the following quadratic equation:

$$\lambda^2 - s\lambda + (C_{11} + C_{22} + C_{33}) = 0. \tag{221}$$

Equations (218), (219), and (220) yield:

$$C_{11} + C_{22} + C_{33} = c_1 c_2 + c_1 c_3 + c_2 c_3 - c_1^2 - c_2^2 - c_3^2 + s^2 - \frac{3}{4} s^2 + 3\eta^2 \tag{222}$$

$$C_{11} + C_{22} + C_{33} = -\frac{1}{2} \left[(c_1 - c_2)^2 + (c_1 - c_3)^2 + (c_2 - c_3)^2 \right] + \frac{1}{4} s^2 + 3\eta^2. \tag{223}$$

Both roots of (221) are negative if and only if the following conditions are satisfied:

$$s < 0 \tag{224}$$

$$\frac{s^2}{4} \geq C_{11} + C_{22} + C_{33} > 0. \tag{225}$$

With the help of (223) and (225- it can be seen that the non-vanishing eigenvalues of C are negative if and only if the trace of C is negative and the following condition is satisfied:

$$\eta^2 \leq \frac{1}{6} \left[(c_1 - c_2)^2 + (c_1 - c_3)^2 + (c_2 - c_3)^2 \right] < \eta^2 + \frac{1}{12} s^2. \tag{226}$$

If C is symmetric and all diagonal elements of C are equal, (226) is satisfied. Sufficiently small deviations from symmetry can be compensated by differences between the diagonal elements which are sufficiently big, but not too big. Suppose, for example, that the differences among the diagonal elements are not too big in the sense that

$$(c_i - c_j)^2 < \frac{s^2}{6} \quad \text{for} \quad i,j = 1,2,3 \tag{227}$$

holds. If this is the case, the middle term in (226) is smaller than the right-hand term. Thus, (226) will be satisfied if at least two diagonal elements differ from one another and if the deviations from symmetry are small enough.

Obviously, the non-negativity of both non-vanishing eigenvalues of C is a restrictive condition; however, the part of the parameter space where the condition is satisfied is not a set of lower dimension.

6. Anticipated Anticipations

The anticipatory learning process assumes that anticipations in the opponent population are not anticipated. This is quite natural in a model of limited rationality. Nevertheless, it is of interest to anwer the question whether a learning process with anticipated anticipations would have different stability properties. We shall restrict our attention to the local stability of completely mixed regular equilibrium points. Since the problem is not of central importance for the theory presented here, no attempt will be made to explore the more general case, which probably will not yield additional insights.

In the anticipatory learning process the change of strategies from one period to the next may be decomposed into a *first order effect* due to the observed strategy on the other side and a *second order effect* due to the anticipated change of the opponent strategy. To these effects one may add a *third order effect* due to the anticipated effect of the opponents anticipation. Of course, this is not yet the end of the story; for any k we may add a *k-th order effect* due to the opponents anticipation of the (k-1)-th order effect. Finally, one may wish to consider the limiting case of full anticipation, where effects of any order k are taken into account.

The preliminary model relies on first order effects. The first order effects do not stabilize genuinely mixed equilibrium points, but in favorable cases they may also fail to amplify deviations from equilibrium. Under these conditions the second order effects have a chance to exert a stabilizing influence in the anticipatory learning process. The higher order effects turn out to be much weaker for sufficiently slow learning and therefore cannot reverse the influence of the second order effect. Of course, these remarks do not intend to provide more than an intuitive interpretation of the results to be obtained below.

6.1 The Learning Process of Order k

Let (A,B) be a quadratic game with h pure strategies 1,...,h for each of both players and let (r,s) be a regular completely mixed equilibrium point of (A,B).

Define

$$\tilde{R} = \begin{pmatrix} 0 & \alpha \tilde{A} \\ \beta \tilde{B}^T & 0 \end{pmatrix} . \tag{228}$$

The *learning process of order* k is defined as follows:

$$\begin{bmatrix} p_+ \\ q_+ \end{bmatrix} = (\underline{D} + \tilde{R} + \tilde{R}^2 + \ldots \tilde{R}^k) \begin{bmatrix} p \\ q \end{bmatrix} . \tag{229}$$

Here $\underline{D}$ is the h×h-unit matrix:

$$\underline{D} = \begin{bmatrix} 1 & & & 0 \\ & \cdot & & \\ & & \cdot & \\ 0 & & & 1 \end{bmatrix} . \tag{230}$$

In view of (56) and (110) the learning process of order 1 is the linear system connected to (r,s) in the preliminary model and the learning process of order 2 is the linear system connected to (r,s) in the anticipatory learning process. Define:

$$\begin{bmatrix} p_>^k \\ q_>^k \end{bmatrix} = (\underline{D} + \tilde{R} + \ldots + \tilde{R}^{k-1}) \begin{bmatrix} p \\ q \end{bmatrix} . \tag{231}$$

We call $p_>^k$ and $q_>^k$ the *anticipated strategies of order* k. In the case k = 1 the matrix $\underline{D} + \ldots + \tilde{R}^{k-1}$ is interpreted as $\underline{D}$. With the help of (231) the learning process of order k can be described as follows:

$$\begin{bmatrix} p_+ \\ q_+ \end{bmatrix} = \begin{bmatrix} p \\ q \end{bmatrix} + \tilde{R} \begin{bmatrix} p_>^k \\ q_>^k \end{bmatrix} . \tag{232}$$

Players adapt to anticipated strategies which would be next period's opponent strategies if the learning process were of order k-1 instead of order k.

6.2 Full Anticipations

Let β be the maximum of the absolute values of elements of $\tilde{R}$. For sufficiently small $\alpha\beta$ we have

$$\mu < \frac{1}{h} . \tag{233}$$

Assume that $\alpha\beta$ is sufficiently small in this sense. It can be seen immediately that the maximum of the absolute values of elements of $\tilde{R}^k$ is at most μ^k. Consequently the infinite sequence

$$Y = \underline{D} + \tilde{R} + \tilde{R}^2 + \ldots \tag{234}$$

converges. Moreover,

$$\underline{D} + \tilde{R}Y = Y \tag{235}$$

yields

$$(\underline{D} - \tilde{R})Y = D \tag{236}$$

$$Y = (\underline{D} - \tilde{R})^{-1}. \tag{237}$$

The *learning process with full anticipations* is defined as follows:

$$\begin{bmatrix} p_+ \\ q_+ \end{bmatrix} = (\underline{D} - \tilde{R})^{-1} \begin{bmatrix} p \\ q \end{bmatrix}. \tag{238}$$

Full anticipations are the limit of anticipations of order k for $k \longrightarrow \infty$. Within a theory of bounded rationality, it is not reasonable to assume that full anticipations are formed. Nevertheless, it is interesting to look at this limiting case.

6.3 Local Stability Condition

In the same way as in 3.8 and 4.8, the learning process of order k can be replaced by a *reduced system* (see (67) and (114)). One receives:

$$\begin{bmatrix} \bar{p}_+ \\ \bar{q}_+ \end{bmatrix} = (\underline{D} + R + \ldots + R^k) \begin{bmatrix} c + \begin{bmatrix} \bar{p} \\ \bar{q} \end{bmatrix} \end{bmatrix}. \tag{239}$$

The notational conventions are the same as in (3.8). The *reduced system* of the learning process with full anticipations assumes the following form:

$$\begin{bmatrix} \bar{p}_+ \\ \bar{q}_+ \end{bmatrix} = (\underline{D} - R)^{-1} \begin{bmatrix} c + \begin{bmatrix} \bar{p} \\ \bar{q} \end{bmatrix} \end{bmatrix}. \tag{240}$$

It can be seen immediately that the learning process of order k is locally stable if and only if the following condition holds for every eigenvalue λ of R:

$$|1 + \lambda + \ldots + \lambda^k| < 1. \tag{241}$$

The eigenvalues of (240) have the form $1/(1-\lambda)$ where λ is an eigenvalue of R. Therefore, the learning process with full anticipation is locally stable if and only if we have:

$$1 < |1 - \lambda|. \tag{242}$$

The learning process of order k and the learning process with full anticipations are defined as linear systems. There is no difference between local and global stability in linear systems. However, the definition of these processes does not make sense unless one restricts it to a sufficiently small neighborhood of (r,s). Therefore, it seems to be better to speak of local stability in order to emphasize the local character of the analysis.

<u>Theorem 7:</u> Let (A,B) be a quadratic game with h pure strategies 1,...,h for each of both players and let (r,s) be a completely mixed regular equilibrium point of (A,B). Then numbers $\alpha_0 > 0$ and $\beta_0 > 0$ can be found such that for $0 < \alpha < \alpha_0$ and $0 < \beta < \beta_0$ the following is true for the learning process of order k with k at least equal to 2 and for the learning process with full anticipations: (r,s) is (locally) stable if and only if all non-vanishing eigenvalues of the interaction covariance matrix C of (r,s) are negative.

<u>Proof</u>: The theorem is a generalization of Lemma 11, which covers the case k = 2. Let λ be an eigenvalue of R. We use the notational conventions of the proof Lemma 11. In particular, we express λ as a function of α and β as in (178).

In view of

$$1 + \lambda + \ldots + \lambda^k = \frac{1-\lambda^{k-1}}{1-\lambda} \tag{243}$$

the stability condition (241) can be expressed as follows:

$$|1 - \lambda^{k+1}| < |1 - \lambda|. \tag{244}$$

Let η be the absolute value of λ:

$$\eta = |\lambda| = \rho\sqrt{\alpha\beta} . \tag{245}$$

In view of (178) we have

$$|1 - \lambda| = \sqrt{(1-\eta\cos\phi)^2+\eta^2\sin^2\phi} \tag{246}$$

$$|1 - \lambda| = \sqrt{1-2\eta\cos\phi+\eta^2} \quad . \tag{247}$$

Similarly, we obtain

$$|1 - \lambda^{k+1}| = \sqrt{1-2\eta^{k+1}\cos(k+1)\phi+\eta^{2(k+1)}} \quad . \tag{248}$$

It follows that (244) holds if and only if we have:

$$-2\eta^{k+1}\cos(k+1)\phi + \eta^{2(k+1)} < -2\eta\cos\phi + \eta^2 \tag{249}$$

or equivalently

$$-2\eta^{k}\cos(k+1)\phi + \eta^{2k+1} < -2\cos\phi + \eta. \tag{250}$$

Suppose that at least one eigenvalue λ of R has a non-vanishing real part. In this case it follows by (i) in Lemma 10 that at least one eigenvalue of R has a positive real part. Assume that λ has a positive real part. Then for $\alpha\beta \to 0$ the right-hand side of (250) converges to the negative constant $-2\cos\phi$ whereas the left-hand side converges to zero. It follows that (250) does not hold for sufficiently small α and β. In view of (iii) and (iv) in Lemma 10, we can conclude that the negativity of all non-vanishing eigenvalues of C is a necessary condition for the local stability with respect to the learning process of order k.

In the full anticipation case local stability requires (242). With the help of (247), this condition can be expressed as follows:

$$0 < -2\eta\cos\phi + \eta^2 \tag{251}$$

or, equivalently

$$0 < -2\cos\phi + \eta. \tag{252}$$

This shows that the negativity of all non-vanishing eigenvalues of C is also a necessary condition for local stability with respect to the learning process with full anticipation.

Now assume that all non-vanishing eigenvalues of C are negative. Then all eigenvalues of R are imaginary and we have $\cos\phi = 0$. Condition (250) assumes the following form

$$\eta^{2k+1} < \eta. \tag{253}$$

Clearly this is the case for sufficiently small α and β. It is also clear that (252) is satisfied: It follows that for the learning process of order k with k at least equal to 2 and for the learning process with full anticipation the non-negativity of all non-vanishing eigenvalues of C is not only necessary, but also sufficient for the local stability of (r,s) if α and β are sufficiently small.

6.4 Comment

Modifications of the anticipatory learning process which consider higher order anticipations do not change the conclusions on the local stability of completely mixed regular equilibrium points. Even if we avoided the full specification of the modified processes for the whole space of mixed strategy pairs including its boundaries, the result suggests that the introduction of higher order anticipations does not yield a fundamentally different theory.

7. Concluding Remark

The theory presented here exhibits anticipation as a potentially stabilizing force in game equilibrium learning. The anticipatory learning process can serve as a point of departure in the analysis of laboratory experiments. Of course, adjustments have to be made for the finiteness of laboratory populations. Frequency distributions over pure strategies cannot be identified with mixed strategies. Moreover, one must expect that the parameters α and β vary from subject to subject. In principle, this does not pose insurmountable obstacles for the theoretical analysis of specific data sets, even if the direct application of the results of this paper may not be possible.

It would be naive to expect that the picture of individual behavior portrayed by the anticipatory learning process is an exact description of reality. Probably modifications will be necessary in the light of experimental evidence. Nevertheless, the author hopes that this paper will not be completely without significance for the development of empirically validated game learning theories.

154

References

Bomze, I.M. (1986). Non-Cooperative Two-Person Games in Biology: A Classification. International Journal of Game Theory **15**: 31-57.

Brown, G. (1951). Iterative Solution of Games by Fictitious Play. In: T. Koopmanns (Ed.): "Activity Analysis of Production and Allocation". pp. 374-376. New York: Wiley.

Bush, R.R., and F. Mosteller (1955). Stochastic Models for Learning. New York: Wiley.

Crawford, V.P, (1985). Learning Behavior and Mixed Strategy Nash Equilibria. Journal of Economic Behavior and Organization **6**: 69-78.

Cross, J.G. (1983). A Theory of Adaptive Economic Behavior. Cambridge: Cambridge University Press.

Darlington, R.B. (1975). Radicals and Squares. New York: Logan Hill Press.

Gabisch, G., and H.W. Lorenz (1987). Business Cycle Theory. Lecture Notes in Economics and Mathematical Systems **283**. Berlin: Springer-Verlag.

Harley, C.B. (1981). Learning the Evolutionarily Stable Strategy. Journal of Theoretical Biology **89**: 611-633.

Harsanyi, J.G. (1973). Games with Randomly Disturbed Payoffs. A New Rationale for Mixed-Strategy Equilibrium Points. Internat. Journal of Game Theory **2**: 235-250.

Maynard Smith, J. (1982). Evolution and the Theory of Games. Cambridge: Cambridge University Press.

Maynard Smith, J., and G.A. Price (1973). The Logic of Animal Conflict. Nature **246**: 15-18.

Milinski, M. (1979). An Evolutionarily Stable Feeding Strategy in Sticklebacks. Zeitschrift für Tierpsychologie **51**: 36-40.

Miyasawa, K. (1961). On the Convergence of the Learning Process in a 2×2 Non-Zero-Sum Two-Person Game. Economic Research Program, Research Memorandum No. 33. Princeton, N.J.: Princeton University.

Nash, J. (1951). Non-Cooperative Games. Annals of Mathematics **54**: 286-295.

O'Neill, B. (1987). Nonmetric Test of the Minimax Theory of Two-Person Zerosum Games. Proceedings of the National Academy of Sciences, U.S.A., Vol. **84**: 2106-2109.

Robinson, J. (1951). An Iterative Method of Solving a Game. Annals of Mathematics **54**: 296-301.

Rosenmüller, J. (1971). Über Periodizitätseigenschaften spieltheoretischer Lernprozesse. Zeitschr. Wahrsch. Verw. Gebiete **17**: 259-308.

Selten, R. (1980). A Note on Evolutionarily Stable Strategies in Asymmetric Animal Conflicts. Journal of Theoretical Biology **83**: 93-101.

Selten, R., and P. Hammerstein (1984). Gaps in Harley's Argument on Evolutionarily Stable Learning Rules and in the Logic of "Tit for Tat". The Behavioral and Brain Sciences 7: 115-116.

Selten, R., and R. Stoecker (1986). End Behavior in Sequences of Finite Prisoner's Dilemma Supergames. Journal of Economic Behavior and Organization 7: 47-70.

Shapley, L.S. (1964). Some Topics in Two-Person Games. In: M. Dresher, L. Shapley, and A. Tucker, "Advances in Game Theory." Annals of Mathem. Studies, No. **52**. pp. 1-28. Princeton, N.J.: Princeton University Press.

Taylor, P.D., and L.B. Jonker (1978). Evolutionarily Stable Strategies and Game Dynamics. Mathem. Biosci. **40**: 145-156.

THE ORIGIN OF ISOGAMOUS SEXUAL DIFFERENTIATION

by

Rolf F. Hoekstra, Yoh Iwasa and Franz J. Weissing

Abstract

It is a general biological rule that sexual reproduction invariably involves the fusion of gametes of two different types. Most evolutionary biologists presume that the differentiation of gametes into mating types started within a primordial population consisting entirely of undifferentiated gametes. However, none of the population genetical models for the evolution of mating types investigated thus far yields a satisfactory explanation for the stable establishment of differentiated gametes. It is the aim of this paper to extend these models by incorporating more realistic assumptions concerning the mating kinetics and the effects of inbreeding. A general class of models for the competition between mating types is developed and analyzed both in dynamical and game—theoretical terms. The results obtained in the general framework are used to investigate several specific models which take account of the inhomogeneous spatial structure of the gamete pool. Surprisingly, the refined models do not qualitatively affect the rather stringent conditions for the stable establishment of differentiated mating types. We have to conclude that a plausible scheme for the evolution of sexual differentiation is still missing.

1. Introduction

In its most basic form, sex involves the union of two specialized cells (syngamy), followed by the fusion of their nuclei (karyogamy) and — often with some delay — by meiosis. There are no exceptions known to the rule that sex is always between gametes of different type. Often the gametes differ in size (anisogamy), but also when the gametes do not differ macroscopically (isogamy) there are characteristic physiological and behavioural differences between the two gametes which form a zygote.

Functional explanations of the origin of sexual differentiation have been explored in theoretical models by Hoekstra (1982, 1987). These models consider the evolution of two mating types in an initially undifferentiated population in which every gamete can mate with any other gamete. The model organism is an aquatic unicellular isogamous 'alga' with a haplontic life cycle. The differences between the original undifferentiated gametes and the differentiated mating types are represented by different configurations of recognition and/or adhesion molecules in the cell membrane of the gametes, as shown in Figure 1.

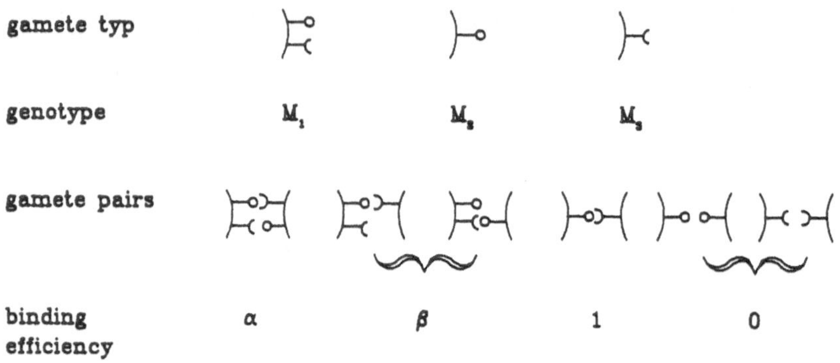

156

gamete typ

genotype M_1 M_2 M_3

gamete pairs

binding
efficiency α β 1 0

FIGURE 1: Different gamete types. M_1 : original undifferentiated type which is able to mate with every type; M_2, M_3 : differentiated mating types, unable to mate with their own type.

A general conclusion emerging from the analysis of these models is that mating types may invade the initially undifferentiated population under fairly broad conditions, but that the removal of the undifferentiated type requires very strong selective forces. Thus so far we do not have a fully convincing scenario for the evolution of mating types.

The basic model (Hoekstra, 1982) is very simplistic in its assumptions with respect to the *mating kinetics* (a term coined by Iwasa & Sasaki, 1987). The population is supposed to be well mixed, and gametes form pairs randomly. Then all pairs consisting of two gametes of the same differentiated mating type are effectively removed from the population, since they are unable to fuse and these gametes are not allowed to search for another mating partner.

Iwasa & Sasaki (1987) have shown that the results in this type of model may crucially depend on the assumptions concerning the mating kinetics. It is therefore important to investigate how strongly the outcomes of the basic model depend on its underlying assumptions. The first assumption (random fusion with respect to gamete type) is bound to be very unrealistic if asexual reproduction is much more frequent than sexual reproduction (like in present–day protists) and if the mobility of the cells is low. Under these conditions a clonal distribution of gametes is expected to result which, kinetically, would favour the undifferentiated gamete type. In fact, the differentiated mating types would have a lower chance of finding a mating partner since the gametes derived from a single clone are much more homogeneous with respect to mating type than the overall population. The second assumption (no second – and third, etc. – chance to find a suitable mating partner) seems unrealistic if the density of the gamete population is high and unmated gametes are not too short–lived. Clearly, relaxation of this assumption would make the conditions more favourable for the differentiated mating types.

Beside the mating kinetics, there are other reasons why the possibility of a clonal distribution of gametes should be incorporated into a mating type model. Probably, the evolution of mating types is closely related to the avoidance of *inbreeding depression*, and the effects of inbreeding are much more transparent in a non–uniformly distributed population. We are still far from a full understanding of the functional significance of sexual reproduction (see, e.g., Stearns, 1987; Michod & Levin, 1988; Eshel, 1990). If sexual reproduction is advantageous, however, its advantage is certainly based on the increase in genetic variability due to the combination of genetic material from different origins. As a consequence, the same factors that favour sexual reproduction itself should also disfavour the mating of gametes with an identical genetic make–up. Undifferentiated gametes of a clonal origin have good chances to find a mating partner in their close vicinity, but they are also in danger of close inbreeding with a gamete from the same origin. Differentiated gametes, on the other hand, are disfavoured kinetically, but the chance is much lower that they mate with a member of their own mother clone.

2. A General Model for the Evolution of Mating Types

Our basic model assumptions are similar in spirit to those in the original mating type model (Hoekstra, 1982). As in the original model we imagine aquatic unicellular haploid organisms, reproducing mainly asexually by mitosis. Under suitable conditions all the cells produce (again mitotically) gametes. There are three different types of gametes as shown in Figure 1. After the process of zygote formation, all zygotes divide meiotically to produce the next generation of vegetative individuals. All genotypes have the same rate of asexual reproduction.

In the original model, selection on mating types was based on gamete recognition differences. In a broader context it is useful to focus more generally on the *zygote formation capacities* of the different mating types. The zygote formation capacity of a mating type (and – correspondingly – its evolutionary success) depends on the mating type distribution in the gamete pool. Therefore, selection on mating types is frequency dependent, and it may be characterized by a *fitness function* which describes how the fitness of a mating type changes with a change in the composition of the gamete pool.

In the mating type models developed in Section 3, the fitness function will be derived from the specific assumptions concerning the clonal distribution and the mating kinetics. The basic characteristics of all these models fit into a general framework which will be established first (Section 2.1). A small set of fairly general assumptions contains enough information in order to classify the evolutionarily stable strategies of the associated *mating type game* (Section 2.2). In contrast to other models (see, e.g., Weissing 1990), the evolutionarily stable strategies of this game correspond perfectly to the global attractors of the discrete and the continuous replicator dynamics (Section 2.3).

2.1 Basic Model Assumptions

In all our models we consider the competition between three different gamete types: an undifferentiated type (called M_1) which is able to mate with every type and two differentiated types (called M_2 and M_3) which are not able to mate with their own type (see Figure 1). Let x_i, $i = 1,2,3$, denote the relative frequency of type M_i. The vector $x := (x_1, x_2, x_3)$ of gamete type frequencies belongs to the simplex

$$\Delta := \{ x \in \mathbb{R}^3 \mid x_i \geq 0, \; \textstyle\sum_i x_i = 1 \}. \tag{1}$$

The gametes in the gamete pool move around and try to form zygotes. A zygote formed by the fusion of two gametes of types M_i and M_j will be indicated by the symbol $M_i M_j$. We make the plausible assumption that there are no position effects, i.e., that two zygotes of types $M_i M_j$ and $M_j M_i$ are indistinguishable from another. Accordingly, up to six different zygote types can be distinguished, three homogametic types ($M_1 M_1$, $M_2 M_2$, $M_3 M_3$) and three heterogametic types ($M_1 M_2$, $M_1 M_3$, $M_2 M_3$).

The general framework for our models is characterized by four basic assumptions which are all very natural. Perhaps with the exception of assumption 4, all these assumptions are almost true by definition or the interpretation of the situation.

ASSUMPTION 1: Zygote formation capacities.

Let $\varphi_{ij}(x)$ denote the probability that a gamete of type M_i finds a gamete of type M_j and forms a successful zygote with it. The notation indicates that we assume that this probability only depends on the gamete types M_i and M_j and on the gamete type frequency distribution x. $\varphi_{ij}(x)$ incorporates the probability to meet and recognize a gamete of type M_j , the adhesion efficiency of the two gametes, and the relative fitness of the resulting zygote $M_i M_j$.
For mathematical convenience, all the functions φ_{ij} are assumed to be continuously differentiable on the simplex Δ.

ASSUMPTION 2: Consistency.

If N denotes the number of gametes in the gamete pool, $N \cdot x_i$ corresponds to the number of gametes of type M_i . According to the interpretation of $\varphi_{ij}(x)$, the number of successfully formed zygotes of type $M_i M_j$ is given by $N \cdot x_i \cdot \varphi_{ij}(x)$. Similarly, the number of successfully formed zygotes of type $M_j M_i$ is given by $N \cdot x_j \cdot \varphi_{ji}(x)$.
If there are no position effects, the two zygote types $M_i M_j$ and $M_j M_i$ are indistinguishable from another. As a consequence, their number has to be equal. This leads to the consistency requirement:

$$x_i \cdot \varphi_{ij}(x) = x_j \cdot \varphi_{ji}(x) \quad \text{for all i,j and all } x \in \Delta. \tag{2}$$

ASSUMPTION 3: Self–incompatibility.

Gametes ot the two differentiated types M_2 and M_3 are assumed to be self–incompatible in the sense that they cannot successfully form zygotes with gametes of their own type. Therefore, we have

$$\varphi_{22}(x) = \varphi_{33}(x) = 0 \quad \text{for all } x \in \Delta. \tag{3}$$

On the other hand, we assume that gametes of different types are always compatible with another, which may be expressed as

$$\varphi_{ij}(x) > 0 \quad \text{for } i \neq j \text{ and } x_j > 0. \tag{4}$$

ASSUMPTION 4: Symmetry.

Gametes of the two differentiated gamete types M_2 and M_3 are treated symmetrically by assuming that they do not differ in their zygote formation capacities in interactions with gametes of type M_1 :

$$\varphi_{21}(x) = \varphi_{31}(x) \quad \text{for all } x \in \Delta. \tag{5}$$

All the mating type models considered in this paper are fully specified by the frequency dependent *fitness matrix*

$$\Phi(x) := (\,\varphi_{ij}(x)\,)_{i,j} , \; x \in \Delta , \tag{6}$$

of zygote formation capacities. Under the model assumptions described above, the (relative) fitness $F_i(x)$ of a gamete of type M_i is given by the sum of the three zygote formation capacities $\varphi_{ij}(x)$, $j = 1,2,3$:

$$F_i(x) := \sum_j \varphi_{ij}(x) , \quad x \in \Delta . \tag{7}$$

Correspondingly, the fitness of mating type M_i may be obtained by summing up the elements in the i'th row of the fitness matrix $\Phi(x)$.

The fitness matrix of a mating type model is not necessarily symmetric. The consistency requirement (2), however, implies some restricted form of symmetry:

$$\varphi_{ij}(x) = \varphi_{ji}(x) \quad \text{if } x_i = x_j . \tag{8}$$

The compatibility requirements (3) and (4) show that two of the diagonal elements of $\Phi(x)$ are equal to zero whereas – for $x > 0$ – all non–diagonal elements of $\Phi(x)$ are positive. Condition (5), finally, implies that the two sub–diagonal elements in the first column of $\Phi(x)$ are identical.

Notice that the compatibility requirement (4) implies that each mating type has a positive fitness if all mating types are present in the population:

$$F_i(\mathbf{x}) > 0 \quad \text{if} \quad \mathbf{x} > 0. \tag{9}$$

Some other consequences of the consistency requirement (2) will be of some importance for the analysis of a mating type model:

LEMMA 1: Consequences of consistency.

1. The zygote formation capacity $\varphi_{ij}(\mathbf{x})$ is zero if no gametes of type M_j are present:

$$\varphi_{ij}(\mathbf{x}) = 0 \quad \text{if} \quad x_j = 0 \quad \text{and} \quad j \neq i. \tag{10}$$

2. If mating types M_2 and M_3 are present in equal frequencies, they contribute equal amounts to the fitness of mating type M_1:

$$\varphi_{12}(\mathbf{x}) = \varphi_{13}(\mathbf{x}) \quad \text{if} \quad x_2 = x_3. \tag{11}$$

PROOF:

1. The consistency requirement (2) implies $\varphi_{ij}(\mathbf{x}) = (x_i/x_j) \cdot \varphi_{ji}(\mathbf{x}) = 0$ for $x_j = 0$ and $x_i > = 0$. By continuity of φ_{ij} we also get $\varphi_{ij}(\mathbf{x}) = 0$ for $x_i = x_j = 0$.

2. From (2) and (5) we get $\varphi_{12}(\mathbf{x}) = (x_2/x_1) \cdot \varphi_{21}(\mathbf{x}) = (x_3/x_1) \cdot \varphi_{13}(\mathbf{x}) = \varphi_{13}(\mathbf{x})$ for $x_1 > 0$ and $x_2 = x_3$. By continuity, this extends to the case $x_1 = 0$ and $x_2 = x_3$.

# # #

The consistency requirement (2) is also very useful in order to characterize the fitness difference between the two differentiated mating types:

LEMMA 2: Fitness differences between mating types.
The fitnesses of the two differentiated mating types M_2 and M_3 are related by the following two equations:

$$x_2 \cdot F_2(\mathbf{x}) - x_3 \cdot F_3(\mathbf{x}) = (x_2 - x_3) \cdot \varphi_{21}(\mathbf{x}), \tag{12}$$

$$F_2(\mathbf{x}) - F_3(\mathbf{x}) = \left(\frac{x_3 - x_2}{x_3}\right) \cdot \varphi_{23}(\mathbf{x}) \quad (\text{if } x_3 > 0). \tag{13}$$

PROOF:

(7) and (3) show that $F_2(x)$ and $F_3(x)$ are given by

$$F_2(x) = \varphi_{21}(x) + \varphi_{23}(x), \quad F_3(x) = \varphi_{31}(x) + \varphi_{32}(x). \tag{14}$$

The consistency requirement (2) implies $x_2 \cdot F_2(x) - x_3 \cdot F_3(x) = x_2 \cdot \varphi_{21}(x) - x_3 \cdot \varphi_{31}(x)$. In view of the symmetry assumption (5), this immediately yields (12).

The symmetry assumption also implies $F_2(x) - F_3(x) = \varphi_{23}(x) - \varphi_{32}(x)$. This is equivalent to (13) since $x_3 > 0$ implies $\varphi_{32}(x) = (x_2/x_3) \cdot \varphi_{23}(x)$.

# # #

The following consequence of Lemma 2 and the compatibility requirement (4) will be of particular importance:

THEOREM 3: Fitness interdependence of differenciated mating types.

The fitness of mating type M_2 is larger than that of mating type M_3 if and only if the complementary mating type M_3 has a higher frequency in the gamete pool than M_2. Similarly, M_3 has a higher fitness than M_2 if and only if M_2 is more frequent than M_3:

$$\text{sgn}(\ F_2(x) - F_3(x)\ ) = \text{sgn}(\ x_3 - x_2\ ). \tag{15}$$

In particular, the fitness of the two differentiated mating types M_2 and M_3 is equal if and only if they occur in equal frequencies in the gamete pool:

$$F_2(x) = F_3(x) \quad \text{if and only if} \quad x_2 = x_3. \tag{16}$$

PROOF:

In view of (4), $\varphi_{23}(x) > 0$ if $x_3 > 0$. Therefore, (15) is a direct consequence of (13) if the frequency of M_3 is positive. Since the two differentiated mating types are treated symmetrical in our models, the same holds true if $x_2 > 0$.

In the remaining case $x_2 = x_3 = 0$, the frequency vector x corresponds to the corner

$$e^1 := (1,0,0) \tag{17}$$

of the simplex. (10) yields that $\varphi_{23}(e^1) = \varphi_{32}(e^1) = 0$. In view of (14) and (5) we get:

$$F_2(e^1) = F_3(e^1) = \varphi_{21}(e^1), \tag{18}$$

which shows that (15) is also true for $x_2 = x_3 = 0$.

# # #

2.2 Game—theoretical Analysis of a Mating Type Game

It has been shown above that the fitness of a mating type is frequency dependent: the mating success of a gamete not only depends on its own mating type but also on the mating type distribution in the gamete pool. John Maynard Smith (1982) has demonstrated that quite often much insight can be gained if a situation with frequency dependent selection is described as an *evolutionary game*. In our 'mating type game', the payoff to the three 'pure strategies' — the three mating types — is given by their frequency dependent fitness $F_i(x)$. In the game context, the frequency distribution of mating types, x, is interpreted as a mixed strategy, the 'population strategy'.

As a consequence, $F_i(x)$ might be interpreted as the payoff to pure strategy i in a contest with an opponent playing the mixed strategy x. In our case, however, the evolutionary game is not induced by pairwise contests but by a situation which Maynard Smith (1982) calls 'playing the field'. In fact, the special models developed in Section 3 will show that F_i is usually not linear in x as it usually is for pairwise contest. This implies that the usual definition of an *evolutionarily stable strategy* (ESS) does not apply for our mating type games.

There are several ways to generalize the concept of an ESS to non—linear frequency dependent selection (see, e.g., Weissing, 1983; Pohley & Thomas, 1983; Vickers & Cannings, 1987; Hines, 1987; Lessard, 1990). Weissing (1983; see also Hofbauer & Sigmund, 1988; Weissing, 1990) argues that the 'spirit' of the concept is best preserved by the following line of reasoning: Consider a population the population strategy of which is given by x^*. Any disturbance of this population strategy which leads to a new distribution of pure strategies, x, may be written in the form

$$x = (1-\epsilon) \cdot x^* + \epsilon \cdot y , \tag{19}$$

where $0 \leq \epsilon \leq 1$ is a real number and y is another mixed strategy. The representation (19) makes it plausible to interpret a disturbance from x^* to x as the 'invasion' of a 'mutant population' with population strategy y into the original x^*—population. ϵ corresponds to the overall frequency of the mutants. Notice that the mean fitness of the members of a subpopulation with population strategy y in a larger population with a population strategy x is given by:

$$F(y,x) := \sum_i y_i \cdot F_i(x) . \tag{20}$$

Correspondingly, the members of the original x^*—population will have a selective advantage with respect to the mutants if

$$F(x^*,x) > F(y,x) , \tag{21}$$

i.e., if the members of the x^*—population have a larger mean fitness than the members of the mutant population. Therefore, a x^*—population will be called 'immune against invasion by a y—population' if

$$F(x^*,(1-\epsilon)x^*+\epsilon y) > F(y,(1-\epsilon)x^*+\epsilon y) \ , \qquad (22)$$

provided that the mutant frequency, ϵ, is low. The x^*—population will be called 'evolutionarily stable' if it is immune against invasion by all possible mutant populations which occur in frequencies that are smaller than a certain (uniform) 'invasion barrier' ϵ_0 . The population strategy of an evolutionarily stable population will be called an 'evolutionarily stable strategy':

DEFINITION 1: Evolutionarily stable strategies.

A mixed strategy x^* is called an 'evolutionarily stable strategy' or simply an ESS if there exists an $\epsilon_0 > 0$ such that (22) holds true for all alternative mixed strategies $y \neq x^*$ and all mutant frequencies which are smaller than ϵ_0 (i.e., $0 < \epsilon < \epsilon_0$).

An evolutionary game is completely characterized by its *payoff function* (or: *fitness function*) $F : \Delta \ x \ \Delta \longrightarrow \mathbb{R}$. Notice that the fitness function of a mating type game is linear in its first component (see (20)) and non—linear in its second component. The linearity of $F(y,x)$ in y leads to the following characterization of evolutionarily stable strategies (see, e.g., Hofbauer & Sigmund, 1988):

PROPOSITION 4: Characterization of evolutionary stability.

If the fitness function of an evolutionary game is linear in its first component, a mixed strategy x^* is an ESS if and only if there is a neighbourhood U of x^* such that

$$F(x^*,x) > F(x,x) \quad \text{for all } x \in U, \ x \neq x^* \ . \qquad (23)$$

The continuity of a fitness function in its second component implies that every evolutionarily stable strategy is a 'Nash strategy' (Weissing, 1983, 1990):

PROPOSITION 5: ESS's are Nash strategies.

Every ESS x^* is a 'best reply to itself', i.e., every ESS is a Nash strategy:

$$F(x^*,x^*) \geq F(x,x^*) \quad \text{for all } x \in \Delta \ . \qquad (24)$$

The proposition follows immediately by letting the ϵ in (22) converge to zero. The name 'Nash strategy' derives from the fact that the strategy pair (x^*,x^*) is a symmetric Nash equilibrium point of the symmetric normal form game which is induced by the fitness function F (see, e.g., Selten, 1983). It is a standard result of game theory that — for a payoff function which is linear in its first component — the Nash strategies can be characterized by the so—called 'fundamental lemma of game theory' (see, e.g., Harsanyi & Selten, 1988):

LEMMA 6: The 'fundamental lemma' of game theory.

If the fitness function $F : \Delta \times \Delta \longrightarrow R$ of an evolutionary game is linear in its first component, the mixed strategy x^* is a Nash strategy if and only if

$$F_i(x^*) = F(x^*,x^*) \quad \text{for all } i \text{ with } x_i^* > 0, \tag{25a}$$

$$F_i(x^*) \leq F(x^*,x^*) \quad \text{for all } i \text{ with } x_i^* = 0. \tag{25b}$$

Let us now apply these general results to a mating type game that is characterized by Assumptions 1 to 4 in Section 2.1. As a direct consequence of Lemma 3 and Lemma 6 we get:

PROPOSITION 7: A property of Nash strategies.

All Nash strategies of a mating type game have the property that the two differentiated mating types occur in equal frequencies. Accordingly, all Nash strategies of a mating type game belong to the line segment of the simplex which is given by

$$\Delta_{equ} := \{ x \in \Delta \mid x_2 = x_3 \}. \tag{26}$$

PROOF:

Let us assume the x^* is a Nash strategy of a mating type game.

If x_2^* and x_3^* are both equal to zero, we have $x^* = e^1$, which belongs to Δ_{equ}. If x_2^* and x_3^* are both positive, (25a) yields $F_2(x^*) = F_3(x^*)$ which — in view of (16) — also implies $x^* \in \Delta_{equ}$.

Let us therefore assume that one of the differentiated mating types occurs with positive frequency whereas the other is not present in the population. Without loss of generality, we may assume $x_2^* = 0$ and $x_3^* > 0$. The fundamental lemma implies $F_3(x^*) \geq F_2(x^*)$ whereas (15) shows that $F_2(x^*) > F_3(x^*)$. This contradiction proves that every Nash strategy belongs to Δ_{equ}.

\# \# \#

Proposition 7 implies that a mating type game admits three types of Nash strategies:
- the 'completely undifferentiated' pure strategy $e^1 = (1,0,0)$ where only M_1 is present,
- the 'completely differentiated' strategy

$$x^1 := (0, 1/2, 1/2) \tag{27}$$

where the undifferentiated type M_1 is absent, and/or
- a 'completely mixed' strategy x^0 where all three types are present in positive frequencies. We have seen that such a strategy belongs to Δ^0_{equ}, the interior of Δ_{equ}.

In view of $F_2(x) = F_3(x)$ for all $x \in \Delta_{equ}$, Lemma 6 implies

PROPOSITION 8:   Characterization of Nash strategies.

1. e^1 is a Nash strategy if and only if $F_1(e^1) \geq F_2(e^1)$ which is equivalent to

$$\varphi_{11}(e^1) \geq \varphi_{21}(e^1) . \tag{28}$$

2. x^1 is a Nash strategy if and only if $F_2(x^1) \geq F_1(x^1)$ or, equivalently,

$$\varphi_{23}(x^1) \geq \varphi_{11}(x^1) + 2 \cdot \varphi_{12}(x^1) . \tag{29}$$

3. $x^0 \in \Delta^0_{equ}$ is a Nash strategy if and only if $F_1(x^0) = F_2(x^0)$ which is equivalent to

$$\varphi_{23}(x^0) = \varphi_{11}(x^0) + [\, 2 - (x_1^0/x_2^0) \,] \cdot \varphi_{12}(x^0) . \tag{30}$$

PROOF:

(28), (29), and (30) are immediate consequences of (2), (3), (10), and (18) which imply:

$$F_1(x) = \varphi_{11}(x) + 2 \cdot \varphi_{12}(x) \quad \text{for} \quad x \in \Delta_{equ} , \tag{31}$$

$$F_2(x) = F_3(x) = \varphi_{21}(x) + \varphi_{23}(x) \quad \text{for} \quad x \in \Delta_{equ} , \tag{32}$$

$$F_1(e^1) = \varphi_{11}(e^1) , \quad F_2(e^1) = F_3(e^1) = \varphi_{21}(e^1) , \tag{33}$$

$$F_2(x^1) = F_3(x^1) = \varphi_{23}(x^1) . \tag{34}$$

Let us call a Nash strategy x^* 'regular' if all inequalities in (25b) are strict, i.e., if all pure strategies that are not used by x^* yield a lower payoff than those for which $x_i^* > 0$ (see, e.g., Taylor & Jonker, 1978; Selten, 1983). The following assumption guarantees that all Nash strategies of a mating type game are regular:

<u>ASSUMPTION 5</u>: Regularity.

The two functions F_1 and F_2 intersect at most once on the line segment Δ_{equ}. If they intersect on Δ_{equ}, they intersect transversally and they do not intersect in the border points e^l and x^l.

In many explicit models (see, e.g., Section 3), the functions F_1 and F_2 are linear on Δ_{equ} (even if they are not linear on the whole simplex Δ). If this is the case, it is clear that F_1 and F_2 intersect at most once on Δ_{equ} and Assumption 5 is satisfied for generic parameter constellations: In the linear case, Assumption 5 may be viewed as a non–degenericity requirement.

We are now able to characterize the evolutionarily stable strategies of a mating type game:

<u>THEOREM 9</u>: Characterization of evolutionarily stable strategies.

Consider a mating type game for which Assumptions 1 to 5 are satisfied.

1. e^l is an ESS if and only if it is a regular Nash strategy, i.e., if

$$\varphi_{11}(e^l) > \varphi_{21}(e^l) . \tag{35}$$

2. x^l is an ESS if and only if it is a regular Nash strategy, i.e., if

$$\varphi_{23}(x^l) > \varphi_{11}(x^l) + 2 \cdot \varphi_{12}(x^l) . \tag{36}$$

3. $x^0 \in \Delta^0_{equ}$ is an ESS if and only if x^0 is a Nash strategy and neither e^l nor x^l are Nash strategies.

<u>PROOF</u>:

Let x^* denote a Nash strategy of the mating type game. We have to show that there exists a neighbourhood U of x^* such that $F(x^*,x) > F(x,x)$ for all $x \in U$, $x \neq x^*$. Let us fix such an x and define

$$\Delta F := F(x^*,x) - F(x,x) . \tag{37}$$

Since M_2 and M_3 are treated symmetrically, there is no loss in generality if we assume that $x_3 > 0$. Let us represent the vector x in the form

$$(x_1, x_2, x_3) = (x_1^* - \eta_1, x_2^* - \eta_2, x_3^* - \eta_3) , \tag{38}$$

where η_1 , η_2 and η_3 are real numbers which sum up to zero. Since the fitness function of a mating type game is linear in its first component, ΔF is given by

$$\Delta F = \eta_1 \cdot F_1(x) + \eta_2 \cdot F_2(x) + \eta_3 \cdot F_3(x) . \tag{39}$$

We have to show that $\Delta F > 0$ if the η_i are small enough. In view of (13) and $\eta_1 = -(\eta_2 + \eta_3)$, (39) may be represented in the form

$$\Delta F = \eta_1 \cdot [F_1(x) - F_2(x)] + \eta_3 \cdot (\eta_3 - \eta_2) \cdot \varphi_{23}(x)/x_3 . \tag{40}$$

Notice that — if the η_i are small — the second term on the right—hand side of (40) is dominated by the first. If $x^* = e^1$ is a regular Nash strategy, we have $F_1(e^1) - F_2(e^1) > 0$ and correspondingly $F_1(x) - F_2(x) > 0$ for all x which are close to x^*. Since η_1 has to be positive for $x^* = e^1$, the first term on the right—hand side of (40) is positive which implies $\Delta F > 0$ for all x which are sufficiently close to x^*.

If $x^* = x^1$ is a regular Nash strategy, $F_1(x^1) - F_2(x^1) < 0$ and $F_1(x) - F_2(x) < 0$ for all x from a small neighbourhood of x^*. For $x^* = x^1$, η_1 has to be negative. Accordingly, the first term on the right—hand side of (40) is positive and we can conclude again that $\Delta F > 0$ for all x which are sufficiently close to x^*.

If x^* is a completely mixed Nash strategy, we have $F_1(x^*) - F_2(x^*) = 0$, and the first term on the right—hand side of (40) no longer dominates the second. If e^1 and x^1 are no Nash strategies, Assumption 5 implies that $F_1(x) - F_2(x) < 0$ if $x_1 > x_1^*$ and $F_1(x) - F_2(x) > 0$ if $x_1 < x_1^*$. This implies that the term $\eta_1 \cdot [F_1(x) - F_2(x)]$ is always positive if $\eta_1 \neq 0$. If, on the other hand, $\eta_1 = 0$ (which implies $\eta_3 = -\eta_2$) ΔF is given by

$$\Delta F = 2 \cdot \eta_3^2 \cdot \varphi_{23}(x)/x_3 > 0 . \tag{41}$$

Combining these two cases, a continuity argument yields $\Delta F > 0$ provided that x is close enough to x^*.

# # #

2.3 Dynamic Analysis of a Mating Type Model

We will now complement the game theoretical analysis of a mating type game by a dynamical analysis of the discrete and the continuous replicator dynamics which are associated with such a game. In contrast to other evolutionary games (see, e.g., Weissing, 1990), the results of the dynamical analysis are completely in line with the predictions of evolutionary game theory. As will be shown below, the evolutionarily stable strategies of a mating type game correspond perfectly to the (globally) stable attractors of the mating type dynamics.

The *replicator equations* represent the simplest dynamical description of a selection process between replicating entities (like the gametes in our mating type models). For discrete, nonoverlapping generations, selection may be characterized by the 'discrete replicator dynamics', a system of recurrence equations given by

$$x_i' = x_i \cdot F_i(\mathbf{x})/\overline{F}(\mathbf{x}) . \tag{42}$$

Here, the vector $\mathbf{x}$ describes the replicator frequency distribution in the parent generation, $\mathbf{x}'$ describes the corresponding distribution in the offspring generation, $F_i(\mathbf{x})$ denotes the (frequency dependent) fitness of replicator i, and

$$\overline{F}(\mathbf{x}) := F(\mathbf{x},\mathbf{x}) = \sum_i x_i \cdot F_i(\mathbf{x}) \tag{43}$$

corresponds to the 'population fitness', i.e., to the mean fitness of the parent population. For overlapping generations, the selection among replicators is more adequately described by the 'continuous replicator dynamics', a system of differential equations which is characterized by

$$\dot{x}_i = x_i \cdot [F_i(\mathbf{x}) - \overline{F}(\mathbf{x})] . \tag{44}$$

(Here and in the sequel the dot denotes a time derivative: $\dot{x}_i := dx_i/dt$). For the motivation of the replicator dynamics as a description of a selection process and a discussion of the underlying assumptions see Hofbauer & Sigmund (1988) and Weissing (1983, 1990).

In the rest of this section, we will assume that selection between different mating types is governed by the discrete or the continuous replicator equation. If the fitness functions F_i of the three mating types M_i satisfy Assumptions 1 to 4 in Section 2.1, the corresponding replicator dynamics will be called a (discrete or continuous) 'mating type dynamics'. We will show that the four basic assumptions contain sufficient information in order to get a complete picture of the global asymptotic behaviour of a mating type dynamics.

Let M_i and M_j denote two replicators which occur in frequencies x_i and x_j. It will be useful to see how the difference $x_i - x_j$ and the quotient x_i/x_j (for $x_j > 0$) changes in the course of time. In the discrete case, this change is characterized by

$$(x_i/x_j)' = (x_i/x_j) \cdot [F_i(\mathbf{x})/F_j(\mathbf{x})] , \tag{45}$$

and

$$(x_i - x_j)' = [x_i F_i(\mathbf{x}) - x_j F_j(\mathbf{x})]/\overline{F}(\mathbf{x}) . \tag{46}$$

The corresponding differential equations for the continuous case are of the form

$$(x_i/x_j)^{\cdot} = (x_i/x_j) \cdot [F_i(\mathbf{x}) - F_j(\mathbf{x})] , \tag{47}$$

and

$$(x_i - x_j)^{\cdot} = [x_i F_i(\mathbf{x}) - x_j F_j(\mathbf{x})] - (x_i - x_j) \cdot \overline{F}(\mathbf{x}) . \tag{48}$$

In view of Lemma 2, these equations take a specific form for the two differentiated mating types M_2 and M_3:

$$(x_2/x_3)' = (x_2/x_3)\cdot\left[1-(x_2-x_3)\cdot\frac{\varphi_{23}(x)}{x_3 F_3(x)}\right],\qquad(49)$$

$$(x_2-x_3)' = (x_2-x_3)\cdot[\varphi_{21}(x)/\overline{F}(x)],\qquad(50)$$

$$(x_2/x_3)^{\cdot} = [1-(x_2/x_3)]\cdot\varphi_{32}(x),\qquad(51)$$

$$(x_2-x_3)^{\cdot} = (x_2-x_3)\cdot[\varphi_{21}(x)-\overline{F}(x)].\qquad(52)$$

It follows immediately from (50) and (52) that the line segment Δ_{equ} is invariant under both mating type dynamics. We will see that the qualitative features of the dynamics on the simplex are completely specified by the dynamics on Δ_{equ} and the three border edges of the simplex which are given by

$$\Delta_{-i} := \{x\in\Delta \mid x_i = 0\}.\qquad(53)$$

Let us first consider the fixed points of a mating type dynamics. It is easy to see that the fixed points of the discrete and the continuous replicator dynamics coincide and that every Nash strategy corresponds to a fixed point of the replicator dynamics. The fundamental lemma implies that a frequency distribution x^* from the interior of the simplex is a fixed point if and only if it is a Nash strategy. As a corollary of Proposition 7 we get that all interior fixed points of a mating type dynamics belong to the line segment Δ_{equ}. The next proposition characterizes the interior and the border fixed points.

PROPOSITION 10: Fixed points of a mating type dynamics.

All fixed points of a mating type dynamics belong either to the line segment Δ_{equ} or to one of the two border edges Δ_{-2} and Δ_{-3}. The fixed points on Δ_{equ} are given by e^1, x^1, and the Nash strategies of the corresponding mating type game. The three 'trivial fixed points'

$$e^1 = (1,0,0),\quad e^2 = (0,1,0),\quad\text{and}\quad e^3 = (0,0,1)\qquad(54)$$

left aside, the fixed points on Δ_{-3} correspond to the intersection points of the functions F_1 and F_2 on this edge whereas the fixed points on Δ_{-2} correspond to the intersection points of the funtions F_1 and F_3 on Δ_{-2}.

PROOF:

Let x^* denote a fixed point with $x_2^* > 0$ and $x_3^* > 0$. This implies that the quotient x_2^*/x_3^* is a fixed point of (49) resp. (51). It is easy to see that this is only possible for $x_2^* = x_3^*$, i.e., $x^*\in\Delta_{equ}$. The other assertions of Proposition 10 are immediate consequences of (45) and (47).

For simplicity, we will tacitly assume in the sequel that F_1 and F_3 intersect at most once on Δ_{-2} and that F_1 and F_2 intersect at most once on Δ_{-3}. The intersection points — if they exist — will be denoted by x^2 and x^3, i.e.,

$$F_1(x^3) = F_2(x^3), \quad F_1(x^2) = F_3(x^2), \quad x^i \in \Delta_{-i}. \tag{55}$$

$x^1 = (0, 1/2, 1/2)$, x^2 and x^3 are the only nontrivial border fixed points of a mating type dynamics.

If x^2 and x^3 exist, they are closer to the corner e^1 than to the corners e^3 and e^2. In fact, their first component is larger than the other nonzero component. For x^3, this can be seen from the relation

$$F_1(x^3) = F_2(x^3) \iff \varphi_{11}(x^3) = [1 - (x_2^3/x_1^3)] \cdot \varphi_{21}(x^3). \tag{56}$$

The following theorem shows that the trivial fixed points e^2 and e^3 and the nontrivial fixed points on the border edges Δ_{-2} and Δ_{-3} (x^2 and x^3) are of limited importance since the line segment Δ_{equ} is a global attractor for the mating type dynamics.

THEOREM 11: Monotone convergence to Δ_{equ}.

Every trajectory of the mating type dynamics that starts in the interior of the simplex or in the interior of the border edge Δ_{-1} converges monotonically to the line segment Δ_{equ}. As a consequence, all stable fixed points of a mating type dynamics belong to Δ_{equ}, and the stability of a fixed point on Δ_{equ} can be determined by analysing its stability with respect to Δ_{equ}.

PROOF:

Let us concentrate on the discrete case. The result for the continuous case follows analogously. Consider a frequency distribution $x \in \Delta$ with $x_2 > 0$ and $x_3 > 0$. We will show that the difference $x_2 - x_3$ does not change sign under the mating type dynamics and that it converges monotonically to zero.

(50) shows that $(x_2 - x_3)'$ is a positive multiple of $(x_2 - x_3)$. Accordingly, $(x_2 - x_3)$ does not change sign. If this sign is positive (i.e., $x_2 > x_3$), (49) shows that the quotient x_2/x_3 — and, of course, also the difference $x_2 - x_3$ — is monotonically decreasing. Similarly, (49) implies that x_2/x_3 and $x_2 - x_3$ are increasing if $x_2 - x_3$ is negative. As a bounded, monotonic sequence, the sequence of the $(x_2 - x_3)$'s converges to a limit point. (49) shows that the only limit point of this sequence is zero.

# # #

Taken together, Proposition 8 and Theorems 9 and 11 imply that the long–term behaviour of a mating type model is governed by its behaviour on the line segment Δ_{equ}. As a consequence, the 'interesting' features of such a model can be derived even if the model is not completely specified:

COROLLARY 12: A reduction principle.

The Nash strategies, the evolutionarily stable strategies, and the stable fixed points of a mating type model can all be determined on the basis of an (incomplete) specification of this model on the line segement Δ_{equ}.

Let us now consider the question whether a border fixed point x^* is immune against invasion by a single mutant which was not yet present at x^*.

PROPOSITION 13: Invasion of border fixed points by pure strategies.

1. The two monomorphic states e^2 and e^3 consisting of one of the differentiated mating types alone can be invaded by any alternative mating type.

2. The monomorphic state e^1 consisting of the undifferentiated mating type alone can be invaded by each of the two differentiated mating types if

$$\varphi_{11}(e^1) < \varphi_{21}(e^1) ; \tag{57}$$

    e^1 is immune against invasion by a single alternative mating type if

$$\varphi_{11}(e^1) > \varphi_{21}(e^1) . \tag{58}$$

3. All nontrivial border fixed points on the border edges Δ_{-2} and Δ_{-3} can be invaded by the mating type that is not yet present.

4. The border fixed point x^1 can be invaded by the undifferentiated mating type M_1 if $F_1(x^1) > F_2(x^1)$, i.e., if x^1 is not a Nash strategy. x^1 is immune against invasion by M_1 if it is an ESS, i.e., if (36) holds true.

PROOF:

1. Theorem 11 shows that every trajectory starting in the interior of the border edge Δ_{-1} converges to x^1. This implies that an M_2–population can be invaded by mating type M_3 and that an M_3–population can be invaded by M_2.
That the undifferentiated mating type M_1 can also invade the monomorphic states e^2 and e^3 is a consequence of

$$F_2(e^2) = 0 < F_1(e^2) \quad \text{and} \quad F_3(e^3) = 0 < F_1(e^3) . \tag{59}$$

2. follows from (33) and (45) or (47) (applied to $i = 1$ and $j = 2$ or $j = 3$).

3. is an immediate consequence of Theorem 11.

4. In view of (32), it is easy to see that the mating type dynamics on the line segment Δ_{equ} is characterized by the equations

$$[\,x_1/(x_2+x_3)\,]' = [\,x_1/(x_2+x_3)\,]\cdot[\,F_1(\mathbf{x})/F_2(\mathbf{x})\,]\,,\tag{60}$$

$$[\,x_1/(x_2+x_3)\,]^{\cdot} = [\,x_1/(x_2+x_3)\,]\cdot[\,F_1(\mathbf{x})-F_2(\mathbf{x})\,]\tag{61}$$

($\mathbf{x}\in\Delta_{equ}$, $x_2 = x_3 > 0$). (60) and (61) show that invasion—proofness of the border fixed point $\mathbf{x}^l$ is governed by the relation between $F_1(\mathbf{x}^l)$ and $F_2(\mathbf{x}^l)$.

# # #

It is now easy to characterize the (hyperbolically) stable fixed points (see, e.g., Weissing, 1990) of a mating type dynamics. The results of the following theorem are illustrated in Figure 2.

THEOREM 14: Stable fixed points of the mating type dynamics.

Consider a mating type model for which Assumptions 1 to 5 are satisfied. Then, a fixed point $\mathbf{x}^*$ of the corresponding mating type dynamics is stable if and only if it is an ESS of the corresponding mating type game. More precisely:

1. A monomorphic M_1–population consisting of the undifferentiated mating type alone is stable against invasion by any combination of the differentiated mating types M_2 and M_3 if and only if it is stable against invasion by single mutants. e^l is stable against single mutants if and only if it is an ESS, i.e., if and only if $\varphi_{11}(e^l) > \varphi_{21}(e^l)$.

2. The population state $\mathbf{x}^l$ consisting of the two differentiated mating types alone is stable if and only if it is stable against invasion by the undifferentiated mating type M_1 . As a consequence, the undifferentiated mating type can go extinct if and only if $\mathbf{x}^l$ is an ESS, i.e., if and only if $\varphi_{23}(\mathbf{x}^l) > \varphi_{11}(\mathbf{x}^l) + 2\cdot\varphi_{12}(\mathbf{x}^l)$.

3. If e^l is stable and $\mathbf{x}^l$ is unstable, e^l is a global attractor, i.e., the two differentiated mating types are always driven to extinction (see Figure 2(a)).

   If $\mathbf{x}^l$ is stable and e^l is unstable, $\mathbf{x}^l$ is a global attractor of the mating type dynamics, i.e., the undifferentiated mating type is driven to extinction as soon as all three mating types are present in the population (see Figure 2(d)).

   If e^l and $\mathbf{x}^l$ are both stable, they are separated by an unstable interior fixed point $\mathbf{x}^0$ which corresponds to a Nash strategy which is not evolutionarily stable. In this case, the undifferentiated mating type may be driven to extinction, but only if the two differentiated mating tpyes reach a rather large threshold frequency (see Figure 2(b)).

   If e^l and $\mathbf{x}^l$ are both unstable, they are separated by an interior fixed point $\mathbf{x}^0$ which corresponds to an evolutionarily stable strategy. $\mathbf{x}^0$ is always stable for the continuous mating type dynamics, and it is stable for the discrete mating type dynamics if selection is weak enough. Accordingly, the differentiated mating types can invade e^l, but the undifferentiated mating type M_1 will not be driven to extinction (see Figure 2(c)).

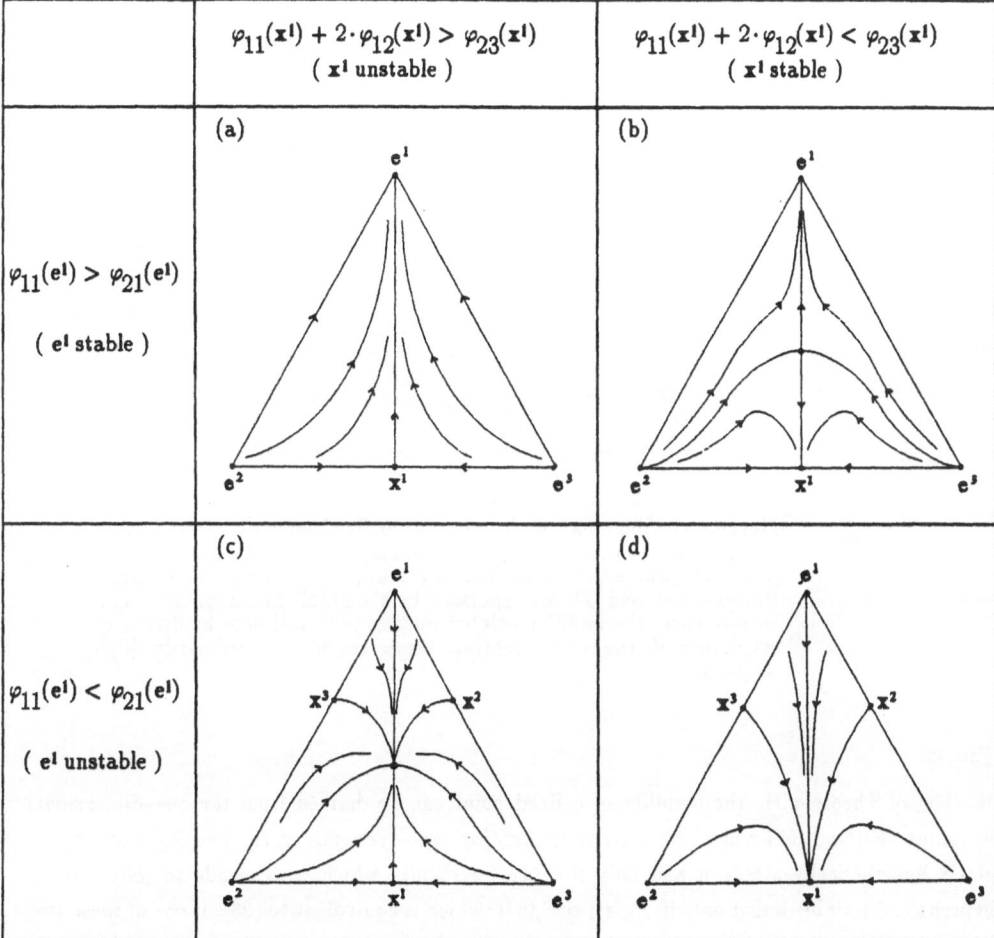

	$\varphi_{11}(\mathbf{x}^{\mathbf{i}}) + 2 \cdot \varphi_{12}(\mathbf{x}^{\mathbf{i}}) > \varphi_{23}(\mathbf{x}^{\mathbf{i}})$ ($\mathbf{x}^{\mathbf{i}}$ unstable)	$\varphi_{11}(\mathbf{x}^{\mathbf{i}}) + 2 \cdot \varphi_{12}(\mathbf{x}^{\mathbf{i}}) < \varphi_{23}(\mathbf{x}^{\mathbf{i}})$ ($\mathbf{x}^{\mathbf{i}}$ stable)
$\varphi_{11}(\mathbf{e}^{\mathbf{i}}) > \varphi_{21}(\mathbf{e}^{\mathbf{i}})$ ($\mathbf{e}^{\mathbf{i}}$ stable)	(a)	(b)
$\varphi_{11}(\mathbf{e}^{\mathbf{i}}) < \varphi_{21}(\mathbf{e}^{\mathbf{i}})$ ($\mathbf{e}^{\mathbf{i}}$ unstable)	(c)	(d)

FIGURE 2: Classification of the mating type dynamics.

Depending on the gamete success rates at the two border fixed points $\mathbf{e}^{\mathbf{i}}$ and $\mathbf{x}^{\mathbf{i}}$ four different scenarios can be distinguished:

(a) $\mathbf{e}^{\mathbf{i}}$ is a global attractor.

(b) $\mathbf{e}^{\mathbf{i}}$ and $\mathbf{x}^{\mathbf{i}}$ are both stable. The outcome of the selection process depends on the starting conditions.

(c) $\mathbf{e}^{\mathbf{i}}$ and $\mathbf{x}^{\mathbf{i}}$ are separated by an interior fixed point which is a global attractor.

(d) $\mathbf{x}^{\mathbf{i}}$ is a global attractor.

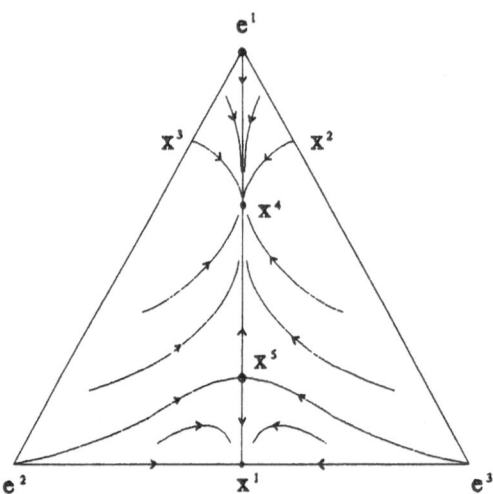

FIGURE 3:   Mating type dynamics with two interior fixed points (x^4 and x^5). The two differentiated mating types may invade e^1 and the fixed point x^1 (where the undifferentiated mating type is absent) is stable. However, e^1 and x^1 are separated by the stable attractor x^4. As a consequence, the undifferentiated mating type will only be driven to extinction if the other mating types reach extraordinarily high frequencies.

PROOF:

In view of Theorem 11, the stability of a fixed point can be derived from the one–dimensional dynamics (60) and (61) which characterize the mating type dynamics on the line segment Δ_{equ}. e^1 is hyperbolically stable if and only if $F_1(e^1) > F_2(e^1)$ which corresponds to (58). x^1 is hyperbolically stable if and only if $F_2(x^1) > F_1(x^1)$ which is equivalent to (36). If one of these two border fixed points is stable whereas the other is unstable, any transverse intersection point of F_1 and F_2 on Δ_{equ} is automatically associated with a second intersection point. Therefore, Assumption 5 implies that F_1 and F_2 cannot intersect at all: e^1 and x^1 are the only fixed points on Δ_{equ}. (Figure 3 shows that this is not necessarily true if Assumption 5 is not satisfied.)

If e^1 and x^1 are both stable or both unstable, Assumption 5 implies that F_1 and F_2 intersect exactly once on Δ_{equ}. Accordingly, there is a unique interior fixed point which will be denoted by x^0. It is easy to see that x^0 is hyperbolically stable with respect to the continuous mating type dynamics if and only if $F_1(x) < F_2(x)$ for all x with $x_1 > x_1^0$ and $F_1(x) > F_2(x)$ for all x with $x_1 < x_1^0$. Notice that this is the case if e^1 and x^1 are both unstable. In the discrete time case, this does not guarantee stability since 'overshooting' may occur (see Weissing, 1990). It can, however, be shown that x^0 is hyperbolically stable with respect to the discrete mating type dynamics if selection is 'weak enough' in the sense that

$$0 < x_1^0 \cdot (1 - x_1^0) \cdot [\, dF_2/dx_1 - dF_1/dx_1 \,](x^0) < F_1(x^0) . \tag{62}$$

3. Some Specific Models

In this section we will analyze a number of specific mating type models. The models differ in their assumptions concerning the mating kinetics. All these models can be treated as special cases of the general model developed in the previous section. The specific assumptions concerning the mating kinetics are then reflected in the success rates $\varphi_{ij}(x)$. Any of these models is thus fully characterized by its fitness matrix

$$\Phi(x) := (\ \varphi_{ij}(x)\ )_{i,j}\ ,\ x \in \Delta\ . \tag{63}$$

In view of the reduction principle, it is only necessary to specify the fitness matrix $\Phi(x)$ for frequency distributions on the line segment Δ_{equ} .

Theorem 14 shows that the analysis of each specific model boils down to the question whether the two differentiated mating types M_2 and M_3 can invade a monomorphic M_1–population, and whether the fixed point x^1 is stable, i.e., whether the undifferentiated mating type M_1 will be driven to extinction. As we have seen, the conditions for invadability and extinction are given by

$$\text{Invadability:} \qquad \varphi_{21}(e^1) > \varphi_{11}(e^1)\ , \tag{64}$$

$$\text{Extinction:} \qquad \varphi_{23}(x^1) > \varphi_{11}(x^1) + 2\cdot\varphi_{12}(x^1)\ . \tag{65}$$

3.1 One–stage Models

The first model we consider is the basic model studied by Hoekstra (1982) with very simple mating kinetics. All gametes are supposed to have one opportunity to combine with another gamete, and this process is random with respect to gamete type. All pairs consisting of the same differentiated type gametes (M_2M_2 and M_3M_3) cannot result in zygotes and are discarded. With the relative fitnesses of the different zygote types as shown in Figure 1, we obtain the following fitness matrix:

$$\Phi_1(x) := \begin{bmatrix} \alpha\cdot x_1 & \beta\cdot x_2 & \beta\cdot x_3 \\ \beta\cdot x_1 & 0 & x_3 \\ \beta\cdot x_1 & x_2 & 0 \end{bmatrix}\ . \tag{66}$$

In the basic model, the conditions for invadability and extinction take the following simple form:

$$\text{Invadability:} \qquad \beta > \alpha\ , \tag{67}$$

$$\text{Extinction:} \qquad \beta < 1/2\ . \tag{68}$$

These conditions are illustrated in Figure 4(a).

176

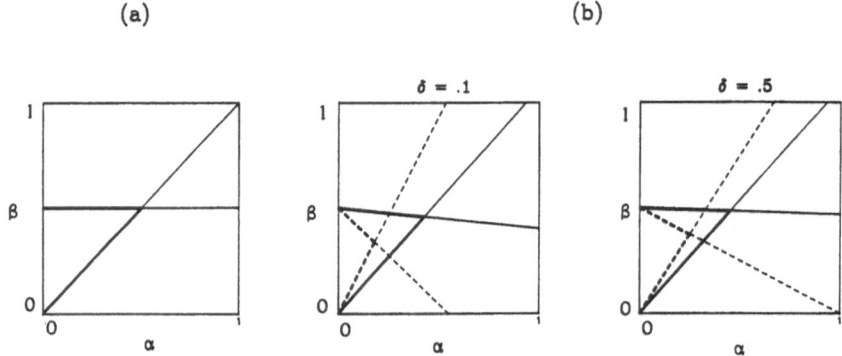

(a)   (b)

FIGURE 4:   Invadability and extinction in one–stage models.

Conditions for the evolution towards two differentiated mating types
starting from an undifferentiated population are fulfilled for those
parameter values which lie between the heavily drawn line segments.
(a)   One–stage random–fusion model.
(b)   One–stage clonal model.
Solid lines: $p = 0.1$, broken lines: $p = 0.5$.

As a second specific model we consider a population with a clonal structure. All individuals
from the same clone are derived by asexual reproduction from a single common ancestor, and are
therefore genetically identical. Different clones may physically overlap in varying degrees. Again
we assume that every gamete will have exactly one opportunity to mate. With probability p, it
meets a gamete from its own clone, with probability $1-p$ it meets a random gamete from the
population gamete pool. The fitness of a zygote resulting from an intraclonal mating is lowered by
an 'inbreeding depression' factor $1-\delta$, where $0 \leq \delta \leq 1$. Actually this parameter δ is better
regarded as the loss of fitness due to *not* having the advantages connected with 'complementation'
occuring in interclonal matings. In the models developed in this chapter we are not concerned with
the nature of this advantage of sex. We just assume that for some reason the occasional occurrence
of sex is advantageous, and that this advantage is less if the two mating gametes are genetically
identical.

The assumptions made above lead to the following fitness matrix:

$$\Phi_2(\mathbf{x}) := \begin{bmatrix} (1-p)\cdot\alpha x_1 + p(1-\delta)\alpha & (1-p)\cdot\beta x_2 & (1-p)\cdot\beta x_3 \\ (1-p)\cdot\beta x_1 & 0 & (1-p)\cdot x_3 \\ (1-p)\cdot\beta x_1 & (1-p)\cdot x_2 & 0 \end{bmatrix}, \tag{69}$$

and the conditions for invadability and extinction are given by:

$$\text{Invadability:} \qquad \beta > \alpha \cdot \left[1 + \frac{p}{1-p} \cdot (1-\delta) \right], \qquad (70)$$

$$\text{Extinction:} \qquad \beta < 1/2 - \frac{p}{1-p} \cdot (1-\delta) \cdot \alpha. \qquad (71)$$

Figure 4(b) shows a graphical representation of these conditions for both high and low values of the parameters p and δ.

3.2 A Two–Stage Model

We now proceed to analyze a specific model in which the assumption of only one mating opportunity for every gamete is relaxed. As an approximation, we consider the process of zygote formation to consist of two subsequent stages. In the first stage, there is opportunity for gametes to form pairs. Suppose that a fraction κ of the gamete population succeeds in finding another gamete to form a pair with. All compatible pairs result in a zygote. However, the gametes from incompatible pairs disengage, and are allowed to become 'free' gametes again. Thus at the end of the first stage we have already a number of zygotes formed, as well as free gametes (either from incompatible pairs or those that have been unsuccessful so far in finding another gamete). In this first stage we assume a clonal population structure — just as in the second model of Section 3.1 — with a probability p for intraclonal pairing and a probability $1-p$ for interclonal pairing.

In the second stage the remaining free gametes are given one additional opportunity to find a mating partner. For reasons of analytical tractability, we assume random pair formation in the second stage. Incompatible pairs are discarded. Therefore, in this second stage we adopt the first model of Section 3.1. The mating success of the gametes in the second stage is assumed to be proportional (with constant of proportionality θ) to that fraction of the original gamete population which is still available for mating.

The fitness matrix $\Phi_3(x)$ for this model can be calculated as follows. The zygote success rates $\varphi_{ij}(x)$ consist of two components: $\varphi_{ij}(x) = \varphi_{ij,I}(x) + \varphi_{ij,II}(x)$. Since in the first stage we are effectively using the second model of Section 3.1, and a fraction κ of the gametes participates in pair formation during the first stage, it follows that

$$\Phi_{3,I}(x) := [\; \varphi_{ij,I}(x) \;]_{i,j} = \kappa \cdot \Phi_2(x), \qquad (72).$$

where $\Phi_2(x)$ is given by (69).

We now proceed to calculate the contributions $\varphi_{ij,II}(x)$ from the second stage. Let ρ_i denote the probability that a gamete of type M_i is still unmated after the first stage. These probabilities are given by

$$\rho_1 = 1 - \kappa \qquad (73a)$$

$$\rho_2 = 1 - \kappa + p\kappa + (1-p)\kappa \cdot x_2, \qquad (73b)$$

$$\rho_3 = 1 - \kappa + p\kappa + (1-p)\kappa \cdot x_3. \qquad (73c)$$

The relative frequencies of the gamete types in the second stage are therefore given by $y_i = \rho_i x_i / \Upsilon$, where $\Upsilon := \Sigma \rho_i x_i$ is a normalizing factor.

Now the probability $\varphi_{ij,II}(x)$ that a gamete of type M_i forms a zygote with a gamete of type M_j in the second stage is equal to the product of the mating success $\theta \cdot \Upsilon$, the probability ρ_i that a gamete of type M_i is still unmated at the start of the second stage, and the frequency y_j of M_j gametes still available in the second stage:

$$\varphi_{ij,II}(x) = \theta \cdot \Upsilon \cdot \rho_i \cdot y_j = \theta \cdot \rho_i \cdot \rho_j \cdot x_j. \qquad (74)$$

Therefore, we obtain the following fitness matrix for the two–stage model:

$$\Phi_3(x) = \kappa \cdot \Phi_2(x) + \theta \cdot \mathrm{diag}(\rho) \cdot \Phi_1(x) \cdot \mathrm{diag}(\rho), \qquad (75)$$

where $\mathrm{diag}(\rho)$ denotes the diagonal matrix with diagonal $\rho = (\rho_1, \rho_2, \rho_3)$; and $\Phi_1(x)$ and $\Phi_2(x)$ are given by resp. (66) and (69).

From the representation (75) we can now derive the conditions for invadability and extinction just as in the preceding models. These conditions are:

Invadability: $$\beta > \alpha \cdot \frac{(1-\delta)\,\kappa p + \kappa(1-p) + \theta(1-\kappa)^2}{\kappa(1-p) + \theta(1-\kappa) - \theta\kappa(1-\kappa)(1-p)}, \qquad (76)$$

Extinction: $$\beta < \frac{\kappa(1-p) + \theta(1-\kappa(1-p)/2)^2 - 2\alpha(1-\delta)\kappa p}{2\kappa(1-p) + \theta(1-\kappa)(2-\kappa(1-p))} \qquad (77)$$

Which regions in the paramter space are most interesting from a biological point of view? A small value of κ (a low probability of finding a mating partner during the first stage) is quite naturally interpreted as indicating a low gamete density. Similarly, a high κ indicates a high population density. High values of p (the probability of intraclonal interactions) would be expected in a population structure with little overlap between different clones, while a small p indicates strongly overlapping clones. This leads us to consider four different special cases, illustrated in Figure 5.

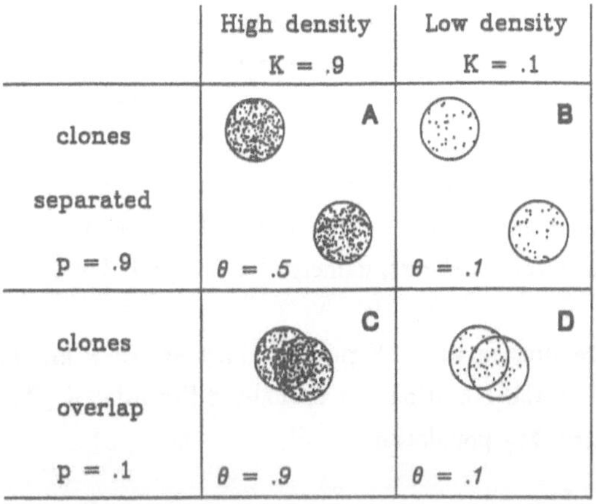

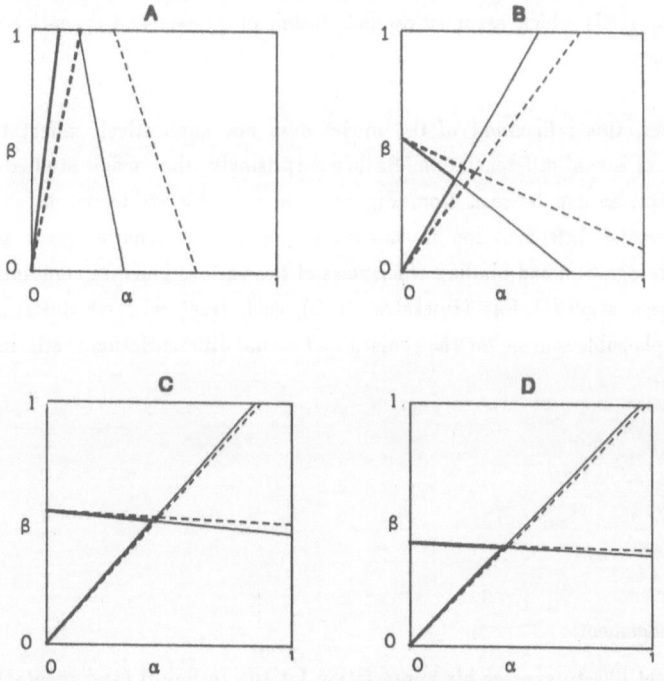

	High density K = .9	Low density K = .1
clones separated p = .9	A $\theta = .5$	B $\theta = .1$
clones overlap p = .1	C $\theta = .9$	D $\theta = .1$

FIGURE 5:   Four special cases of the two–stage model of Section 3.2.
The circles represent clones, while the density of dots indicates the
density of the gamete population.

FIGURE 6:   Conditions for the evolution of two differentiated mating tpyes are
fulfilled for paramter values between the heavily drawn line
segments.
Cases A, B, C and D correspond to those of Figure 5.
Solid lines: $\delta = 0.1$; broken lines: $\delta = 0.5$.

What are likely values of θ in these four special cases? In cases B and D (low population density) it should be hard for the remaining gametes in the second stage to find a mating partner. This implies a low value of θ, and we set $\theta = 0.1$ in these cases. When density is high and clones are also overlapping, we expect θ to be high: $\theta = 0.9$. In the remaining case (A), where density is high but clones are separated, an intermediate value of θ ($\theta = 0.5$) seems the most realistic choice. We analyze all cases for two different values of δ (the disadvantage associated with intraclonal versus interclonal sex), namely $\delta = 0.1$ and $\delta = 0.5$.

All these cases are illustrated in Figure 5 which shows a graphical representation of the conditions which lead to the evolution of a sexually differentiated (M_2, M_3) population starting from an undifferentiated M_1-population.

4. Conclusions

The theoretical model developed in this paper enables us to be much more subtle about the 'mating kinetics' of a gamete population. The incorporation of both a clonal distribution and the second chance for gametes from incompatible pairs seems to add realism to the original model (Hoekstra, 1982) which assumes random fusion of gamets and (genetic) death of incompatible gamete pairs.

However, this refinement of the model does not qualitatively affect the conditions for the evolution of sexual differentiation. Rather surprisingly, the models studied do not differ much in this respect, as can be seen from Figures 4 and 6. We are forced to conclude that evolution towards sexual differentiation in isogamous populations requires great selective differences in terms of recognition and binding efficiencies of the various pairwise combinations of gamete types. As has been argued before (Hoekstra, 1982), such great selective differences do not seem very likely. A plausible scheme for the evolution of sexual differentiation is still missing.

Acknowledgements:

F.W. would like to express his appreciation for the technical (and mental) support provided by Anke Huckriede. Without her help this paper had not been completed in time.

References:

Eshel, I. (1990): Game Theory and Population Dynamics in Complex Genetical Systems: The Role of Sex in Short Term and in Long Term Evolution. This volume.

Harsanyi, J.C. & R. Selten (1988): A General Theory of Equilibrium Selection in Games. Cambridge, Ms: The MIT Press.

Hines, W.G.S. (1987): Evolutionarily Stable Strategies: A Review of Basic Theory. Theor. Pop. Biol. 31: 195–272.

Hoekstra, R.F. (1982): On the Asymmetry of Sex: Evolution of Mating Types in Isogamous Popualtions. J. theor. Biol. 98: 427–451.

Hoekstra, R.F..(1987): The Evolution of Sexes. In: S.C. Stearns (ed.): The Evolution of Sex and its Consequences. Basel: Birkhäuser Verlag, pp. 59–91.

Hofbauer, J. & K. Sigmund (1988): The Theory of Evolution and Dynamical Systems. Cambridge: Cambridge University Press.

Iwasa, Y. & A. Sasaki (1987): Evolution of the Number of Sexes. Evolution 41: 49–65.

Lessard, S. (1990): Evolutionary Stability: One Concept, Several Meanings. Theor. Pop. Biol. 37: 159–170.

Maynard Smith, J. (1982): Evolution and the Theory of Games. Cambridge: Cambridge University Press.

Michod, R.E. & B.R. Levin (eds) (1988): The Evolution of Sex. Sunderland: Sinauer Ass.

Pohley, H.J. & B. Thomas (1983): Nonlinear ESS–models and Frequency Dependent Selection. Biosystems 16: 87–100.

Selten, R. (1983): Evolutionary Stability in Extensive Two–Person Games. Mathem. Social Sciences 5: 269–363.

Stearns, S.C. (ed.) (1987): The Evolution of Sex and its Consequences. Basel: Birkhäuser Verlag.

Taylor, P.D. & Jonker, L.B. (1978): Evolutionarily Stable Strategies and Game Dynamics. Math. Biosci. 40: 145–156.

Vickers, G.T. & C. Cannings (1987): On the Definition of an Evolutionarily Stable Strategy. J. theor. Biol. 129: 349–353.

Weissing, F.J. (1983): Populationsgenetische Grundlagen der Evolutionären Spieltheorie. Bielefeld: Materialien zur Mathematisierung der Einzelwissenschaften, Vols 41 & 42.

Weissing, F.J. (1990): Evolutionary Stability and Dynamic Stability in a Class of Evolutionary Normal Form Games. This volume.

THE EVOLUTIONARY STABILITY OF BLUFFING IN A CLASS OF EXTENSIVE FORM GAMES[1]

by

Roy Gardner and Molly Morris

I.   INTRODUCTION

It is common in many different organisms for contests to be settled through the use
of agonistic displays rather than physical fights. For example, in contests where males
compete for females, Red deer will settle the contest based on roaring tempo (Clutton-
Brock and Albon 1979), toads on the pitch of a males croak (Davies and Halliday 1978)
and African buffalo on ritualized head-on charges (Sinclair 1977).   The traditional
view of communication in agonistic encounters suggests that displays should contain
information about who would win an escalated contest (i.e. Cullen 1972; Parker 1974).
Exchanging this information would be benefical to both individuals, as they could settle
the contest without the high cost of a fight. The recent concept of communication as
exploitive (Krebs and Dawkins 1984) and the continuing application of game theory to
aggressive encounters (Maynard Smith and Parker 1976; Bishop and Cannings 1978; Maynard
Smith 1982; Enquist 1985), have stirred interest in the possiblity of bluffing (i.e. one
individual gives false information in order to win a nonescalated contest). Maynard
Smith (1982) reasoned that if bluffing did not have a high cost, it would invade,
rendering a signal uninformative and eventually ignored by opponents. Along the same
lines of reasoning, it has been argued that a display that affects the outcome of a
cont st should transmit information about resource holding power (RHP) rather than
information about intention (Maynard Smith 1974; Zahavi 1977, 1979). A signal
correlated to RHP would be more costly to bluff and therefore contain more reliable
information as compared to a signal of intention. The general predictions, therefore,
hav been that displays will not contain information about intention and that bluffing
will not be part of a stable strategy.

Many displays used to settle contests appear to be correlated with resource holding
power and also appear to be costly to bluff (eg. roaring in Red deer, Clutton-Brock and
Albon 1979; pitch of croak in the European toad, Davies and Halliday 1978). However,
there are examples of displays that do not appear to follow this pattern (e.g. status
badges in sparrows, Rohwer 1985; ritualized displays in Blue Tits, Stokes 1962). In
addition, examples of signals that announce intent are beginning to appear in the animal
behavior literature (e.g. Poole 1989, Piersma and Veen 1988). The question remains as

[1]The authors wish to thank Franjo Weissing for his helpful comments. Financial
support from the National Science Foundation and the Center for Interdisciplinary Research,
University of Bielefeld helped make this research possible.

to how such systems remain stable against the invasion of bluffing.

We examined the possiblity that bluffing can be part of an evolutionarily stable strategy (ESS) by developing a simple extensive form model (Gardner and Morris 1989) based on the best documentation of bluffing in the literature; the display behavior of the mantis shrimp <u>Gonodactylus</u> <u>bredini</u> (Steger and Caldwell 1983). Extensive form games can incorporate assessment, which is an important component of all communication in animals (Parker 1974; Parker and Rubenstein 1981). In general, our model allows one player to assess its opponent before choosing to attack or flee. We also allow for information asymmetries by considering an information index, which reflects the information an individual possesses about his own role in relation to his oppon nt.

Two conclusions were drawn from the results of this model. First, when the cost of bluffing is high in relation to the resource, a completely informative signal is the ESS. However, when the cost of losing a fight and bluffing are both low, bluffing and informative signals are both behaviors found in the candidate ESS. There is no ESS in this regime. Instead, the population cycles, with the candidate ESS providing a reasonable estimate of the average behavior of the population.

The purpose of the present paper is to determine the robustness of the results from the mantis shrimp model, both in relation to bluffing and the dynamics of the system. The model based on the mantis shrimp contains assumptions in the parameter values that are specific to the biology of the system. For example, we assigned a higher cost to bluffing than an honest display, and we did not assign a cost to fleeing for resid nts that retreat after their bluff is called. In addition, we simplified the analysis of the model by only considering cases in which the strong resident won all escalated contests. Now, we look at a dynamical system representing the evolutionary dynamics of an 8-dimensional parameter space, and explore what happens when the parameter value specifications of our original model are relaxed.

II. Evolutionary Dynamics of an Extensive Form Game

In this section, we study the evolutionary dynamics on the extensive form gam depicted in figure 1. There are two players, denoted 1 and 2. Moreover, there ar two types of player 1 present in the population, denoted by 1-A and 1-B, with frequ ncies X and 1-X respectively. At the beginning of the contest, Player 2 does not know Play r 1's type, however both players know their own type. Player 1 is the first to mov . If player 1 moves left (displays), then it is player 2's turn to move. If player 1 m ves right (flees), the game is over. In this case, Player 1 of either type gets th pay ff 0 and player 2 gets the payoff a_1. We interpret a move to the right by either player as a move which terminates the interaction, and all payoffs are normalized so that the play r that terminates the game gets a payoff of 0. If player 2 terminates the interaction by moving right, the payoff to player 1 is type dependent (a_4 if player 1-A of frequ ncy X and a_7 if player 1-B of frequency 1 - X). Otherwise, player 2 continues the interaction by also moving left, which we interpret as an attack. This results in the payoffs to player 1 (respectively player 2) of a_2 (a_3) or a_5 (a_6) depending on player 1's type. One thus has a 9-dimensional parameter space on which evolutionary dynamics are defined; 8 dimensions for payoff parameters and 1 dimension for the player type parameter.

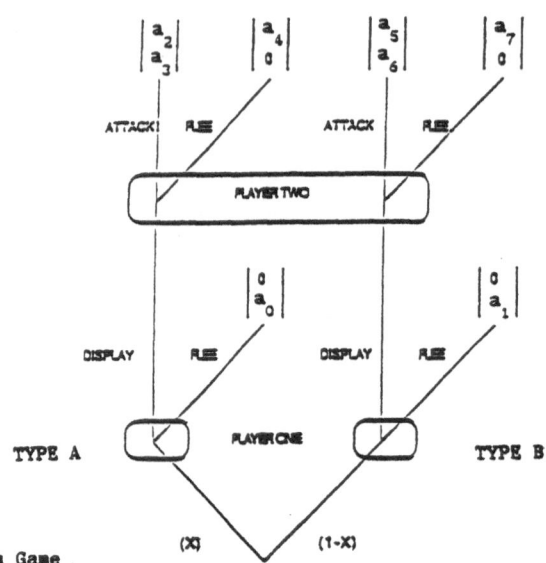

Figure 1: Extensive Form Game

W will study here the evolutionary dynamics introduced by Taylor and Jonker (1978), applied to the local strategies of the extensive form. The intuition behind these dynamics is that if a strategy played by a certain player or type of player has an above-av rage payoff, then that strategy will be represented to a greater extent in the population in the next generation. Underlying this intuition are the customary hypoth s s of genetic inheritance of the behavior of an individual player. We will assum here that there are no genetic constraints which would prevent any possible population from actually arising. Using this intuition, one arrives at the following dynamic system:

$$\dot{x}_i = x_i(1-x_i)F_i(x) \qquad\qquad i = 1,2,3 \qquad\qquad (1)$$

where

$$F_1(x) = x_3 a_2 + (1 - x_3)a_4 \qquad\qquad (2)$$

$$F_2(x) = x_3 a_5 + (1 - x_3)a_7 \qquad\qquad (3)$$

$$F_3(x) = Xx_1 a_3 + (1 - X)x_2 a_6. \qquad (4)$$

where x_1 denotes the percentage of players 1-A who move left, x_2 denotes the percentage of players 1-B who move left, and x_3 denotes the percentage of players 2 who move left. $\dot{x}_1$, $\dot{x}_2$, and $\dot{x}_3$ denote the rate of change in the population of x_1, x_2 and x_3 respectively. The reasoning behind equation (1) is as follows. Take a player 1-A who has moved left. Then with probability x_3, he is matched against a player 2 who is also programmed to move left, in which case player 1 gets the payoff a_2. With probability 1 $- x_3$, he is matched against a player 2 who is programmed to move right, in which case player 1 gets the payoff a_4. So, this type player 1 can expect the payoff $x_3 a_2 + (1 - x_3)a_4$. At the same time, this type player 1 who moves right gets the payoff 0. Hence, the average payoff to all players 1 of this type is

$$\text{average} = x_1 (x_3 a_2 + (1 - x_3) a_4) + (1 - x_1) 0.$$

In the evolutionary dynamics, the rate of change of x_1 in the population, $\dot{x}_1$, is given by

$$\dot{x}_1 = x_1 [x_3 a_2 + (1 - x_3) a_4 - \text{average}]$$

which reduces to equation (1) for $i = 1$. When $F_1(x)$ is positive, players 1-A are getting an above average payoff. The arguments leading to (1) for $i = 2$ and 3 are similar. Notice that the payoff parameters a_0 and a_1 do not appear in the dynamic system (1): they are strategically irrelevant.

W now turn to the evolutionarily stable strategies (ESS) which are possible in various regions of the parameter space for the dynamical system (1).

III. Evolutionarily Stable Strategies

An ESS is a behavior in a population which cannot be invaded by a mutant behavior (Selten, 1980). If $x^* = (x_1^*, x_2^*, x_3^*)$ is an ESS, then x^* is a stable dynamic equilibrium of the system (1). By a result of Selten (1980), every ESS of an asymmetric game is a strict Nash equilibrium strategy. It is easy to see that a strict Nash equilibrium corresponds to a dynamic equilibrium which is hyperbolically stable with respect to the dynamics (1). Selten's result applied to the present asymmetric gam s m ans that ESS x^* can only occur on the corners of the unit cube. Concerning thes eight corners, the following can be established:

Case 1. $x^* = (1, 1, 1)$ is the ESS where all players move left. The parameter configuration sufficient for this to be an ESS is that

$$a_2 > 0, \ a_5 > 0 \ \text{and} \ Xa_3 + (1 - X) a_6 > 0. \tag{5}$$

To see this, consider the three possible types of mutants to the population. First, consider a mutant player 1-A that moves right. Since $x_1 = 1 - \epsilon$, for ϵ small (ϵ is the mutant frequency in the population of players 1-A), one has

$$\dot{x}_1 = \epsilon(1 - \epsilon) \ (a_2) > 0.$$

The b havior by the established player 1-A (display) is getting an above average payoff, hence selection works against the mutant. Next consider a mutant player 1-B that mov s right. Since $x_2 = 1 - \epsilon$, for ϵ small, one has

$$\dot{x}_2 = \epsilon(1 - \epsilon) \ (a_5) > 0.$$

The behavior by the established player 1-B (display) is getting an above average payoff and s lection works against the mutant. Finally, consider a mutant player 2. Since $x_3 = 1 - \epsilon$ for ϵ small, one has

$$\dot{x}_3 = \epsilon \ (1 - \epsilon) \ (Xa_3 + (1 - X)a_6) > 0,$$

and again s lection operates against the mutant.

An alternative argument uses the Jacobian of the replicator dynamics. The Jacobian $J(x)$ of (1) at population x is given by:

$$J(x) = \begin{matrix} (1 - 2x_1)F_1(x) & 0 & x_1(1 - x_1)(a_2 - a_4) \\ 0 & (1 - 2x_2)F_2(x) & x_2(1 - x_2)(a_5 - a_7) \\ x_3(1 - x_3)Xa_3 & x_3(1 - x_3)(1 - X)a_6 & (1 - 2x_3)F_3(x) \end{matrix} \tag{6}$$

Evaluating the Jacobian (6) at $x^* = (1,1,1)$ one has the diagonal matrix $J(x^*)$.

$$J(x^*) = \begin{matrix} -a_2 & 0 & 0 \\ 0 & -a_5 & 0 \\ 0 & 0 & -Xa_3 - (1 - X)a_6 \end{matrix} \tag{7}$$

Th ignvalues of $J(x^*)$ are the diagonal entries, namely $-a_2$, $-a_5$ and $-Xa_3 - (1 - x)a_6$. For hyp rbolic dynamic stability, all the eignvalues must be negative. This establish s (5).

Using material similar to (7), one verifies the following cases 2 through 6 as w ll. <u>Case 2.</u> $x = (1, 1, 0)$ is an ESS. Both types of players 1 move left (display) and players 2 move right (flee). The parameter configuration sufficient for this is that

$a_4 > 0$, $a_7 > 0$, and $Xa_3 + (1 - X)a_6 < 0$.

<u>Case 3.</u> $x^* = (1, 0, 1)$. Players 1-A move left (display) and players 2 move left (display). Players 1-B move right (flee). The parameter configuration sufficient for this is that

$a_2 > 0$, $a_5 < 0$, and $a_3 > 0$.

<u>Case 4.</u> $x^* = (1, 0, 0)$. Players 1-A move left (display), while all others move right (flee). The parameter configuration for this is

$a_3 < 0$, $a_4 > 0$, $a_7 < 0$.

<u>Case 5.</u> $x^* = (0, 1, 1)$. Players 1-A move right (flee), while all others move left (display). The parameter configuration for this is

$a_2 < 0$, $a_5 > 0$, $a_6 > 0$.

<u>Case 6.</u> $x^* = (0, 1, 0)$. Players 1-B move left (display), while all others move right (flee). The parameter configuration for this is

$a_4 < 0$, $a_6 < 0$, $a_7 > 0$.

As we shall see in a moment, these six cases exhaust the pure ESS possibilities.

It is worth pointing out that there may be multiple ESS. One can check that there are non mpty interestions of the following cases: 1 and 4; 1 and 6; 2 and 3; 2 and 5; 3 and 6; and 4 and 5. Consider for instance the following parameter configuration:

$$a_2 = a_4 = a_5 = 1 \qquad a_3 = a_7 = -1 \qquad a_6 = 4 \text{ and } X = .5.$$

This c nfiguration answers to both case 1 and case 4. Thus, both $x^* = (1,1,1)$ and $x^* = (1,0,0)$ ar ESS. These are very different behaviors. As usual in such a case, the ESS which ultimately evolves depends on the history of the dynamic system (1). One can furth r check that these are the only patterns of ESS possible with two ESS, and that ther is no possible pattern with more than two ESS.

In addition to the ESS possibilities, two cases of limit ESS in the sense of Selten (1983, 1988) are possible. To fix ideas, consider the corner $x^* = (0,0,1)$. The Jacobian at x^*, $J(x^*)$, is

$$J(x) = \begin{array}{ccc} a_2 & 0 & 0 \\ 0 & a_5 & 0 \\ 0 & 0 & 0 \end{array} \qquad (8)$$

whose eigenvalues are a_2, a_5, and 0. The 0 eignvalue shows that there cannot be an ESS here. Suppose that the population is in an ϵ neighborhood of x^*, say $x = (\epsilon, \epsilon, 1-\epsilon)$. Selten's idea in this case is that if $\dot{x}_1 < 0$, $\dot{x}_2 < 0$ and $\dot{x}_3 > 0$ when $x = (\epsilon, \epsilon, 1-\epsilon)$, then the limit of the population as ϵ goes to 0 is a limit ESS. The interpretation of ϵ in such case is that of mistake probabilities inherent in the behavioral programming that keep the population even x^*, even if x^* itself is not an ESS.

Suppose then that one has a_2, a_0 and $a_5 < 0$. Evaluating (1) at $(\epsilon, \epsilon, 1-\epsilon)$, one has

$$\dot{x}_1 = \epsilon(1-\epsilon)[(1-\epsilon)a_2 + \epsilon a_4]$$

$$\dot{x}_2 = \epsilon(1-\epsilon)[(1-\epsilon)a_5 + \epsilon a_7]$$

$$\dot{x}_3 = \epsilon(1-\epsilon)[X\epsilon a_3 + (1-x)\epsilon a_6]$$

$$= \epsilon^2(1-\epsilon)[Xa_3 + (1-X)a_6]$$

Now for ϵ small enough, $\dot{x}_1 < 0$ and $\dot{x}_2 < 0$, from the negativity of a_2 and a_5. If in addition one has $Xa_3 + (1-X)a_6 > 0$, then $\dot{x}_3 > 0$ and $x^* = (0,0,1)$ is a limit ESS.

Summarizing this discussion:

Case 7. (Limit ESS). $x^* = (0,0,1)$ is a limit ESS when

$$a_2 < 0, \ a_5 < 0, \text{ and } Xa_3 + (1-X)a_6 > 0.$$

By a similar argument, one can show:

Cas 8. (Limit ESS). $x^* = (0,0,0)$ is a limit ESS when a_4, a_7, and $Xa_3 + (1-X)a_6$ are all negative. Clearly, at most one limit ESS is possible. It is also possible for there to be a pattern with both an ESS and a limit ESS, namely case 2 and case 7.

Now it is clear that there are parameter regions where neither an ESS nor a limit ESS xists. These involve candidate ESS where $x_i = 0$ for all i at x^* not on the corners of th unit cube. Hence, it turns out that two new dynamic phenomena are possible, a saddle point x^* or a cycle around x^*. These two possibilities turn out to be rather intimately related.

Consid r the following case:

Case 9. (Candidate ESS). $x^* = (1, x^*_2, x^*_3)$ is a candidate ESS when $0 < x^*_2 < 1$ and $0 < x_3 < 1$, at which $\dot{x}_i = 0$, all i.
In particular

$$x_2 = \frac{-Xa_3}{(1-x)a_6} \qquad\qquad x^*_3 = \frac{a_7}{a_7 - a_5}$$

The Jacobian evaluated at x^*, $J(x^*)$ is given by

$$J(x^*) = \begin{matrix} -x^*_3 a_2 - (1-x^*_3)a_4 & 0 & 0 \\ 0 & 0 & x^*_2(1-x^*_2)(a_5-a_7) \\ x^*_3(1-x^*_3)Xa_3 & x^*_3(1-x^*_3)(1-X)a_6 & 0 \end{matrix} \qquad (9)$$

whose eignvalues are

$$\lambda = -x^*_3 a_2 - (1-x^*_3)a_4$$

and $\qquad (10)$

$$\lambda^2 = x^*_3(1-x^*_3)(x^*_2)(1-x^*_2)(1-X)a_6(a_5-a_7)$$

In order for $x^*_1 = 1$ to be maintained in the population it is sufficient that $a_2 > 0$, $a_4 > 0$; this makes the eigenvalue in the x_1 direction, $-x^*_3 a_2 - (1-x^*_3)a_4 < 0$. The nature of the remaining eigenvalue depends on the sign of a_6 $(a_5 - a_7)$, since all other quantities in (10) are positive.

If $\lambda^2 > 0$ because $a_6(a_5-a_7)$ is positive, then x^* is a saddle point on the face of the cube where $x^*_1 = 1$. If $\lambda^2 < 0$, however, $a_6(a_5-a_7)$ is negative, then x^* is a center of closed orbits on the interior of this face. The cyclic case in particular appears to have significant biological implications, which we pursue in the discussion below.

IV. Biological Interpretations

In this section we discuss the ESSs in relation to possible biological interpretations of the parameter space. We assume that payoffs consist of the valu of the contested resource and two costs; displaying for players 1 and the cost of an escalated contest (player 2 attacks) for both players. The cost of an escalated encounter, the cost of a display and the value of the resource can vary depending on play rs role or type.

Th most common situations to which this game can be applied are those in which player 1 is the resident on a territory or in possession of a resource that player 2, the intruder, wants. Consider that players 1-A, are more likely to win an escalat d contest than players 1-B. If the display is going to carry information about who would win an escalated contest, then a player 1-B that displays rather than flees is bluffing. Within this framework we can also define honest signals as a display by players 1-A. When both types of player 1 use the same behavior, either display or flee, the behavior of player 1 does not contain any information relative to type, and therefore discussion of bluffing is irrelevant. For the pure and limit ESSs, we can examine the conditi ns under which an honest signal would remain stable as compared to the conditions und r which bluffing would invade, rendering the signal uninformative. Bluffing is found as part of a mixed strategy in only the candidate ESS.

Some of the cases are paradoxical given the assumptions stated above. For example, in the case 1. where player 2 always attacks (case 1), bluffing invades because the payoff to a bluffer that is caught, a_5, is positive. However, since we have assumed that type B players 1 are more likely to loose an escalated contest, it is difficult to think of a biological example of this case. Cases 5 and 6 are also paradoxical as n both player 1-A flees even though we have assumed an asymmetry giving 1-A a higher probability of winning an escalated contest. While it is possible to imagine circumstances in which such paradoxical ESSs could occur, they are not as likely to occur in nature (Maynard Smith and Parker 1976) and therefore we will focus our attention on the other three pure ESSs.

In case 2 both types of player 1 use the same behavior, and therefore no information about type is available to player 2 (players 1 always display and players 2 always flee). Players 1-A receive a positive payoff when they display and players 2 retreat, and players 1-B also receive a positive payoff when they bluff successfully. Therefore, the costs of displaying are not high in relation to the value of the resource. However, there is a very high cost for players 2 that attack. Dependent on the frequency of the types of players 1 in the population, it does not pay player 2 to attack with the hopes of catching a bluffer. The intruder only needs to see that another individual is on the territory and he flees.

In those cases where the signal is not informative (1, 2 and 7) the residents ith r win all encounters, or immediately flee. Dawkins and Krebs (1978) discuss several situations in which prior residency is used as a cue to settle contests. They suggest that in some contests of this kind an arbitrary rule of "resident wins" or "intrud r wins" is used so that the contest can be settled quickly with very little cost to either player. The payoff structure in our model, however, dictates the strategies for each player and therefore they are not arbitrary. The model is set up in such a way that allows for assesment of player 1 by player 2, and yet an informative signal would not remain stable within these payoff regimes.

In the pure ESS cases 3 and 4, the signal is completely informative. Player 2 knows wh n he has encountered a type A individual as they always display and type B always fle . The intruder always attacks in case 3. The parameter structure in which this ESS is found gives both players 1 type A and players 2 positive payoffs when player 1 displays and player 2 attacks. This would be reasonable assuming that the cost of an escalated encounter for both players is low in relation to the value of the resource. In addition, the intruder should win some proportion of these contests, so that both play rs have a probability of winning the resource. Bluffing does not invade in this system due to the negative payoff to a Type B player 1 that displays and is caught. Therefore, it is not necessarily a high cost to the dishonest display to prevent the invasion of bluffing, but a high probability of being detected and suffering the cost of losing an escalated encounter.

A possible example for this case can be found in systems where individuals signal their status in aggressive encounters. Status signals based on plumage appear to correlate with fighting ability in some birds (e.g. Rowher and Rowher 1978; Ketterson 1979; Fugle et al. 1984). However, mimicking dominant plumage should be both energetically and evolutionarily easy. There is good evidence that plumage badges in the yellow warbler (<u>Dendroica petechia</u>) are reliable signals of aggressive effort (Studd and Robertson 1985). In this system, brighter birds are more aggressive, but in addition elicit more aggressive responses from other birds. Therefore, a bluff would carry the cost of more aggressive encounters, without the necessary RHP for defense.

Case 4 also has a completely informative signal, however, here the intruder always flees. This case differs from case 3 in that player 2 receives a negative payoff if he attacks a strong player 1 and the type B player 1 does not gain by bluffing even if player 2 does not attack. The biological interpretation of this case is straightforward, and fits the Maynard Smith's assertion as to how displays remain honest (for examples and discussion, see Maynard Smith 1982). The signals remain honest because a dishonest display, even without the consequences of an escalated encounter, entails a high cost. In addition, the cost of an escalated encounter for player 2 in relation to the value of the resource or the probability of winning against a type A player 1 is too low to warrant attacking a resident that displays.

Finally, we have determined the payoff configuration necessary to obtain a candidate ESS that consists of both an honest signal and bluffing. To obtain a stable cycle of these behaviors, as illustarated in our original model of bluffing in the stomatopods (Gardner and Morris 1989), only two parameter requirements are necessary. These requirements are more general than the configuration used to describe contests in the stomatopods, and can be interpreted biologically. First,

$$0 < -Xa_3/(1-X)a_6 < 1$$

can be interpreted as it does not pay to bluff all the time. If it paid to bluff all of the time, this value would be equal to or greater than one. Second,

$$a_6 (a_5 - a_7) < 0$$

implies two things; 1) that getting caught bluffing is worse than not getting caught ($a_5 - a_7 < 0$), and 2) that the resource is worth something to an intruder ($a_6 > 0$). The honest signal is maintained in the populations by limiting the frequency of bluffs and invoking a cost to getting caught bluffing greater than the benefits of getting away with bluffing.

V. Discussion

There has been much confusion about the role of signalling in encounters between animals in conflict situations. We contend that much of the analysis of agonistic encounters between animals has been handicapped by not including assessment. While assessment has been recognized as an important component of contests (Parker 1974, Parker and Rubenstein 1981), the assessment phase of encounters have typically been modeled in such a way that players can not alter their behavior based on information gained during the contest. In this paper, we apply the theory of extensive form games to signalling behavior in conflict situations. The extensive form, in which opponents move sequentially, allows for assessment, and therefore allows us to examine conditions under which a honest signal would evolve. In addition, we have taken this extensive form model and explored the limitations of the parameter.configurations that produce honest signals as well as the configuration that produce candidate and limit ESSs consisting of honest signals and bluffing.

Maynard Smith (1982) suggested that signalling an opponent honestly about one's own resource holding power (RHP, Parker 1974) is a stable strategy so long as the cost of bluffing a signal indicating greater RHP is large. Signalling about ones intention or motivational state would not be as stable since bluffing of this kind of information would be much less costly. In the parameter space of our model there were only two cases in which a completely informative (players 1-A and 1-B behaved differently) and honest signal (player 1-A displayed) was an ESS. In one case the display remains hon st du to a high cost of the bluffing display. However in the other case, bluffing does not invade due to the cost involved in getting caught bluffing, or the inevitable loss if confronted in an escalated encounter. We suggest that this second case could explain why some signals that appear to be easily bluffed, such as status signals, remain honest.

The biological interpretation of the model we have presented here also demonstrates that in asymmetric games, it will be difficult to separate information into two types; information about RHP and information about intentions. Signals intended to carry information about fighting ability (RHP) alone can carry information about intentions (Enquist 1985). When the resident displays rather than flees, it can mean "I am strong", but in the long run it means, "because I am strong, I plan to stay and fight". Mod ls that have examined the transmission of information about intention considered symm tric games, in which there were no real differences between the opponents that would decide the outcome of the contest (see Maynard Smith 1974). We suggest that an asymmetric game is a more realistic representation of most contests, therefore, th distinction of information type may not always be possible nor necessarily reflect the probability of invasion by bluffing.

When the overall parameter space is considered, we note that the conditions for getting an ESS in a signalling model are very strong. In a rough sense, it appears that the odds of encountering an ESS are small. However, the biological literature is full of examples in which displays appear to contain information. One possibile explanation is that many of the honest signals encountered are not stable. In the parameter space with no ESSs, there are many possible evolutionary configurations. Two that we have uncovered are saddle points and stable cycles. The behaviors that one would encounter in such parameter regions would all appear similar, regardless of the underlying turbulence. Displays would be continually changing with new displays invading and old displays loosing their reliability. Andersson (1980) has offered just such an explanation for why many species have several different threat displays. Phylogenetic comparisons of displays can be used to explore this possibility (Maynard Smith 1984). Threat displays of related species should differ more than other kinds of displays due to their more rapid evolution.

References

Andersson, M. (1980). Why Are There So Many Threat Displays? J. Theor. Biol., 86: 773–781.

Bishop, D.T. & Cannings, C. (1978). A Generalized War of Attrition. J Theor. Biol, 70: 85–124.

Clutton-Brock, T.H. & Albon, S.D. (1979). The Roaring of Red Deer and the Evolution of Honest Advertisement. Behaviour, 69: 145–170.

Cullen, J.M. (1972). Some Principles of Animal Communication. In: Non-verbal Communication (ed. Hinde, R.A.). Cambridge University Press, Cambridge.

Davies, N.B. & Halliday, T.M. (1978). Deep Croaks and Fighting Assessment in Toads Bufo Bufo. Nature, Lond., 274: 683–685.

Dawkins, R. & Krebs, J.R. (1978). Animal Signals: Information or Manipulation: In: Behavioural Ecology: An evolutionary approach (eds. Krebs, J.R. & Davies, N.B.) Blackw 11 Scientific Publications, Oxford.

Dingle, H. (1969). A Statistical and Information Analysis of Aggressive Communication in th Mantis Shrimp Gonodactylus bredini Manning. Anim. Behav., 17: 561–575.

Enquist, M. (1985). Communication During Aggressive Interaction with Particular Reference to Variation in Choice of Behavior. Anim. Behav., 33: 1152–1161.

Fugle, G.N., Rothstein, S.I., Osenberg, C.W. & McFinley, M.A. (1984). Signals of Status in Wintering White-Crowned Sparrows, Zonotrichia leucophrys gambelii. Anim. Behav., 32: 86–93.

Gardner, R. & Morris, M.R. (1989). The Evolution of Bluffing in Animal Contests: An ESS Approach. J. Theor. Biol., 137: 235–243.

194

Hofbauer, J. & Sigmund, K. (1988). The Theory of Evolution and Dynamical Systems. Cambridge: Cambridge University Press.

Ketterson, E.D. (1979). Status Signaling in Dark-Eyed Juncos. Auk, 96: 94-99.

Krebs, J.R. & Dawkins, R. (1984). Animal Signals: Mind-reading and Manipulation. In: Behavioural Ecology 2nd Edition (ed. by Krebs, J.R. & Davies, N.B.), pp. 380-402.

Maynard Smith, J. (1982). Evolution and the Theory of Games. Cambridge University Press, Cambridge.

Maynard Smith, J. & Parker, G.A. (1976). The Logic of Asymmetric Contests. Anim. Behav., 24: 159-175.

Parker, G.A. (1974). Assessment Strategy and the Evolution of Fighting Behavior. J. Theor. Biol. 47: 223-243.

Parker, G.A. & Rubenstein, D.I. (1981). Role Assessment, Reserve Strategy, and Acquisition of Information in Asymmetric Animal Conflicts. Anim. Behav., 29: 221-240.

Piersma, T. & Veen, J. (1988). An Analysis of the Communication Function of Attack Calls in Little Gulls. Anim. Behav., 36: 773-779.

Poole, J.H. (1989). Announcing Intent: The Aggressive State of Musth in African Elephants. Anim. Behav., 37: 140-152.

Rohwer, S.A. & Rohwer, F.C. (1978). Status Signalling in Harris' Sparrows: Experimental D ceptions Achieved. Anim. Behav., 26: 1012-1022.

Selten, R. (1980). A Note on Evolutionarily Stable Strategies in Asymmetrical Animal Conflicts. J. Theor. Biol., 84: 93-101.

Selten, R. (1983). Evolutionary Stability in Extensive Two-Person Games. Math. Soc. Sci., 5: 269-363.

Selten, R. (1988). Evolutionary Stability in Extensive Two-Person Games--Correction and Further Development. Math. Soc. Sci., 16: 223-266.

Sinclair, A.R.E. (1977). The African Buffalo. A Study of Resource Limitation of Populations. University of Chicago Press, Chicago.

Steger, R. & Caldwell, R.L. (1983). Intraspecific Deception by Bluffing: A Def ns Strategy of Newly Molted Stomatopods (Arthropoda: Crustacea). Science, 221:558-560.

Stokes, A.W. (1962). Agonistic Behaviour Among Blue Tits at a Winter Feeding Station. B haviour, 19: 118-138.

Studd, M.V. & Robertson, R.J. (1985). Evidence for Reliable Badges of Status in Territorial Yellow Warblers (Dendroica petchia). Anim. Behav., 33: 1102-1113.

Taylor, P. and Jonker, L. (1978). Evolutionarily Stable Strategies and Game Dynamics. Math. Biosciences, 40: 145-156.

Zahair, A. (1979). Reliability in Communication Systems and the Evolution of Altruism: In: Evolutionary Ecology (ed. B. Stonehouse & C.M. Perrins) pp. 253-259. Macmillan, London.

Zahair, A. (1979). Ritvalizations and the Evolution of Movement Signals. Behaviour 72, 77-81.

POLLINATOR FORAGING AND FLOWER COMPETITON IN A GAME EQUILIBRIUM MODEL

by

Reinhard Selten and Avi Shmida

1. Introduction

Why do flowers produce nectar? A simple answer suggests itself. It is the purpose of
nectar to attract pollinators. However, this is not a convincing explanation of the
phenomenon. Pollinators are attracted to flowers which provide the right visual and
olfactory stimuli. Once a pollinator visits a flower, pollination services are
supplied whether nectar is found or not. The magnificent shapes and colors of flowers
are advertising efforts, but it is not clear why the product should be delivered
after payment has been obtained by successful advertising. In fact, many species of
orchids heavily invest in advertising without offering any resource to pollinators.

Pollinators who visit flowers may be compared to customers who go to shops.
However, the analogy is not a close one. The customer pays by his contribution to the
plant's reproductive success, but the shop is not forced to offer anything in
exchange. The shopkeeper smears the customer with pollen in the hope that it will be
carried to other shops of the same kind to be dropped there. Ordinarily a retailer
would not like to induce his customers to do business with his competitors but the
pollination market is different.

It is the aim of this paper to examine the impact of optimal pollinator foraging
on flower competition for effective transfer of pollen to other members of the same
species. A mathematical model will be presented which is sufficiently simple to be
analytically tractable. Of course, a simple abstract model cannot capture all the
intricacies of pollinator-plant interaction reported in the literature (Proctor and
Yeo 1973, Vogel 1983, Real 1983, Barth 1984). One cannot hope to achieve more than a
logically stringent presentation of a possible explanation for the phenomenon of
resources offered to pollinators by flowers. Analytical tractability requires extreme
simplifications in the description of reality. Much more complexs models and computer
simulations rather than analytical techniques would be needed for applications to
specific ecological systems.

Remarks on the Empirical Background

Plants produce a variety of resources offered to pollinators, mainly nectar and pollen. In some cases also other kinds of resources are offered, e.g. sleeping places, fragrants used by male bees to attract females and special nutritive tissue (Faergi and Van der Pijl 1966, Dafni 1984). In this paper, we shall leave the resource offered to the pollinators unspecified. However, our model assumes that the resource is not replenishable; therefore it seems to be more appropriate to think of pollen rather than nectar as the resource.

Many plants are hermaphrodite in the sense that male and female sex organs are found in every flower. Our model deals with hermaphrodite insect pollinated plants. Figure 1 illustrates the process of pollination for such plants.

We are interested in the question of why resources are offered to pollinators. It is by no means obvious that this should be the case. It is known that nectar is costly in terms of energy consumption (Vogel 1983, Pleasants and Chaplin 1983). A plant which invests in resources offered to pollinators has less energy to allocate to growth and reproduction (Charnov and Bull 1986). If pollination services are supplied whether the resource is offered or not, it is advantageous to avoid these energy costs. In fact, non-rewarding insect pollinated flowers exist in plants, but they are rare. They are found especially in mimetic plants (Dafni 1984) or deceptive species (Boyden 1982) or in plants who rely on "pollination by chance" (Cochran 1986). In the literature there are at least two examples of nectarless populations of usually rewarding species (Feinsinger 1978, Brown and Kodric-Brown 1979). In the autumn population of Colchicum in Israel we also observe non-rewarding individuals (Shmida, unpublished data).

Many species of orchids invest in advertising without offering any resource to pollinators (Dafni 1984). The goal of this paper is not to explain how non-rewarding plants can coexist by mimicry or pollination by chance (Eastabrock 1974, Heinrich 1983, Bell 1986, Cochran 1986), but rather to ask another question: How can a population of rewarding plants be in stable equilibrium? What prevents a less rewarding mutant to invade and to take over? This important question was not new to some of the eminent scientists in pollination biology. Heinrich (1983, p. 212) wrote: "In a resource-limited wild population of plants having flowers with little nectar should be an advantage, provided they get pollinated because their neighbors produce nectar."

Modelling Considerations

Empirical studies show that pollinators tend to search for the next flower in the neighborhood of the flower visited last, as long as the yield is high enough. A low yield induces them to search further away (Waddington 1983). This is a reasonable search strategy if yield levels are the result of a stochastic process with a

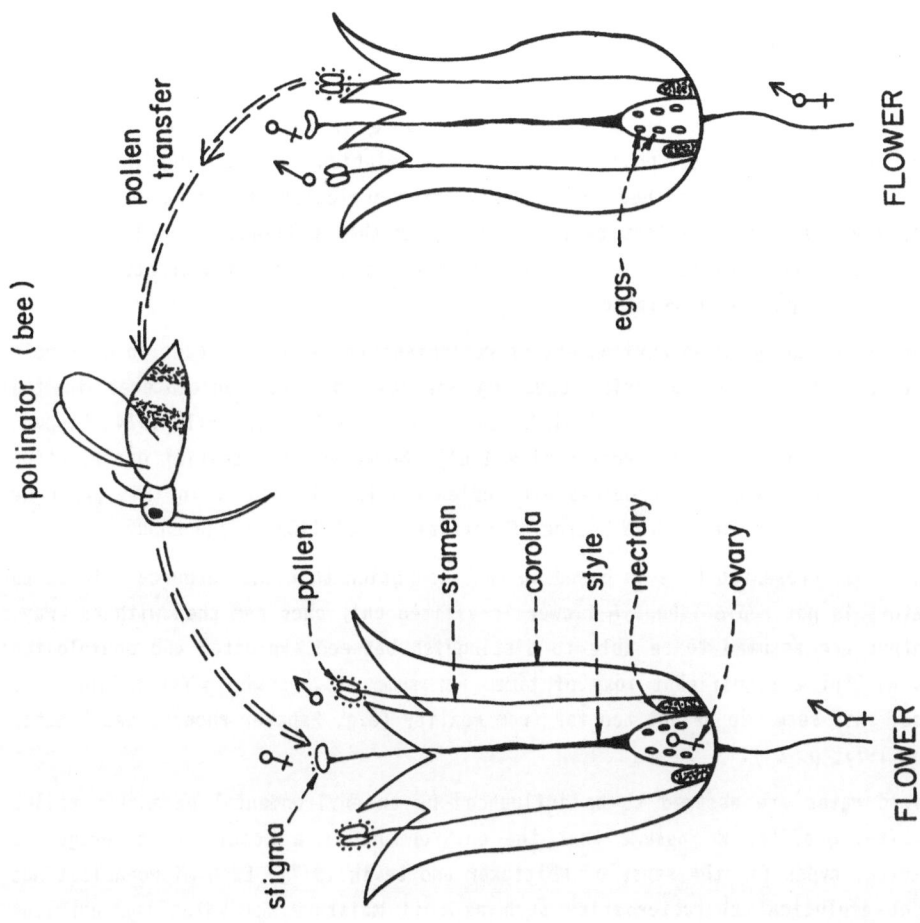

Figure 1: Pollen transfer in insect pollinated hermaphroditic plants

positive *spatial correlation* For the sake of simplicity, our model reduces the pollinator's decision on interflower flight distance to a choice between "near search" and "far search".

Different assumptions can be made on the source of spatial correlation. Maybe the most obvious cause of positive spatial correlation is the foraging process itself. If nectar is replenished continuously after depletion, yield rates depend on the time elapsed since the last exploitation by another pollinator. Since the last pollinator is likely to have visited several flowers in the neighborhood, one can expect positive spatial correlation.

Optimal foraging in an environment of replenishable resources leads to difficult mathematical problems of Bayesian updating similar to those encountered in the economic literature on price search (Kohn and Shavell 1974, Rothschild 1974, Lippman and McCall 1979, Harstad and Postlewaite 1981). Moreover, the spatial distribution of yields is endogenous if resources are replenishable. Therefore, in this paper we look at a simpler case where the source of spatial correlation is exogenous.

The model presented here is based on the assumption that the resource offered to pollinators is not replenished. A flower is visited only once and then withers away. Pollinators are assumed to be able to distinguish between exploited and unexploited flowers without a significant loss of time. For some flowers who offer pollen as a resource, this seems to be not too far from reality (e.g. Papaver rhoeas, see Proctor and Yeo, 1973, p. 54).

Yield rates are assumed to be influenced by an environmental parameter called *microhabitat quality*. We assume that the environment is a mosaic of heterogenous microhabitat types (in the sense of Whittaker and Levin 1977). Each microhabitat has different ecological characteristics such as soil moisture, pH-value and nutrient composition which jointly determine microhabitat quality.

The higher the microhabitat quality is, the more favorable are the conditions for the energetic budget of the plant. In the model presented here, the percentage of the energy budget devoted to resources offered to pollinators depends on an investment decision which is thought of as fixed genetically before a response to environmental conditions is possible. The amount of resources offered to pollinators is proportional to the product of the investment level and microhabitat quality.

Microhabitat qualities of neighboring plant sites are positively correlated since they are likely to belong to the same microhabitat. This leads to a positive spatial correlation of yield levels. Therefore, the yield level observed by a pollinator supplies information about the yield levels of neighboring plants.

In the model presented here the search problem of a pollinator is reduced to a decision between "near search" and "far search". In near search, a neighboring flower is visited next whereas in far search the flight distance is sufficiently great to make the effects of spatial correlation negligible. It will be shown that the optimal search policy of a pollinator is characterized by a critical yield level. The decision to search near or far depends only on the last flower visit. The pollinator chooses near search above the critical level and far search below the critical level. Presumably the optimal critical yield level is learnt rather than inherited (Bertsch 1987). The learning process could be similar to the process described by Houston and McNamara (1980). Learning is assumed to be sufficiently quick to justify the neglect of initial deviations from optimality.

The search behavior of the pollinators is the background for flower competition. For the sake of simplicity it is assumed that each plant has only one exploitable flower at any given point of time. Multiple exploitable flowers would greatly increase the complexity of the analysis, even if an extension of the model in this direction seems to be feasible.

In the model presented here, the yield level is the product of microhabitat quality and the level of investment in resources offered to pollinators. The investment is thought of as fixed genetically before a phenotypic response to microhabitat conditions is possible.

The model describes a hypothetical ecological system with two species of flowers A and B and one species of a pollinator who visits both species of flowers and no others. Apart from the fact that crossbreeding is excluded, the two species of flowers are assumed to be equal in all respects. The *production technologies* are equal in the sense that the relationship between investment level and microhabitat parameter on the one hand and yield level on the other hand is the same one in both species.

Both species of flowers have the same overall frequency but species abundance shows positive spatial correlation. Therefore, in near search it is highly probable that the next flower will be of the same species as the last one. The model assumes that the effect of spatial correlation is negligible in far search. There is an even chance for the next flower being of each of both species in far search.

The strategic variable of a plant is the investment level. This variable does not influence female fitness. Once a flower is visited, it is fertilized. In view of the absence of strategic influences on female fitness, the model can concentrate on male fitness.

Male fitness is related to the amount of own pollen transferred to other members of the same species. This variable is named *pollen flow*. The level of investment in resources offered to pollinators is connected with fitness costs. It is assumed that total fitness depends linearly on male fitness and the investment level. Female fitness enters the linear function as a constant which need not be considered explicitly.

It is advantageous for a flower if the next flower visited is of the same species. If this happens, more pollen is effectively transported to other flowers of the same species than in the opposite event. Therefore, plants have an interest to induce pollinators to engage in near search rather than in far search.

A high yield level ordinarily indicates a high microhabitat quality. Therefore, it is advantageous for the pollinator to engage in near search after a high yield level at the last visited flower. The yield level depends on the investment level and microhabitat quality. A higher investment level induces near search for a wider range of microhabitat qualities and thereby increases the expected pollen flow. This is the advantage of producing resources offered to pollinators. Our model shows how under appropriate conditions this advantage stabilizes a positive equilibrium investment level.

The model presented here explains resources offered to pollinators as an adaptation to the optimal foraging of pollinators. The resources are not produced to attract pollinators but to convey favorable information on environmental conditions to pollinators. Favorable information induces pollinators to engage in near search rather than far search. Near search offers a greater chance than far search for the next flower being of the same species. Therefore near search results in a higher effective pollen transfer to flowers of the same species than far search does. This is the reason why it is advantageous for an individual plant to offer resources to pollinators.

Summary of the Model Assumptions

In the following we loosely describe the basic assumptions of our model, most of which have already been discussed above.

1. At each point of time each plant has at most one flower.

2. The resource offered to pollinators is non-replenishable, this means that each flower is exploited only once.

3. The number of pollinators per exploitable flower is constant over time.

4. Pollinators can distinguish exploited and unexploited flowers from outside, but the amount of resources offered by an unexploited flower is known only after a visit to the flower.

5. After a visit to a flower a pollinator has to decide between *near search* and *far search*. In the case of near search, the pollintor moves to an unexploited flower nearby whereas in the case of far search the pollinator flies to a new neighborhood much further away from the last visited flower.

6. The amount of resources offered to pollinators depends multiplicatively on the *investment level* in resources offered to pollinators and on *microhabitat quality*.

The investment level is the strategic variable of the plant. It is genetically determined before microhabitat quality exerts its influence.

7. Microhabitat quality (favorability in the sense of Terborgh 1973) is subject to spatial random variation. It is described by a parameter x. The model is based on the idea of a patchwork of randomly distributed microhabitats of random extension covering the space where pollinator - flower interactions take place. A pollinator engaged in near search stays in the same microhabitat with probability p from one flower visit to the next and moves to a different microhabitat with probability q = 1-p. Far search always leads to a new microhabitat. For the sake of simplicity it is assumed that there is no correlation between the environmental parameters of neighboring microhabitat patches.

8. There are two flower types (different species) in the model. They are equal in all morphological aspects, such as color and shape. There is only one type of pollinator. The pollinators do not distinguish between both types of flowers.

9. Both types of flowers have the same overall density, but there is a positive spatial correlation of local densities within each species. Therefore, in near search the probability α of meeting a flower of the same type as the last visited is greater than 1/2. After far search the probability of meeting a flower of the same type as the last visited one is exactly 1/2.

10. At each visit of a flower, a pollinator drops a fraction τ of the pollen in its *transferable pollen load* (some pollen may be not transferable since it is safely closed in scopa). The pollen lost is replaced by pollen of the flower visited.

11. Male fitness is proportional to pollen flow divided by average pollen flow in the population. Pollen flow is defined as the amount of pollen of a flower transferred to other individuals of the same species. The number of flowers visited in one foraging trip is assumed to be sufficiently large to justify the neglect of trips back to the nest in the computation of the pollen flow.

12. Total fitness depends linearly on male fitness and the level of investment in resources offered to pollinators.

These assumptions will be discussed in more detail in the chapters were they are first used.

Aim of the Analysis

It is the purpose of the analysis to explore the conditions of existence and the structural properties of a special kind of game equilibrium of the model. In this equilibrium the pollinator's decision between near search and far search optimizes the amount of resource collected per time. All flowers choose the same investment level which maximizes fitness for each plant individually, given the optimal behaviour of pollinators and the investment decisions of the other plants.

The model does not exclude the possibility of equilibria which are polymorphic with respect to investment levels, but this possibility will not be explored here. Optimal pollinator foraging may be viewed as the result of a learning process, but the equilibrium examined here neglects initial non-optimal behavior. This may be justified if learning is sufficiently quick. The structure of pollinator behavior is not specified in advance, but derived as a consequence of optimal foraging.

The main emphasis of the analysis will be on structural properties and conditions of existence. In addition to this, the influence of parameter changes will be explored.

Structure of the Paper

Section 2 describes the decision problem of a pollinator. After a heuristic discussion of optimal foraging in Section 3, formal results on the structure of optimal foraging are proved in Section 4. Parameter influences on the critical level are discussed in Section 5. The investigation of plant competition begins with Section 6. The transfer of pollen to other flowers is investigated there. An equation for the "pollen flow", the amount of pollen transferred to flowers of the same species is derived on the basis of previous results on pollinator foraging. Section 7 connects pollen flow and fitness and defines the investment game among the flowers. The notion of a strong evolutionarily stable strategy is explained (a strategy of this kind can be described as a pure evolutionarily stable strategy without alternative best replies). Necessary conditions for strong evolutionary stability are derived in Section 8 and sufficient conditions are given in Section 9. Influences of parameters are examined in Section 10. Section 11 is devoted to the discussion of the results of this paper.

2. The Decision Problem of a Pollinator

2.1 Introduction

Pollinators like female solitary bees collect nectar and pollen to build up a stock of nutrition for their offspring (Linsley 1958, Eickwort and Ginsberg 1980). This suggests the idea that the expected number of offspring is proportional to the size of the stock piled up during the pollinator's lifetime. Therefore, it is assumed in the model presented here that pollinators aim at the maximization of the amount of resource collected per time unit (Bertsch 1987, Schmid-Hempel et al. 1985, Kacelnik et al. 1987).

Pollinators interrupt foraging for trips back to their nest as soon as the load of collected food becomes to heavy. A heavy load increases time and energy requirements for foraging. This leads to an optimization problem which has been investigated in the literature (Kacelnik et al. 1986). The model presented here neglects trips back to the nest and the influence of load weight on foraging behavior. It is assumed that the number of flowers visited in one trip is sufficiently great to justify the description of foraging by an infinite stochastic control process as a tolerable mathematical approximation.

The problem of optimal foraging will be examined under the assumption that the investment in the production of resources offered to pollinators is the same for all flowers of both species. In this paper, attention is restricted to equilibrium situations of this kind. (One cannot exclude the possibility of different types of equilibrium situations.)

2.2 The Model

Consider an individual flower. Let x be the microhabitat quality at the location of the flower and let c be the investment of the plant in resources offered to pollinators. c is measured in terms of fitness. We assume that the yield level y is proportional to the microhabitat quality x and to the investment level c. Therefore, we have:

$$y = acx \tag{1}$$

where a is a positive constant which is the same for all flowers. In equilibrium, c is equal to an *equilibrium investment level* $c*$ which is the same for all flowers. Therefore, in equilibrium the yield level of a flower is proportional to the microhabitat quality at its location

$$y = ac*x. \tag{2}$$

The microhabitat quality x is a non-negative continuous variable, whose probability density at a randomly selected site is $f(x)$. The corresponding cumulative distribution is denoted by $F(x)$:

$$F(x) = \int_0^x f(s)ds. \tag{3}$$

The support of a density function is the closure of the set of all points at which it is positive. It is assumed that the support of f is a closed interval $[x',x'']$. In view of (2), in equilibrium the cumulative distribution F on x induces a cumulative distribution H on y:

$$H(y) = H(ac*x) = F(x) = F(\frac{y}{ac*}). \tag{4}$$

204

The corresponding density h(y) of y is as follows:

$$h(y) = \frac{1}{ac*} \cdot f(\frac{y}{ac*}).$$ (5)

Positive spatial correlation of the environmental parameter x is modelled in the simplest possible way. Consider a path of a pollinator engaged in *near search*. It is assumed that from one flower to the next the microhabitat quality remains unchanged with probability p; with probability q = 1-p the next microhabitat type is chosen randomly according to the distribution f.

Far search is assumed to move the pollinator sufficiently far away to make the influence of spatial correlation negligible. Immediately after far search the microhabitat x is chosen randomly according to the distribution f.

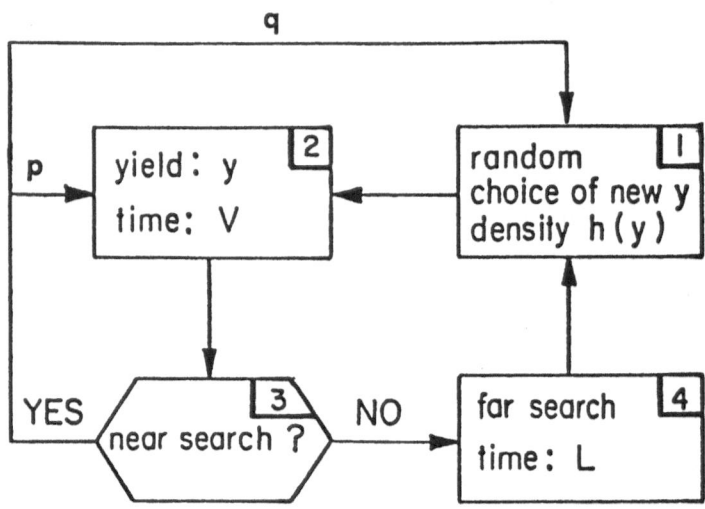

Figure 2: The decision problem of the pollinator

Explanation of the symbols:

y   - yield level
h(y)   - probability density of yield level
V   - time for visiting and handling one flower, including local travel in near search
L   - travel time for far search
p   - probability of unchanged microhabitat conditions in near search
q=1-p - probability of a change of microhabitat conditions

The parameters V, L, p and q are assumed to be positive. The support of h is an interval [y',y"] with 0 ≤ y' < y".

The decision problem of a pollinator is described by the flow diagram of Figure 2. After each visit to a flower, the pollinator has to decide between near search and far search. In Figure 2, the point of decision is represented by rhomboid 3. If the pollinator chooses YES, which means near search, then rectangle 2 is reached with probability p and rectangle 1 is reached with probability q = 1-p. Rectangle 2 represents the visit of a flower. A visit takes the time V, which includes handling time and time for local travel. If rectangle 2 is reached directly from rhomboid 3, then the yield y is the same as that of the last visited flower. If rectangle 1 is reached, a new yield level y is randomly chosen according to the distribution h. From rectangle 1 the process moves to rectangle 2, where the newly determined yield y is collected and the time V is spent.

If the pollinator chooses NO, which means far search, then the process moves to rectangle 4, which represents travel to a new area. This travel requires the time L. From there rectangle 1 is reached, where a new yield y is chosen randomly according to h.

It is assumed that the pollinator has no way to determine the yield level before the visit of a flower, but during exploitation the yield level is exactly measured. The pollinator can make his decision at rhomboid 3 dependent on the yield level observed at the last visited flower.

Figure 2 concentrates attention on the situation of one individual pollinator. However, many pollinators are assumed to be present; all of them belong to the same species and face the same decision problem. The competition among the pollinators need not be modelled explicitly, since we look at a stationary state where the number of pollinators per unexploited flower is constant over time. New flowers open at the same rate as unexploited ones are exploited. The intensity of competition among the pollinators is reflected by the parameter V which includes the time of travel from one unexploited flower to the next. An increase of the number of pollinators per flower would decrease the density of unexploited flowers and thereby increase the distance from one unexploited flower to the next. The time for local travel included in V is proportional to this distance. The model presented here treats V as an exogenously fixed parameter and thereby avoids an explicit theory about the mechanism which determines the number of pollinators per unexploited flower. As far as the optimization problems of an individual pollinator is concerned, V can be considered as fixed. The same is true for the optimization problem of a single flower. Therefore, the results of our analysis can be expected to remain valid for an extended theory which endogenizes the number of pollinators per flower.

In principle, the search behavior of a pollinator could depend on all the yield levels observed in the past (Houston et al. 1982, Kacelnick and Cuthill 1987, Staddon 1987). Presumably, this would be the case in an explicit theory of learning (Staddon 1987, Kacelnick et al. 1987). However, optimal behavior of the foraging bee does not need more information than that contained in the yield level of the last visited flower. The distribution of future yields reached by near search depends only on the yield of the last visited flower. The distribution of future yields reached by far search is completely independent of past observations. Therefore, the optimal decision at rhomboid 3 depends only on the yield level of the last visited flower. The words *Markoff-property* will be used for this feature of the decision problem described by Figure 2.

The Markoff-property is due to the extremely simple assumptions on positive spatial correlation embodied in the model presented here. Without the Markoff-property it would be much more difficult to characterize *optimal search policies*. The decision problem of Figure 2 has an additional feature which facilitates analysis. Regardless of whether near search or far search is chosen, the process will eventually reach rectangle 1. Therefore, it is natural to think of rectangle 1 as the starting point of a "cycle" of the process which ends as soon as rectangle 1 is reached again.

In view of the Markoff-property, attention will be restricted to pollinator search policies which depend only on the yield y of the last visited flower. Formally, a search policy is described by the set S of all $y \geq 0$ where the policy chooses near search if y has been observed as the yield of the last visited flower. We refer to S as the *near search set* of the policy, but we also speak of policy S where we actually mean the policy with the near search set S. The complement of S in the set of all non-negative real numbers is denoted by $\bar{S}$ and is called the *far search set*. The policy chooses far search if the yield of the last visited flower is in the far search set.

2.3 Admissible Search Policies

Attention will be restricted to search policies with the following properties:

(a) The near search set is closed and the far search set is open (relative to the set of all non-negative numbers).

(b) Each of both sets S and $\bar{S}$ is a union of finitely many intervals of positive length or empty.

(c) Let $[y',y'']$ be the support of h. An $\varepsilon > 0$ exists such that the following is true: either $[0,y'+\varepsilon] \subseteq S$ or $[0,y'+\varepsilon) \subseteq \bar{S}$, similarly either $[y''-\varepsilon,\infty) \subseteq S$ or $[y''-\varepsilon,\infty) \subseteq \bar{S}$.

Search policies with the properties (a), (b), and (c) are called *admissible*. Property (a) is an arbitrary convention which removes cumbersome formal distinctions between essentially equal policies. Property (b) avoids uninteresting mathematical technicalities. Property (c) is the only one which needs some comment in view of its substantial significance.

If the problem of optimal foraging were the only concern of this paper, it would not be necessary to worry about decisions after observations of yield levels outside the support of h. However, later it will be necessary to look at the fitness of rare mutant flowers. A deviation from equilibrium investment in resources offered to pollinators results in yield levels outside the support of h with positive probability.

Property (c) is based on the idea that a pollinator who meets an unusual yield level outside the range of yield levels encountered before will react to it in the same way as to the nearest yield level covered by previous experience. This *extrapolation principle* seems to be a reasonable assumption on the underlying learning process. Obviously, optimality as the result of learning cannot be assumed outside the range of previous experience.

It is necessary to specify an exact optimality criterion for the selection among *admissible search policies*. The easiest way to formalize the concept of resources collected per time unit is based on the expected yield and the expected time per cycle. For every admissible search policy S let Y_S be the total expected yield and let T_S be the expected time for a cycle starting at rectangle 1 and lasting until return to rectangle 1. The *yield rate*

$$u_S = \frac{Y_S}{T_S} \tag{6}$$

is the optimality criterion applied to pollinator search in this paper.

It is not difficult to determine Y_S and T_S for a given search policy S. Let $Y(y)$ and $T(y)$ be the expected yield and the expected time for the cycle under the condition that the yield level y is the realization of the random choice at the beginning of the cycle. Consider the case $y \in S$. With the help of Figure 2 it can be seen that we have:

$$Y(y) = y + pY(y) \tag{7}$$

$$T(y) = V + pT(y) \tag{8}$$

for $y \in S$. Obviously, for $y \notin S$ we have $Y(y) = y$ and $T(y) = V+L$. This, together with (7) and (8), yields:

$$Y(y) = \begin{cases} \frac{y}{q} & \text{for } y \in S \\ y & \text{for } y \notin S \end{cases} \tag{9}$$

and

$$T(y) = \begin{cases} \dfrac{V}{q} & \text{for } y \in S \\ V+L & \text{for } y \notin S \end{cases}. \tag{10}$$

With the help of (9) and (10) one obtains:

$$Y_S = \frac{1}{q} \int_S yh(y)dy + \int_S yh(y)dy \tag{11}$$

$$T_S = \frac{V}{q} \int_S h(y)dy + (V+L) \int_S h(y)dy. \tag{12}$$

As before, let $[y',y'']$ be the support of h. In view of the boundedness of the support it is clear that the mean $\bar{y}$ of y with respect to h exists:

$$\bar{y} = \int_{y'}^{y''} yh(y)dy. \tag{13}$$

Obviously, we have

$$\frac{V}{q} - V = \frac{p}{q} V. \tag{14}$$

With the help of (13) and (14), Eqs. (11) and (12) can be rewritten as follows:

$$Y_S = \bar{y} + \frac{p}{q} \int_S yh(y)dy \tag{15}$$

$$T_S = V + L + (\frac{p}{q} V - L) \int_S h(y)dy. \tag{16}$$

In view of (6), (15) and (16) we obtain:

$$u_S = \frac{\bar{y} + \frac{p}{q} \int_S yh(y)dy}{V+L+(\frac{p}{q} V-L) \int_S h(y)dy}. \tag{17}$$

If we speak of an optimal policy, we mean an admissible search policy which maximizes the right hand side of (17) over the set of all admissible search policies. It will be shown in Section 4 that there is a uniquely determined optimal policy.

3. Heuristic Discussion of Optimal Pollinator Foraging

A heuristic discussion sometimes reveals insights which may easily remain hidden behind formal proofs. The analysis of the decision problem of Figure 2 will yield an optimality criterion which compares two estimates of total yields obtained by *near search* and *far search* during equal periods of time. The heuristic derivation of this criterion throws light on the structure of optimal pollinator foraging in the environment described by Figure 2.

The criterion leads to the conclusion that near search is indicated above a critical level for the yield of the last visited flower and far search below this critical level. An equation for the critical level can be derived from the optimality criterion. Exact proofs for the conclusions reached heuristically in this section will be given in the next section.

Consider a pollinator who has to decide betwen near search and far search on the basis of a specific yield level y observed at the last visited flower. In order to see which of both decisions is preferable from the point of view of the pollinator, we look at the total yield obtained by a period of near search starting at rhomboid 3 and lasting until rectangle 1 is reached. Unlike in the previous section, the starting point is now rhomboid 3 and not rectangle 1, but otherwise the situation is similar. The total yield for near search, starting at rhomboid 3 and lasting until rectangle 1 is reached, is as follows

$$Y_1(y) = py + p^2y + p^3y + \ldots = \frac{p}{q} y. \tag{18}$$

In order to obtain this total yield, the pollinator has to spend the total time T_1:

$$T_1 = pV + p^2V + p^3V + \ldots = \frac{p}{q} V. \tag{19}$$

What is the expected total yield obtained in the same time T_1 if far search is chosen? We shall look at this question under the assumption:

$$T_1 > L. \tag{20}$$

If L is larger than T_1, then far search never can be optimal, since rectangle 1 is reached more quickly by near search, which offers the additional advantange that resources can be collected. Assume that (20) holds and consider the total yield obtained by far search in the time T_1. No yield is obtained during the travel time L, but in the remaining time T_1-L resources can be collected. Suppose that u_S is the yield rate of an optimal policy S. It is reasonable to assume that in the time T_1-L resources can be collected at the rate u_S. This yields the following *heuristic estimate* Y_0 for the total expected yield obtainable by far search in the time T_1:

$$Y_0 = (T_1 - L)u_S. \qquad (21)$$

The comparison of $Y_1(y)$ and Y_0 suggests an *optimality criterion: optimal behavior requires near search for $Y_1(y) > Y_0$ and far search for $Y_1(y) < Y_0$.*

A search policy is called *distinguishing* if each of both sets S and $\check{S}$ contains an interval of positive length within the support $[y',y'']$ of h. Other search policies are *non-distinguishing*. There are only two admissible non-distinguishing policies, the policy $S = [0,\infty)$ which chooses near search everywhere and the policy $S = \emptyset$ which chooses near search nowhere.

An admissible search policy which chooses near search for $Y_1(y) > Y_0$ must choose near search in the border case $Y_1(y) = Y_0$ too. This is a consequence of the convention that S is closed, which is part of the admissibility definition.

Let S be a distinguishing admissible search policy which satisfies the heuristically derived optimality criterion. Obviously, S is connected to a *critical level* y^* above which near search is chosen and below which far search is chosen. In view of (18), (19) and (21), the critical level is as follows:

$$y^* = (V - \frac{q}{p} L)u_S , \qquad (22)$$

The policy S has the form $S = [y^*,\infty)$ where y^* satisfies (22). This means that the following equation holds:

$$S = \{y | y \geq (V - \frac{p}{q} L)u_S\}, \qquad (23)$$

A search policy of the form $[z,\infty)$ will be called a *critical yield policy*. It follows by condition (c) in the definition of admissibility that a critical yield policy $[z,\infty)$ is admissible if and only if we either have $z = 0$ or $y' < z < y''$, where $[y',y'']$ is the support of h. Since by assumption $S = [y^*,\infty)$ is distinguishing and admissible, we must have

$$y' < y^* < y''. \qquad (24)$$

Of course, optimal search may require a non-distinguishing policy. Therefore, we want to explore the question under which conditions an y^* can be found which satisfies (22) and (24) with $S = [y^*,\infty)$. For this purpose, we evaluate u_S for $S = [y^*,\infty)$ with the help of (17):

$$u_S = \frac{y + \frac{p}{q} \int_{y^*}^{y''} yh(y)dy}{\frac{V}{q} - (\frac{p}{q} V - L)H(y^*)}. \qquad (25)$$

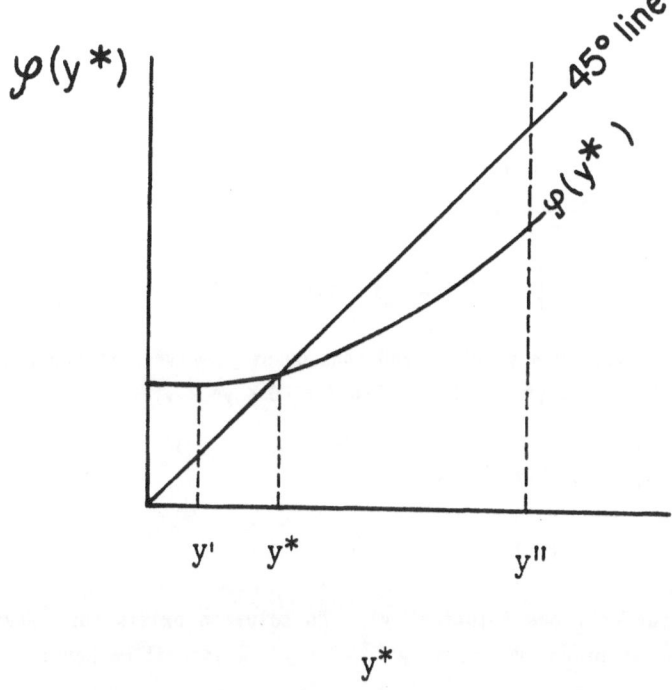

Figure 3: Determination of the critical yield level.

The right-hand side of (25) can be inserted for u_S in (22). In this way, one obtains an equation for the critical yield level $y*$:

$$y* = \frac{\frac{q}{p}\bar{y} + \int_{y*}^{y''} yh(y)dy}{\frac{V}{pV-qL} - H(y*)} .$$

(26)

Multiplication by the denominator, division by $V/(pV-qL)$, and a rearrangement of terms yield an equivalent transformation of (26):

$$y* = p(1 - \frac{qL}{pV}) \left[\frac{q}{p}\bar{y} + y*H(y*) + \int_{y*}^{y''} yh(y)dy \right].$$

(27)

In order to examine the conditions under which Eq. (27) has a solution $y*$, we introduce the notation $\varphi(y*)$ for the right-hand side of (27) as a function of $y*$:

$$\varphi(y^*) = p(1 - \tfrac{qL}{pV})\left[\tfrac{q}{p}\,\bar{y} + y^*H(y^*) + \int_{y^*}^{y''} yh(y)dy\,\right]. \tag{28}$$

In the diagram of Figure 3, a solution of (27) corresponds to an intersection of $\varphi(y^*)$ with the $45°$-line. In order to examine the shape of $\varphi(y^*)$, we look at the derivative $\varphi'(y^*)$ with respect to y^*:

$$\varphi'(y^*) = p\left[1 - \tfrac{qL}{pV}\right]H(y^*) \quad \text{for} \quad 0 \le y^* \le y''. \tag{29}$$

Since $H(y^*)$ vanishes for $0 < y^* < y'$ and consequently $\varphi(y^*)$ is constant there, (29) holds not only for $y' < y^* < y''$ but also for $0 \le y^* < y'$.

Obviously, $\varphi(y^*)$ is equal to $\varphi(y')$ for $0 < y^* < y'$ and to $\varphi(y'')$ for $y^* > y''$. It follows by (29) that we have:

$$\varphi'(y^*) < 1 \quad \text{for} \quad 0 \le y^* < y''. \tag{30}$$

Therefore, (27) has at most one solution y^*. No solution exists for $\varphi(y') < y'$. A uniquely determined solution y^* with $y' < y^* < y''$ exists if we have:

$$\varphi(y') > y' \tag{31}$$

and

$$\varphi(y'') < y''. \tag{32}$$

These inequalities are equivalent to the following ones:

$$\left[1 - \tfrac{qL}{pV}\right]\bar{y} > y' \tag{33}$$

$$\left[1 - \tfrac{qL}{pV}\right](q\bar{y} + py'') < y''. \tag{34}$$

It follows by the continuity of h that $\bar{y}$ is smaller than y''. Therefore, $q\bar{y} + py''$ is smaller than y''. This has the consequence that (34) is always satisfied under our assumptions on the parameters and the distribution h. However, (33) imposes restrictions on the parameters and the distribution h. As we shall see in the next section, far search does not pay, if (33) fails to be satisfied.

The conclusion that the optimal policy is of the form (23) has been drawn with the help of the heuristic estimate Y_0 and therefore cannot claim to be exact. However, the reasoning based on (22) and (23), which has led to (27), is correct. The results obtained in this connection will be used in the next section.

4. Derivation of Results on Optimal Foraging

In Section 2.3, an admissible search policy has been defined by three conditions (a), (b), (c). An *optimal yield policy* is an admissible policy which maximizes the yield rate u_S given by (17) over all admissible search policies. In this section, a theorem on the existence and uniqueness of a distinguishing optimal policy and its structural properties will be proved. Auxiliary results will be stated as lemmas.

All results will be based on the same *standard assumptions* on parameters and the distribution h: p,q,V, and L are positive parameters; p+q = 1; the distribution h is continuous; the support of h is an interval [y',y"] with $0 \leq y' < y"$. These standard assumptions are always made without being explicitly mentioned in the assertions of the lemmas and theorems of this paper.

A *critical yield policy* has been defined as a search policy S of the form S = [z,∞). Admissibility requires either z = 0 or y' < z < y". The first lemma will be concerned with all critical yield policies, including the inadmissible ones.

Lemma 1: For every z with z ≥ 0 let u_z be the yield rate u_S of the critical policy [z,∞). If

$$\left[1 - \frac{qL}{pV}\right]\bar{y} \leq y' \tag{35}$$

holds, then u_z assumes its maximum over the interval z ≥ 0 at z = 0. If

$$\left[1 - \frac{qL}{pV}\right]\bar{y} > y' \tag{36}$$

holds, then u_z assumes its maximum over the interval z ≥ 0 at z = y* and nowhere else; here y* is the uniquely determined solution of (27) with y' < y* < y".

Proof: It follows by (17) that we have:

$$u_z = \begin{cases} \dfrac{\bar{y}}{V} & \text{for } 0 \leq z \leq y' \\[2ex] \dfrac{\bar{y} + \frac{p}{q}\displaystyle\int_z^{y''} yh(y)dy}{\frac{V}{q} - (\frac{p}{q}V-L)H(z)} & \text{for } y' \leq z \leq y'' \\[2ex] \dfrac{\bar{y}}{V+L} & \text{for } y'' \leq z . \end{cases} \tag{37}$$

214

The derivative of u_z with respect to z can be computed without difficulty. One obtains:

$$\frac{du_z}{dz} = \frac{p}{q} \cdot \frac{h(z)}{\frac{V}{q} - (\frac{p}{q} V-L)H(z)} \left[(V - \frac{q}{p} L)u_z - z \right]. \tag{38}$$

Equation (38) does not only hold for $y' < z < y''$ but for all $z \geq 0$ since u_z is constant for $0 \leq z \leq y'$ and for $z \geq y''$ and since $h(z)$ vanishes there. (38) yields the following necessary condition for a maximum of u_z at a value z with $y' < z < y''$:

$$z = (V - \frac{q}{p} L)u_z. \tag{39}$$

As we have seen in connection with Eqs. (22) to (34) in Section 3, a uniquely determined solution $z = y*$ of (39) with $y' < y* < y''$ exists if (36) is satisfied, but no such solution exists if (35) holds. In view of (37), we can conclude that in the case of (35) the yield rate u_z assumes its maximum for $z = 0$. This shows that the lemma holds as far as (35) is concerned.

In the following we shall assume (36). It follows from (36) and (37) that we have

$$(V - \frac{q}{p} L)u_{y'} > y'. \tag{40}$$

In the interval $y' < z < y''$ the right-hand sinde of (38) vanishes only at $z = y*$. With the help of (37) and $\bar{y} < y''$ it can be seen that the right-hand side of (38) is negative at $z = y''$. It follows by (40) that du_z/dz is positive for $y' < z < y*$ and negative for $y* < z < y''$. This has the consequence that u_z assumes a maximum at $y* = z$.

Lemma 2: Let R and S be two admissible search policies. Then

$$u_S > u_R \tag{41}$$

holds if and only if we have

$$Y_S - Y_R > (T_S - T_R)u_S. \tag{42}$$

Proof: (41) implies

$$Y_R < \frac{Y_S}{T_S} T_R. \tag{43}$$

This yields:

$$Y_S - Y_R > Y_S - \frac{Y_S}{T_S} T_R = (T_S - T_R)u_S. \tag{44}$$

Lemma 3: If (35) holds, then $S = [0,\infty)$ is an optimal policy and no other admissible search policy is an optimal policy.

Proof: (37) shows that $S = [0,\infty)$ has the yield rate $\bar{y}/V$ and that the the yield rate fo $R = \emptyset$ is lower. It remains to show that every distinguishing admissible policy R has a lower yield rate than $S = [0,\infty)$. Let R be a distinguishing admissible search policy. Define

$$D = [y',y'']\backslash R. \tag{45}$$

Since R is distinguishing, it is a consequence of the definition of admissibility that D contains an interval of positive length. Therefore, we have

$$\int_D h(y)dy > 0. \tag{46}$$

We now apply Lemma 2. Equations (15) and (16) yield:

$$Y_S - Y_R = \frac{p}{q} \int_D yh(y)dy > \frac{p}{q} y' \int_D h(y)dy \tag{47}$$

$$(T_S - T_R)u_S = \frac{p}{q} (1 - \frac{qL}{pV})\bar{y} \int_D h(y)dy. \tag{48}$$

In view of Assumption (35), it follows by (47) and (48) that (42) holds. Lemma 2 yields the conclusion $u_S > u_R$.

Lemma 4: If (36) holds, then $S = [y^*,\infty)$ is an optimal policy and no other admissible policy is an optimal policy; here y^* is the uniquely determined solution of (27) with $y' < y^* < y''$.

Proof: Let R be an admissible search policy which is different from S. It follows by Lemma 1 and (37) that the yield rate u_S of S is greater than y/V. Therefore, u_S is greater than the yield rates for both non-distinguishing admissible search policies $[0,\infty)$ and $\emptyset$. It remains to show that $u_S > u_R$ holds for distinguishing admissible search policies R. Assume that R is distinguishing.

Define:

$$C = [y', y^*) \cap R \tag{49}$$

$$D = [y^*, y'') \setminus R . \tag{50}$$

Equations (15) and (16) yield:

$$Y_S - Y_R = \frac{p}{q} \left[- \int_C yh(y)dy + \int_D yh(y)dy \right] \tag{51}$$

$$T_S - T_R = (\frac{p}{q} V - L) \left[- \int_C h(y)dy + \int_D h(y)dy \right]. \tag{52}$$

Since R is different from S and admissible, at least one of both sets C and D contains an interval of positive length. Moreover, y^* is an upper bound of C and a lower bound of D. Therefore, (51) permits the following conclusion:

$$Y_S - Y_R > \frac{p}{q} y^* \left[- \int_C h(y)dy + \int_D h(y)dy \right]. \tag{53}$$

In view of (22) we have:

$$u_S = \frac{y^*}{V - \frac{q}{p} L}. \tag{54}$$

Equations (52) and (54) show that $(T_S - T_R)u_S$ agrees with the right-hand side of (53). Therefore, (42) holds. $u_S > u_R$ follows by Lemma 2.

Theorem 1: The optimal policy is uniquely determined. If (35) holds, then $[0,\infty)$ is the optimal policy; if (36) holds, then $[y^*,\infty)$ is the optimal policy, where y^* is the uniquely determined solution of (27) with $y' < y^* < y''$.

Proof: The theorem follows by Lemma 3 and Lemma 4.

5. The Critical Level of the Microhabitat Quality

It is the purpose of this paper to examine equilibrium situations, in which all flowers invest the same amount c^* in resources offered to pollinators. The proposed explanation of the phenomenon of resources offered to pollinators requires a distinguishing optimal policy.

Equation (2) describes the relationship between yield level y and microhabitat quality x in equilibrium. Assume that (36) holds and that y^* is the critical yield level determined by (27). The *critical level* x^* of the microhabitat quality

x is defined by the following equation:

$$y* = ac*x*. \tag{55}$$

A theorem will show that the critical level x* does not depend on the parameter a and the equilibrium investment c*.

Theorem 2: (36) holds if and only if

$$(1 - \frac{qL}{pV})\bar{x} > x' \tag{56}$$

is satisfied; here [x',x"] is the support of f and x̄ is the expected value of x with respect to f.   If (56) holds, then the critical level x* of the microhabitat quality is uniquely determined by x' < x* < x" and

$$x* = (1 - \frac{qL}{pV})\left[\bar{x} + p \int_{x'}^{x*} F(x)dx\right]. \tag{57}$$

Proof: In view of ȳ = ac*x̄ and y' = ac*x' inequalities (36) and (56) are equivalent. (5) together with dy = ac*dx yields:

$$\int_{y'}^{y''} yh(y)dy = ac* \int_{x'}^{x''} xf(x)dx. \tag{58}$$

With the help of (58), it can be seen that division of (27) by ac* leads to the following equation:

$$x* = p(1 - \frac{qL}{pV})\left[\frac{q}{p}\bar{x} + x*F(x*) + \int_{x*}^{x''} xf(x)dx\right]. \tag{59}$$

It remains to show that (57) and (59) are equivalent. This can be seen as follows: For x* = x' the right-hand sides of (57) and (59) agree. Moreover, it can be seen without difficulty that the right-hand sides of (57) and (59) have the same derivative with respect to x*. This completes the proof of Theorem 2.

Comparative statics of the critical level: In the following, we shall examine the question of how parameter changes influence the critical level x*. Let ψ(x*) be the right-hand side of (57) as a function of x*.   Imagine a diagram which shows ψ(x*) as a function of x*; up to a change of units of measurement this diagram is nothing else than Figure 3 with y', y*, y", and φ(y*) replaced by x', x*, x", and $\psi(x^*)$, respectively. The critical level corresponds to the intersection of ψ with the 45°-line. Every influence which results in an upward shift of ψ increases the critical level x^*.

With the help of (57) it can be seen that an increase of p or V results in an upward shift of ψ. An increase of L results in a downward shift of ψ. An increase of p or V increases the critical level x^* and an increase of L decrease the critical level x^* as long as (56) remains valid. The signs of the influences are shown by Table 1. They demonstrate how the critical level of the microhabitat quality is influenced by the different parameters of the model.

<u>Mean preserving spread</u>: Table 1 also includes the influence of a mean preserving spread of the distribution f. Loosely speaking, the notion of a mean preserving spread formalizes the idea of flattening and stretching a distribution in a roughly uniform way without changing its mean. The concept has been introduced as an instrument of comparative statics which avoids the necessity to restrict attention to specific parametric families of distributions (Rothschild and Stiglitz 1970, Stuart 1979).

Let g(x) be a continuous distribution whose support is an interval of non-negative numbers. We say that g is a *mean preserving spread* of f if we have:

$$\int_0^\infty xf(x)dx = \int_0^\infty xg(x)dx \tag{60}$$

and

$$\int_0^z F(x)dx < \int_0^z G(x)dx \tag{61}$$

for every z in the interior support of f,

p	L	V	mean preserving spread of f
+	−	+	+

<u>Table 1</u>: Signs of influences on the critical level x^* (critical microhabitat quality).

<u>Explanation of symbols</u>:

    p - probability to encounter the same microhabitat quality in near search at the next flower
    L - travel time for far search
    V - time for handling and local travel
    f - density of microhabitat quality

where G is the cumulative distribution of g. This definition is not the most general one, but apart from this it agrees with that given in the literature. Condition (60) requires that both distributions have the same mean. The interpretation of (61) is less obvious. In order to throw light on the meaning of (61), we show that the following two identities hold:

$$\int_0^z (z-x)f(x)dx = \int_0^z F(x)dx \tag{62}$$

$$\int_z^\infty (x-z)f(x)dx = \bar{x} - z + \int_z^\infty F(z)dz. \tag{63}$$

In both cases, it can be seen without difficulty that both sides agree for z = 0 and that both sides have the same derivative. For x < z the difference z - x may be interpreted as a *deficit below* z; similarly for x > z the difference x - z may be interpreted as a *surplus above* z. In this sense, we refer to the left-hand side of (62) as the *expected deficit below* z and to the left-hand side of (63) as the *expected surplus above* z. Equations (62) and (63) show that (61) has the following consequence: for every z in the interior of the support of f the expected deficit below z and the expected surplus above z are both increased by a mean preserving spread.

Inequality (61) permits the conclusion that a mean preserving spread results in an upward shift of the right-hand side of (57) and, therefore, causes an increase of the critical level x*.

The expected deficit below z may be looked upon as a measure of risk. In this sense, a mean preserving spread increases the riskiness of the situation. An increase of x* means that the pollinator changes his response to environmental conditions in favor of far search. Since near search seems to be the safer alternative, it is hard to understand why increased riskiness should favor far search.

In the context of the model presented here, the interpretation of a mean preserving spread as an increase of the riskiness of the situation is misleading. The pollinators are assumed to maximize expected yield per time without any consideration of risk. An adequate interpretation of the increase of the critical level x* caused by a mean preserving spread must focus on the surplus above z. A higher surplus over z indicates better chances to find high yield levels by far search. Not the increase of risk, but the simultaneous increase of the chances to find higher yield levels by far search is the important aspect of a mean preserving spread in the context of the model presented here. In this light, the positive influence of a mean preserving spread on the critical level x* becomes understandable.

220

6. Pollen Flow

The investigation of optimal pollinator foraging in the framework of the model presented here is now complete. This section begins to examine the influence of optimal pollinator foraging on flower competition. It will be assumed that the fitness of a flower depends on the total amount of pollen transferred to other flowers of the same species by a visiting pollinator. The name *pollen flow* will be used for this variable.

It is assumed that the pollinators visit flowers of two species A and B. In equilibrium all flowers invest the same amount $c*$ in resources offered to pollinators. Both species are equally frequent, but there is a positive spatial correlation of local densities. A pollinator who engages in near search has a probability α with $1/2 < \alpha < 1$ to visit a flower of the last visited species as the next one. This probability α is the same for both species and it is independent of the microhabitat quality x.

A pollinator may collect pollen and carry at least part of it in a way which prevents it to be lost on a flower. Here, we are only concerned with the *transferable pollen load* which is potentially lost during visits to flowers. This transferable pollen load is assumed to be constant. It is assumed that on each visit to a flower a fraction τ (or $100\tau\%$) of the transferable pollen load is lost there and replaced by pollen of this flower. Without loss of generality, the amount exchanged on a visit can be normalized to 1, by the appropriate choice of the quantity unit. Pollen from previously visited flowers are represented in the fraction lost in the same proportions as in the transferable load. One may think of the fraction lost as a random sample out of a huge number of pollen. τ is called the *transfer rate*. τ is a parameter with $0 < \tau < 1$.

It is assumed that the expected contributions to the fertilization of a flower visited due to the pollen of a previously visited flower is proportional to the amount of this pollen lost during the visit. This means that the proportion in which pollen of both species are represented in the transferable load is assumed to be without influence. The pollen of the other species does not compete for fertilization, but it may fall on the stigma and thereby diminish the reproductive success of pollen of the same species. It is of course an extremely simplifying assumption that this effect is equally strong as the influence of the competition by other pollen of the same species. This is maybe not the most objectionable distortion of reality for the sake of analytical tractability of the model presented here.

It is the intention of this paper to investigate necessary and sufficient conditions for an evolutionarily stable investment in resources offered to pollinators. Therefore, we look at a situation where every flower with the exception of a rare mutant of species A chooses the same investment $c*$. Let c be the investment chosen by the mutant. Assume that the environmental parameter at the site of a mutant is x. The yield collected by a pollinator who visits the mutant is $y = acx$. Let x^* be the critical level of the microhabitat quality. We have

$$acx = y \geq y^* = ac^*x^* \tag{64}$$

if and only if the following is true:

$$x \geq \frac{c^*}{c} x^*. \tag{65}$$

Define

$$\tilde{x} = \frac{c^*}{c} x^*. \tag{66}$$

We call $\tilde{x}$ the *mutant's critical level*, since it has the same significance for the reaction of a visiting pollinator as x^* does in the case of a normal member of the population. In anthropomorphic terms one can say that the visitor of a mutant is deceived about the true microhabitat quality. For $c > c^*$ the microhabitat quality is overestimated by the visitor and $\tilde{x}$ is smaller than x^*. The pollinator will be induced to engage in near search in the interval $\tilde{x} \leq x < x^*$, where it normally would choose far search. Similarly for $c < c^*$ we have $\tilde{x} > x^*$ and the pollinator is induced to engage in far search in the interval $x^* \leq x < \tilde{x}$, where it normally would choose near search.

It is convenient to introduce the following notation:

$$F^* = F(x^*) \tag{67}$$

$$\tilde{F} = F(\tilde{x}) \tag{68}$$

$$J^* = 1 - F^* \tag{69}$$

$$\tilde{J} = 1 - \tilde{F} \tag{70}$$

$$\beta = 1 - \alpha \tag{71}$$

$$\eta = 1 - \tau \tag{72}$$

$$\rho = 2\alpha - 1. \tag{73}$$

The probability α of staying with the same species in near search and the transfer rate τ satisfy the following inequalities:

$$\frac{1}{2} < \alpha < 1 \tag{74}$$

$$0 < \tau < 1 . \tag{75}$$

The parameter ρ defined by (73) can be interpreted as a measure of spatial correlation of species abundance. In view of (74) this *spatial correlation parameter* satisfies the following inequality

$$0 < \rho < 1. \tag{76}$$

We now turn our attention to the expected pollen flow P of a rare mutant of species A in an equilibrium situation where all non-mutants of both species invest the amount c^*. The mutant invests c. It is sufficient to look at a mutant of species A; in view of the symmetry of the situation, the same reasoning would apply to a mutant of species B.

The flow diagrams of Figures 4 and 5 describe what happens with which probabilities after a visit of a pollinator to a mutant flower. The pollen flow depends on the microhabitat quality x at the site of the mutant flower. Rectangle 1 symbolizes the event that this parameter x is smaller than $\tilde{x}$. This happens with probability $\tilde{F}$. The expected pollen flow under the condition $x < \tilde{x}$ is denoted by P_1. Similarly, P_2 denotes the expected pollen flow under the condition $\tilde{x} \leq x < x^*$; this condition cannot occur unless we have $c > c^*$. If $c > c^*$ holds, the probability for $\tilde{x} < x < x^*$ is $F^*-\tilde{F}$. This environmental situation is described by rectangle 2. The situation symbolized by rectangle 3 arises if x is greater than $\tilde{x}$ and x^*. For $c \leq c^*$ we have $\tilde{x} \geq x^*$, and $x^* \geq \tilde{x}$ holds for $-\geq c^*$. Therefore, the probability for this situation is $\tilde{J}$ for $c \leq c^*$ and J^* for $c \geq c^*$. The expected pollen flow in the case of rectangle 3 is denoted by P_3.

The expected pollen flows P_1, P_2, and P_3 are computed for the possible situations after the microhabitat quality of the mutants site has been determined. The expected pollen flow of the mutant before the microhabitat quality x has been determined is denoted by P. The expected pollen flow is obtained as the average of the P_i weighted by the probabilities of the corresponding situation:

$$P = \begin{cases} \tilde{F}P_1 + \tilde{J}P_3 & \text{for } c \leq c^* \\ \tilde{F}P_1 + (F^*-\tilde{F})P_2 + J^*P_3 & \text{for } c \geq c^* . \end{cases} \tag{77}$$

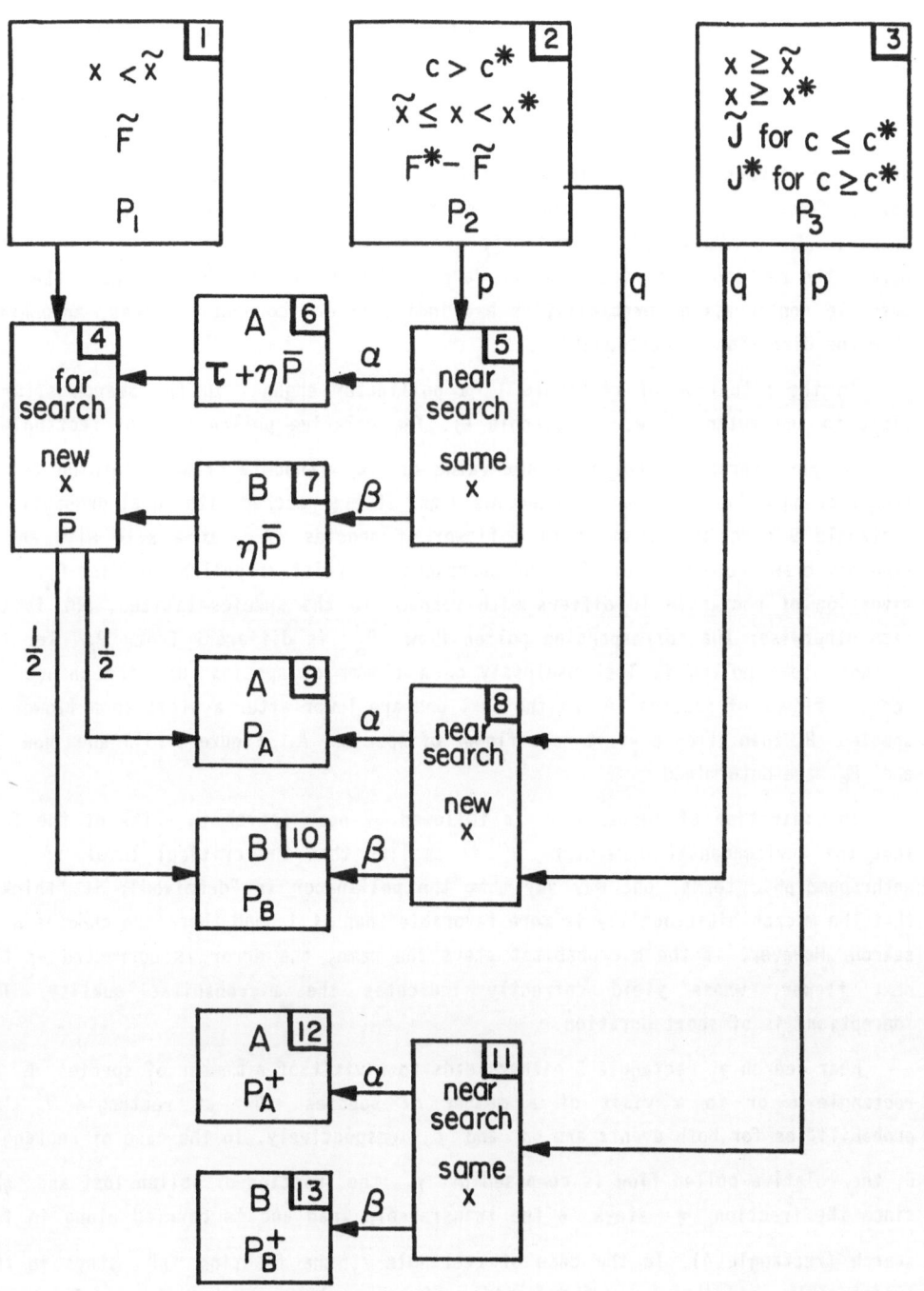

Figure 4: Beginning of the pollen flow from a mutant of species A

Rectangles 4 to 11 in Figure 4 correspond to situations possibly encountered by a pollinator after a visit to a mutant flower. The probability that another mutant flower is encountered is neglected. This is justified since we want to investigate evolutionarily stable strategies in pure strategies. The species of the flower is indicated at rectangles symbolizing visits. $\bar{P}$, P_A, P_B, P_A^+ and P_B^+ denote *relative pollen flows*, i.e., expected amounts of pollen eventually dropped on flowers of species A per unit of the mutant's pollen still in the transferable load just before the rectangle is reached which carries the symbol. If there is more than one possible continuation, probabilities are indicated at connecting lines. Arrowheads show the direction of continuation.

In the situation of rectangle 1, a pollinator engages in far search after a visit to the mutant flower (rectangle 4). The relative pollen flow at rectangle 4 is $\bar{P}$. Far search results in a new value of x randomly drawn according to f. From rectangle 4 each of both rectangles 9 and 10 are reached with equal probability. Rectangle 9 describes a visit to a flower of species A, at a site with an x randomly drawn according to f; the corresponding relative pollen flow is P_A. The situation of rectangle 10 differs with respect to the species visited, but is the same otherwise; the corresponding pollen flow P_B is different from P_A for two reasons: some pollen is lost uselessly on a flower of species B; the chances to reach a flower of species A as the next one are lower after a visit to a flower of species B than after a visit to a flower of species A. Figure 5 will show how P_A and P_B are determined.

The situation of rectangle 2 is followed by near search in spite of the fact that the environmental parameter x is smaller than the critical level x*. In anthropomorphic terms, one may say that the pollinator is "deceived"; it "thinks" that the microhabitat quality is more favorable than it is and therefore chooses near search. However, if the microhabitat stays the same, the error is corrected at the next flower, whose yield correctly indicates the microhabitat quality. The "deception" is of short duration.

Near search at rectangle 5 either leads to a visit of a member of species A at rectangle 6 or to a visit of a member of species B at rectangle 7. The probabilities for both events are α and β, respectively. In the case of rectangle 6, the relative pollen flow is composed of τ, the fraction of pollen lost and $\eta\bar{P}$, since the fraction η stays in the transferable load and is carried along in far search (rectangle 4). In the case of rectangle 7, the fraction $\eta\bar{P}$ stays in the transferable load, but τ, the fraction of pollen lost, is wasted on a flower of species B.

At rectangle 5, the microhabitat quality is the same as at the site of the mutant. With probability q a new random draw of x according to f follows rectangle 2 at rectangle 8. This leads to the rectangles 9 and 10 with probabilities α and β.

Near search with a new x may also follow the situation of rectangle 3. From there, also near search with the same x as at the mutant's site may be reached (rectangle 11). the pollen flows P_A^+ and P_B^+ at rectangles 12 and 13 are different from P_A and P_B, respectively, since here the microhabitat quality was not newly determined.

Figure 4 yields the following equations:

$$P_1 = \bar{P} \tag{78}$$

$$P_2 = p(\alpha\tau + \eta\bar{P}) + q(\alpha P_A + \beta P_B) \tag{79}$$

$$P_3 = p(\alpha P_A^+ + \beta P_B^+) + q(\alpha P_A + \beta P_B) \tag{80}$$

$$\bar{P} = \frac{1}{2} P_A + \frac{1}{2} P_B. \tag{81}$$

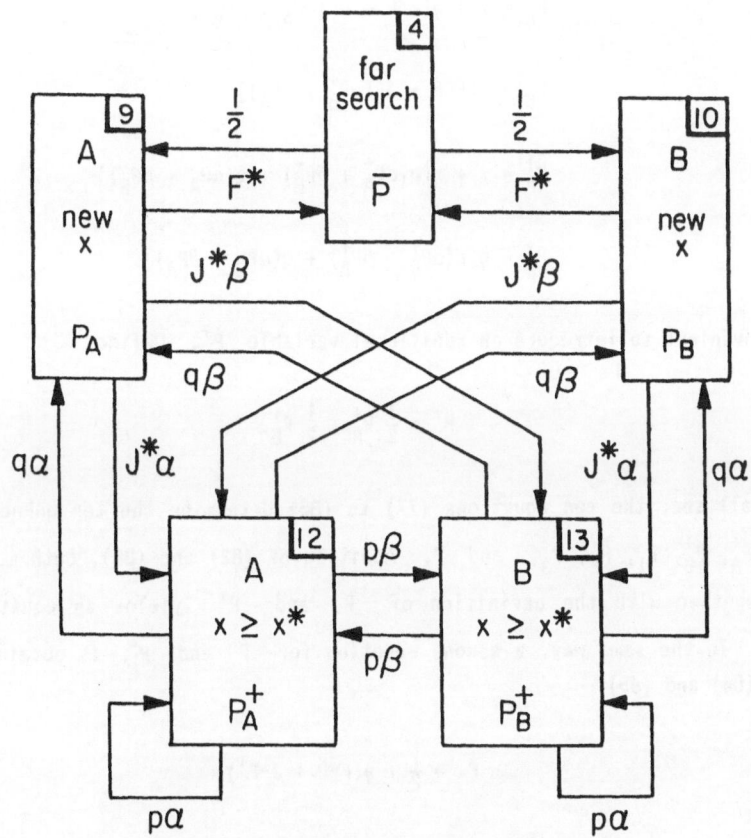

<u>Figure 5</u>: Continuation of the pollen flow

Figure 5 is the continuation of Figure 4. Rectangles 4, 9, 10, 12, and 13 appear there again, graphically arranged in a different way. The graphical conventions are the same as in Figure 4. The same flower is never visited again, but the same type of situation arises again with different flowers at the same rectangle.

With the help of Figure 5, equations for P_A, P_B, P_A^+, and P_B^+ can be derived. Consider rectangle 9 in Figure 5. Rectangle 9 corresponds to a visit to a flower of species A, where the fraction τ of the mutant's pollen transported up to this point is lost. The remaining fraction η is carried along in far search with probability F* or in near search with probability J*; the species of the next visited flower is A with probability α and B with probability β. Since the microhabitat quality and the species of the next flower are determined by stochastically independent processes, rectangle 12 is reached with probability J*α and rectangle 13 with probability J*β. This yields an equation for P_A. Equations for P_B, P_A^+, and P_B^+ are derived analogously:

$$P_A = \tau + \eta[F*\bar{P} + J*(\alpha P_A^+ + \beta P_B^+)] \tag{82}$$

$$P_B = \eta[F*\bar{P} + J*(\alpha P_B^+ + \beta P_A^+)] \tag{83}$$

$$P_A^+ = \tau + \eta[p(\alpha P_A^+ + \beta P_B^+) + q(\alpha P_A + \beta P_B)] \tag{84}$$

$$P_B^+ = \eta[p(\alpha P_B^+ + \beta P_A^+) + q(\alpha P_B + \beta P_A)]. \tag{85}$$

It is convenient to introduce an additional variable P^+. Define

$$P^+ = \frac{1}{2} P_A^+ + \frac{1}{2} P_B^+. \tag{86}$$

As we shall see, the ten Equations (77) to (86) determine the ten unknowns P, P^+, P_A, P_B, P_A^+, P_B^+, P_1, P_2, P_3, and P. Addition of (82) and (83), both multiplied by 1/2, together with the definition of P and P^+ yields an equation for P and P^+. In the same way, a second equation for P and P^+ is obtained with the help of (84) and (85):

$$\bar{P} = \frac{\tau}{2} + \eta(F*\bar{P} + J*P^+) \tag{87}$$

$$P^+ = \frac{\tau}{2} + \eta(pP^+ + q\bar{P}). \tag{88}$$

Since F*-q is equal to p-J*, this yields:

$$P - P^+ = \eta(F^* - q)(P - P^+). \tag{89}$$

In view of the fact that η, F*, and q are positive and smaller than 1, it is clear that $\eta(F^*-q)$ is unequal to 1. Therefore, (89) permits the conclusion that P is equal to P^+. This together with (87) yields:

$$P = P^+ = \frac{1}{2}. \tag{90}$$

The result that P is equal to 1/2 can also be obtained without computation by a symmetry argument outlined in the following. Let X_1, X_2,..., X_n with X_i = A or X_i = B be the species of the n first flowers visited by the pollinator after far search (rectangle 4). Let X_1', X_2',...,X_n' be the *opposite sequence* with X_i' = B for X_i = A and X_i' = A for X_i = B. It is clear that the model generates the same probability for both sequences X_1,...,X_n and X_1',...,X_n'. This has the consequence that the expected amount of pollen dropped on flowers of species A during the first n visits after far search is equal to the expected amount of pollen on flowers of species B during the first n visits after far search. The argument applies to every n and, therefore, shows that P must be equal to 1/2.

The task to determine the remaining unknowns is simplified by the introduction of three auxiliary variables U_A, U_A^+ and U:

$$U_A = P_A - \frac{1}{2} \tag{91}$$

$$U_A^+ = P_A^+ - \frac{1}{2} \tag{92}$$

$$U = pU_A^+ + qU_A. \tag{93}$$

With the help of the definition of P and P^+ it can be seen that (91) and (92) have the following consequences:

$$P_B = \frac{1}{2} - U_A \tag{94}$$

$$P_B^+ = \frac{1}{2} - U_A^+ . \tag{95}$$

In view of (73) this yields:

$$\alpha P_A + \beta P_B = \frac{1}{2} + (2\alpha-1)U_A = \frac{1}{2} + \rho U_A \tag{96}$$

$$\alpha P_A^+ + \beta P_B^+ = \frac{1}{2} + (2\alpha-1)U_A^+ = \frac{1}{2} + \rho U_A^+ . \tag{97}$$

With the help of (93), (96) and (97) Eqs. (82) and (84) can now be rewritten as follows:

$$U_A = \frac{\tau}{2} + \rho\eta J*U_A^+ \tag{98}$$

$$U_A^+ = \frac{\tau}{2} + \rho\eta U_A . \tag{99}$$

The right-hand side of (99) can be substituted for U_A^+ in (98):

$$U_A = \frac{\tau}{2}(1+\rho\eta J*) + \rho^2\eta^2 J*U . \tag{100}$$

In view of (93), addition of (99) multiplied by p and of (100) multiplied by q yields an equation for U whose solution is as follows:

$$U = \frac{1 + q\rho\eta J^*}{1 - \rho\eta(p+q\rho\eta J^*)} \cdot \frac{\tau}{2}. \tag{101}$$

In view of $q = 1-p$ and $\eta = 1-\tau$ as well as $\rho = 2\alpha-1$ and $J* = 1-F*$, Eq. (101) determines U as a function of p,τ,α and $F(x*)$. The auxiliary variables U_A^+ and U_A as well as the pollen flows P_A, P_B, P_A^+, P_B^+, P_1, P_2, P_3, and finally P can now be computed as functions of the key variable U. We are mainly interested in P which depends on the *pollen flow components* P_1, P_2, and P_3. In view of (90), (96) and (97), Eqs. (78), (79), and (80) yield:

$$P_1 = \frac{1}{2} \tag{102}$$

$$P_2 = p\alpha\tau + p\frac{\eta}{2} + q(\frac{1}{2} + \rho U_A) \tag{103}$$

$$P_3 = \frac{1}{2} + \rho(pU_A^+ + qU_A). \tag{104}$$

With the help of (72), (73), and (93), it can be seen that (103) and (104) can be rewritten as follows:

$$P_2 = \frac{1}{2} + p\rho\frac{\tau}{2} + q\rho U_A \tag{105}$$

$$P_3 = \frac{1}{2} + \rho U. \tag{106}$$

In view of (93) and (99) we have

$$qU_A = U - pU_A^+ = U - p(\frac{\tau}{2} + \rho\eta U).$$ (107)

This, together with (105), yields:

$$P_2 = \frac{1}{2} + \rho(1 - p\rho\eta)U.$$ (108)

Equations (102), (108), and (106) can now be used to determine P as a function of U. With the help of (77) one obtains:

$$P = \begin{cases} \frac{1}{2} + \rho U \tilde{J} & \text{for } c \leq c* \\ \frac{1}{2} + \rho U[1-p\rho\eta)(F^*-\tilde{F})+J^*] & \text{for } c \geq c* \, . \end{cases}$$ (109)

This equation, together with (101), describes the pollen flow of a mutant as a function of the investment c. The auxiliary variable U does not depend on c. The investment c determines the mutant's critical level $\tilde{x}$ defined by (66) and thereby $\tilde{F} = F(\tilde{x})$ and $\tilde{J} = 1-\tilde{F}$. These intermediate variables then influence the mutant's pollen flow P.

7. The Investment Game

On the basis of the results of the last section, it is now possible to examine the competition among the flowers with respect to the investment c in resources offered to pollinators. First, the relationship between pollen flow and fitness will be discussed. The competition of the flowers is modelled as a game whose payoff is the fitness obtained for an investment level c against an equilibrium level $c*$ chosen by all flowers with the exception of a rare mutant of one species. Payoffs will be defined for pure strategies only. Therefore, only pure evolutionarily stable strategies without alternative best replies will be considered. The name *strong evolutionarily stable strategy* will be attached to such strategies.

Let $P*$ be the pollen flow of a non-mutant. $P*$ is obtained as the special case $c = c*$ of (109):

$$P* = \frac{1}{2} + \rho U J*.$$ (110)

The expected reproductive success of a flower due to its pollen is assumed to be proportional to the pollen flow P. In order do determine the influence of P on fitness, it is necessary to compare P with $P*$, the average pollen flow in the population. The expected relative reproductive success of a rare mutant due to its pollen flow is the quotient P/P*, which will be referred to as the mutant's male fitness. As has been pointed out in the Introduction, female fitness need not be considered here, since it is not influenced by the investment level c. It is assumed that the male fitness P/P* and investment costs c have additive influences on the flower's overall fitness. Without loss of generality, investment is measured in male fitness units. This means that up to a positive linear transformation the mutant's overall fitness W is as follows:

$$W = \frac{P}{P*} - c.$$ (111)

In game-theoretical terms, flower competition is described as a population game (Selten 1980). A *population game* (Ⅱ,E) consists of a strategy set Ⅱ and a payoff function E. In the *investment game* considered here, Ⅱ is the set of non-negative numbers c (the possible investments) and the payoff E is defined by the following equation:

$$E(c,c*) = \frac{P}{P*} - c.$$ (112)

Mixed strategies are not considered in this paper. Therefore, the payoff function E is defined for pairs of pure strategies only. An evolutionarily stable strategy can be described as a symmetric equilibrium strategy with an additional stability property. A *symmetric equilibrium strategy* c* for (Ⅱ,E) is a strategy c* ∈ Ⅱ which satisfies the following condition:

$$E(c*,c*) \geq E(c,c*) \quad \text{for every } c \in Ⅱ.$$ (113)

Inequality (113) is a necessary condition for evolutionary stability. Our analysis will be based on a slightly different sufficient condition von evolutionary stability:

$$E(c*,c*) > E(c,c*) \quad \text{for every } c \in Ⅱ \text{ with } c \neq c*.$$ (114)

A strategy c* ∈ Ⅱ with (114) will be called a *strong evolutionarily stable strategy*. The word *strong* has been chosen in view of the definition by a strong inequality.

8. Necessary Conditions for Strong Evolutionary Stability in the Investment Game

In order to find investment levels c^* which are strong evolutionarily stable strategies, one has to explore the conditions under which the payoff $E(c,c^*)$, defined by (112), assumes its maximum with respect to c at $c = c^*$. In view of (109), the pollen flow P is not differentiable with respect to c at $c = c^*$. The same is true for $E(c,c^*)$. A strong evolutionarily stable investment level c^* must satisfy the following condition:

$$\left. \frac{\partial E(c,c^*)}{\partial c}\right|_{c^*}^{-} \geq 0 \geq \left.\frac{\partial E(c,c^*)}{\partial c}\right|_{c^*}^{+} \tag{115}$$

where the upper indices "-" and "+" indicate left and right derivatives, respectively. Both derivatives are taken at $c = c^*$ as expressed by c^* as a lower index. The same notational convention is also applied to the left and right derivatives of P at $c = c^*$. In order to compute these derivatives, we first look at the derivatives of $\tilde{x}$ defined by (66) and of $\tilde{F}$:

$$\frac{\partial \tilde{x}}{\partial c} = - \frac{c^*x^*}{c^2} = - \frac{\tilde{x}^2}{c^*x^*} . \tag{116}$$

In view of (68) this yields

$$\frac{\partial \tilde{F}}{\partial c} = f(\tilde{x}) \frac{\partial \tilde{x}}{\partial c} = - \frac{\tilde{x}^2 f(\tilde{x})}{x^*c^*} . \tag{117}$$

With the help of (109) one obtains:

$$\frac{\partial P}{\partial c} = \begin{cases} \frac{\rho U}{x^*c^*} \tilde{x}^2 f(\tilde{x}) & \text{for } c < c^* \\[2ex] (1-p\rho\eta)\frac{\rho U}{x^*c^*} \tilde{x}^2 f(\tilde{x}) & \text{for } c > c^* \end{cases} \tag{118}$$

and therefore

$$\left.\frac{\partial P}{\partial c}\right|_{c^*}^{-} = \frac{\rho U}{c^*} x^*f(x^*) \tag{119}$$

$$\left.\frac{\partial P}{\partial c}\right|_{c^*}^{+} = (1-p\rho\eta) \frac{\rho U}{c^*} x^*f(x^*) . \tag{120}$$

In view of (111) and (112) we have:

$$\frac{\partial E(c,c^*)}{\partial c} = \frac{1}{P^*} \frac{\partial P}{\partial c} - 1 \tag{121}$$

for $c \neq c*$. With the help of (119), (120), and (121), the necessary condition (115) can be rewritten as follows:

$$\frac{\rho U}{p* c*} \; x*f(x*) \geq 1 \geq (1-p\rho\eta) \; \frac{\rho U}{p* c*} \cdot x*f(x*). \tag{122}$$

This yields an inequality for $c*$:

$$(1-p\rho\eta) \; \frac{\rho U}{p*} \; x*f(x*) \leq c* \leq \frac{\rho U}{p*} \; x*f(x*). \tag{123}$$

Inequality (123) is equivalent to (115). Therefore, (123) is a necessary condition for strong evolutionary stability of an investment level $c*$.

The necessary condition (123) permits a whole range of investment levels. This is due to the asymmetric effects of increases and decreases of investment c at $c = c*$ on the pollen flow P in (109). The pollen flow as a function of c has a kink at $c = c*$. The derivative drops to a lower level. Therefore, local necessary conditions do not single out a unique investment level.

The payoff $E(c,c*)$ is not necessarily a concave function of c. Condition (123) does not imply that the local maximum at $c = c*$ is a global one. As we shall see, $E(c,c*)$ has another local maximum at $c = 0$. This leads to an additional global necessary condition which will be derived in the following.

Equation (66) connects every $c > 0$ to the corresponding critical level $\tilde{x}$. Let c' and c'' be the values of c connected to x' and x'' in this way:

$$c' = c* \; \frac{x*}{x'} \tag{124}$$

$$c'' = c* \; \frac{x*}{x''} . \tag{125}$$

In the special case $x' = 0$, Eq. (110) is interpreted as $c' = \infty$. It is assumed that (56) holds. In view of Theorem 2, we have $x' < x* < x''$. The following equation is a more detailed version of (109):

$$P = \begin{cases} \frac{1}{2} & \text{for } 0 \leq c \leq c'' \\ \frac{1}{2} + \rho U \tilde{J} & \text{for } c'' \leq c \leq c* \\ \frac{1}{2} + \rho U[(1-p\rho\eta)(F*-\tilde{F}) + J*] & \text{for } c* \leq c \leq c' \\ \frac{1}{2} + \rho U p\rho\eta & \text{for } c' \leq c. \end{cases} \tag{126}$$

Of course, for $x' = 0$, the case in the last line does not arise. (126) together with (111) and (112), yields a similar equation for $E(c,c^*)$:

$$E(c,c^*) = \begin{cases} \frac{1}{2p^*} - c & \text{for } 0 \leq c \leq c'' \\ \frac{1}{2p^*} + \frac{\rho U}{p^*} \bar{J} - c & \text{for } c'' \leq c \leq c^* \\ \frac{1}{2p^*} + \frac{\rho U}{p^*}[(1-p\rho\eta)(F^*-\bar{F}) + J^*] - c & \text{for } c^* \leq c \leq c' \\ \frac{1}{2p^*} + \frac{\rho U}{p^*} p\rho\eta - c & \text{for } c' \leq c . \end{cases} \qquad (127)$$

Equation (127) shows that for $0 \leq c \leq c''$ the payoff $E(c,c^*)$ is a decreasing function of c. Therefore, $E(c,c^*)$ assumes a local maximum at $c = 0$. For $c' \leq c$, too, $E(c,c^*)$ is a decreasing function of c. Therefore, a local maximum at a value of c with $c \neq 0$ cannot be assumed outside the interval $c'' < c \leq c'$. However, without further assumptions of f one cannot exclude the possibility of several local maxima in this interval. Therefore, the condition

$$E(c^*,c^*) > E(0,c^*) \qquad (128)$$

does not yield more than an additional necessary condition for strong evolutionary stability of c^*. In view of (127), condition (128) is equivalent to the following inequality:

$$c^* < \frac{\rho U}{p^*} J^*. \qquad (129)$$

In order to have a convenient way to express the necessary conditions obtained up to now, we introduce the following notation:

$$c_1^* = (1-p\rho\eta) \frac{\rho U}{p^*} x^* f(x^*) \qquad (130)$$

$$c_2^* = \frac{\rho U}{p^*} x^* f(x^*) \qquad (131)$$

$$c_3^* = \frac{\rho U}{p^*} J^* . \qquad (132)$$

The results obtained in this section are summarized by the following theorem

Theorem 3: Assume that (56) holds. If c^* with $c^* > 0$ is a strong evolutionarily stable strategy of (Π,E), then the following inequalities are satisfied:

$$c_1^* \leq c^* \leq c_2^* \qquad (133)$$

$$c^* < c_3^*. \qquad (134)$$

234

Proof: The proof has been given above.

Comment: Theorem 3 shows that the existence of a strong evolutionarily stable strategy c* cannot be expected for all possible constellations of the parameters p, L, V, α, τ, and the distribution f. Since $p\rho\eta$ is positive, it is clear that c_1^* is smaller than c_2^*. However, c_3^* may be smaller than c_1^*.

If the constellation of p, L, V, α, τ, and f does not permit a strong evolutionarily stable strategy, other types of evolutionarily stable situations may still be possible. It is conceivable that the system described by our model also permits equilibria where one of both species offers resources to pollinators whereas the other species does not do this. Obviously, asymmetric equilibria of this kind would require a different kind of pollinator behavior. Such possibilities cannot be investigated in the limited framework of our analysis.

9. Sufficient Conditions for Strong Evolutionary Stability in the Investment Game

No attempt will be made to derive conditions which are both necessary and sufficient for strong evolutionary stability of an investment level c*. Two different sets of sufficient conditions will be presented. Both of them involve strong assumptions on f. These assumptions are listed below.

(A) Zero densities at the borders:

$$f(x') = f(x'') = 0. \tag{135}$$

(B) Continuous differentiability: In the interval [x',x''] the density f has a continuous first derivative f'. Here f'(x') and f'(x'') are right and left derivatives, respectively.

(C) Continuous second derivative: In the interval [x',x''] the density f has a continuous second derivative f''. Here f''(x') and f''(x'') are right and left derivatives, respectively.

(D) Strict concavity: We have

$$f''(x) < 0 \quad \text{for} \quad x' \le x \le x''. \tag{136}$$

Here f''(x') and f''(x'') are right and left derivatives, respectively.

(E) Exclusion of a sharp relative decrease at x*:

$$x^*f'(x^*) + 2f(x^*) > 0. \tag{137}$$

(F) Exclusion of a sharp relative decrease in the support:

$$xf'(x) + 2f(x) > o \quad \text{for} \quad x' \leq x \leq x''. \tag{138}$$

Here $f'(x')$ and $f'(x'')$ are right and left derivatives, respectively.

(G) Necessary conditions:

$$c_1^* \leq c^* \leq c_2^* \tag{139}$$

$$c^* < c_3^*. \tag{140}$$

Of course, previous assumptions on the parameter p, V, L, α, τ and on the distribution f are also made in this section. For the sake of shortness these "standard assumptions" are not mentioned by the theorems to be proved. As in the last section, it will be assumed explicitly that the precondition (56) of Theorem 2 is satisfied.

One set of sufficient conditions involves Assumptions (A) to (E), and (G), but not (F). Assumptions (B), (F), and (G) form the other set of sufficient conditions.

Assumption (A) is quite natural; it means that f is not only continuous in [x',x''], but also in [0,∞). Assumptions (B) and (C) are technical ones. Assumption (D) is slightly stronger than the condition that f is strictly concave over the support of f. Assumption (D) seems to be stronger than necessary, but it is difficult to replace it by an easily interpretable weaker condition. In order to throw light on Assumptions (E) and (F), we look at the *elasticity* $\varepsilon(x)$ of f(x):

$$\varepsilon(x) = \frac{f'(x)}{f(x)} x. \tag{141}$$

The elasticity $\varepsilon(x)$ can be interpreted as the limit of the quotient of the relative changes $[f(x+\Delta x)-f(x)]/f(x)$ and $\Delta x/x$ for $\Delta x \rightarrow 0$. Conditions (137) and (138) can be expressed as follows:

$$\varepsilon(x^*) > -2 \tag{142}$$

$$\varepsilon(x) > -2 \quad \text{for} \quad x' \leq x \leq x''. \tag{143}$$

This shows in which sense (E) and (F) exclude sharp relative decreases of f. Assumption (G) simply repeats the necessary conditions of Theorem 3.

In the proofs of Theorems 4 and 5, we shall have to look at the first and second derivative of E(c,c*) with respect to c.

With the help of (118), (121), and (127) one obtains:

$$\frac{\partial E(c,c^*)}{\partial c} = \begin{cases} -1 & \text{for } 0 < c < c'' \\ \frac{\rho U}{P^* x^* c^*}\, \tilde{x}^2 f(\tilde{x}) - 1 & \text{for } c'' < c < c^* \\ (1-p\rho\eta)\frac{\rho U}{P^* x^* c^*}\, \tilde{x}^2 f(\tilde{x}) - 1 & \text{for } c^* < c < c' \\ -1 & \text{for } c' < c . \end{cases} \tag{144}$$

Under Assumption (B), we can compute the derivative of $\tilde{x}^2 f(\tilde{x})$ with respect to c with the help of (116):

$$\frac{\partial \tilde{x}^2 f(\tilde{x})}{\partial c} = - \frac{\tilde{x}^3}{c^* x^*}\, (\tilde{x} f'(\tilde{x}) + 2f(\tilde{x})). \tag{145}$$

This, together with (144), yields:

$$\frac{\partial^2 E(c,c^*)}{\partial c} = \begin{cases} 0 & \text{for } 0 < c < c'' \\ - \frac{\rho U \tilde{x}^3}{P^* (x^* c^*)^2}\, (\tilde{x} f'(\tilde{x}) + 2f(\tilde{x})) & \text{for } c'' < c < c^* \\ - (1-p\rho\eta)\, \frac{\rho U \tilde{x}^3}{P^* (x^* c^*)^2}\, (\tilde{x} f'(\tilde{x}) + 2f(\tilde{x})) & \text{for } c^* < c < c' \\ 0 & \text{for } c' < c. \end{cases} \tag{146}$$

Of course, also here it is assumed that (B) holds.

<u>Theorem 4</u>: Assume that (56) holds and that Assumptions (B), (F) and (G) are satisfied. Then c^* is a strong evolutionarily stable strategy of (Π, E).

<u>Proof</u>: It follows by (B) that (146) holds. In view of (146), Assumption (F) implies that $E(c,c^*)$ is a strictly concave function of c in the interval $c' \le c \le c''$. A global maximum at $c = 0$ is excluded by (G). Assumption (G) also implies that $E(c,c^*)$ has a local maximum at $c = c^*$. since $E(c,c^*)$ is strictly concave in the interval $c' \le c \le c^*$, this interval cannot contain another local maximum. As we have already seen, a global maximum outside the interval $c' \le c \le c^*$ must be assumed at $c = 0$. Since this possibility is excluded, we can conclude that the global maximum is assumed at $c = c^*$.

<u>Theorem 5</u>: Assume that (56) holds and that Assumptions (A), (B), (C), (D), (E), and (G) are satisfied. Then c^* is a strong evolutionarily stable strategy of (Π, E).

<u>Proof</u>: As we have seen in Section 8, inequality (140) is equivalent to the condition that $E(0,c^*)$ is smaller than $E(c^*,c^*)$. In view of this fact, it is sufficient to prove that the following assertion holds. With increasing c the function $E(c,c^*)$ first decreases in an interval $0 \le c \le c_{min}$ and then increases in the interval $c_{min} \le c \le c^*$ and finally decreases everywhere for $c \ge c^*$ (see Figure 5).

It follows by (144) that the assertion holds as far as values of c outside the interval $c'' < c < c'$ are concerned. It remains to show that the following is true:

$$\frac{\partial E(c,c*)}{\partial c} < 0 \quad \text{for} \quad c'' \leq c < c_{min} \tag{147}$$

$$\left.\frac{\partial E(c,c*)}{\partial c}\right|_{c_{min}} = 0 \tag{148}$$

$$\frac{\partial E(c,c*)}{\partial c} > 0 \quad \text{for} \quad c_{min} < c < c' \tag{149}$$

$$\frac{\partial E(c,c*)}{\partial c} < 0 \quad \text{for} \quad c* < c < c' . \tag{150}$$

It follows by (135) and (144) that (147) holds at $c = c''$. The discussion of necessary conditions in Section 8 shows that (139) implies (115). Since $\partial E(c,c*)/\partial c$ is continuous in the interval $c'' < c < c*$, we can conclude that $E(c,c*)$ assumes a local minimum in this interval. Of course, we did not yet exclude the possibility of several local minima in this interval. Let c_{min} be the smallest value of c with $c'' < c_{min} < c*$ at which a local minimum is attained. Obviously, this definition of c_{min} implies (147) and (148).

The proof of (149) requires an argument involving the second derivative of $E(c,c*)$. Equation (146) shows that for $c'' < c < c*$ the sign of the second derivative is the opposite sign of $\tilde{x}f'(\tilde{x})+2f(\tilde{x})$. We want to show that this sign changes at exactly one value $\hat{c}$ with $c'' < \hat{c} < c*$. For this purpose, it is sufficient to show that the equation

$$\hat{x}f'(\hat{x}) + 2f(\hat{x}) = 0 \tag{151}$$

has exactly one solution with $x* < \hat{x} < x''$. This solution $\hat{x}$ is the mutant's level corresponding to $\hat{c}$:

$$\hat{x} = \frac{x*c*}{\hat{c}} . \tag{152}$$

In order to prove the assertion on (151), we look at the derivative of the left-hand side:

$$\frac{\partial}{\partial x} (xf'(x) + 2f(x)) = xf''(x) + 3f'(x). \tag{153}$$

238

In view of (D), the distribution f is unimodal. Let x_0 be the mode of f. It follows by (D) that $f'(x)$ is negative for $x_0 < x < x''$. Therefore, $xf'(x)+2f(x)$ is decreasing in the interval $x_0 \le x \le x''$. It follows by (D) that the left-hand side of (151) is positive for $x' < x \le x_0$. It follows by (A) and (D) that $x''f'(x'')+2f(x'')$ is negative. Consequently, $xf'(x)+2f(x)$ is positive for $x' < x \le x_0$ and then decreases for $x_0 \le \hat{x} \le x''$ and finally becomes negative at x''. It follows that Eq. (151) holds at exactly one $\hat{x}$ in the interval $x' < \hat{x} < x''$ and that $\hat{x} > x_0$ holds for this value. Moreover, Assumption (E) implies that $\hat{x} > x^*$ holds. This shows that $E(c,c^*)$ as a function of c has exactly one

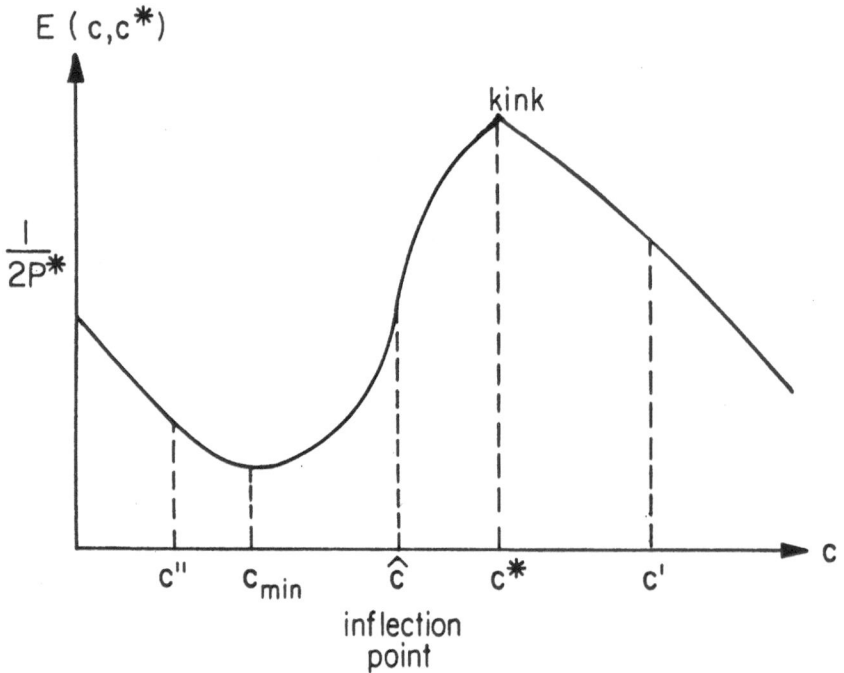

Figure 6: Typical shape of $E(c,c^*)$ as a function of c under Assumptions (A) to (E) and (G).

inflection point in the interval $c'' < c < c'$ and that $c'' < \hat{c} < c^*$ holds for the c-coordinate $\hat{c}$ of this inflection point.

Suppose that $E(c,c^*)$ assumes a local maximum in the interval $c_{min} < c < c^*$. Since (G) implies (115), the derivative $\partial E(c,c^*)/\partial c$ is positive in a left neigborhood of c^*. Therefore, the function $E(c,c^*)$ would have to assume a second local minimum in this interval; a second inflection point would have to be present between the local maximum and the second local minimum. Since this possibility has been excluded, it follows that $E(c,c*)$ assumes no local maximum in the interval $c_{min} < c < c*$. The presence of only one inflection point also excludes the possibility of other stationary points in this interval. It follows by the continuity of $\partial E(c,c*)/\partial c$ that (149) holds.

The presence of exactly one inflection point in the interval $c'' < c < c'$ has the additional consequence that $E(c,c*)$ is strictly concave in the interval $c* < c < c'$. Since (G) implies (115), it follows that (150) holds. This completes the proof of the theorem.

First numerical example: In the case of a rectangular distribution f, the critical level $x*$ is easy to compute since in this case (57) is a quadratic equation. Assume:

$$f(x) = \begin{cases} 1 & \text{for } 0 \leq x \leq 1 \\ 0 & \text{for } x \geq 1. \end{cases} \tag{154}$$

In view of $x' = 0$, condition (56) is satisfied for this density if we have $qL < pV$. Eq. (57) assumes the following form:

$$x* = \left[1 - \frac{qL}{pV}\right]\left[\frac{1}{2} + \frac{p}{2}(x*)^2\right]. \tag{155}$$

Eq. (155) has only one root in the support of f:

$$x* = \frac{1}{p-q\frac{L}{V}} - \sqrt{\frac{1}{(p-q\frac{L}{V})^2} - \frac{1}{p}}. \tag{156}$$

Assume that the parameters p, V, L, α, and τ are as follows:

$$p = .8 \tag{157}$$

$$V = 1 \tag{158}$$

$$L = 2 \tag{159}$$

$$\alpha = .75 \tag{160}$$

$$\tau = .1 . \tag{161}$$

Equation (156) yields:

$$x* = .264. \tag{162}$$

With the help of (101) one obtains:

$$U = .0873. \tag{163}$$

Equation (110) yields the equilibrium pollen flow:

$$P^* = .532. \tag{164}$$

Interestingly, P^* is not much greater than .5, the pollen flow obtained by a mutant with $c = 0$. Equations (130), (131) and (132) yield:

$$c_1^* = .0139 \tag{165}$$

$$c_2^* = .0217 \tag{166}$$

$$c_3^* = .0604. \tag{167}$$

It is clear that the sufficient conditions (B), (F) and (G) of Theorem 4 hold for c^* if c^* satisfies the following inequality:

$$.0139 \leq c^* \leq .0217. \tag{168}$$

All investment levels in the interval (168) are strong evolutionarily stable strategies. It is important for the interpretation of (164) that the investment level c is measured in male fitness units. The male fitness P/P^* of a mutant with $c = 0$ is $.5/P^* = .940$, independently of the value of c^*. The failure to offer resources to pollinators entails a male fitness loss of .060 which is not compensated by the advantage of saving costs of at most .0217.

Second numerical example: The sufficient conditions of Theorem 4 have been used in the first numerical example. We now look at a similar example which satisfies the sufficient conditions of Theorem 5. The density f is as follows:

$$f(x) = \begin{cases} .75x(2-x) & \text{for } 0 \leq x \leq 2 \\ 0 & \text{for } x \geq 2 . \end{cases} \tag{169}$$

Obviously, (A), (B), (C) and (D) are satisfied for this density f. Let the parameters p, V, L, α, and τ be the same as in the first numerical example. (57) becomes a cubic equation. One obtains:

$$x^* = .524 \tag{170}$$

$$U = .0886 \tag{171}$$

$$P^* = .537 \tag{172}$$

$$c_1^* = .0160 \tag{173}$$

$$c_2^* = .0251 \tag{174}$$

$$c_3^* = .0685 . \tag{175}$$

Since x* is below the mode, f'(x*) is positive. Therefore, condition (E) holds. (G) is satisfied if and only if the following inequality holds:

$$.0160 \leq c^* \leq .0251. \tag{176}$$

The strong evolutionarily stable strategies are the investment levels c* in the interval (176). The sufficient conditions of Theorem 5 are satisfied.

The numerical results are similar to those of the first example. This is due to the fact that the parameters p, V, L, α, and τ are the same in both cases. For this parameter constellation, the influence of a change of the density f seems to be small. The values of U are near to each other; this is due to the fact that the coefficient of J* in the numerator and the denominator on the right-hand side of (101) are small.

A numerical example of non-existence: It follows by Theorem 3 that strong evolutionarily stable strategies c* do not exist if we have:

$$c_3^* < c_1^*. \tag{177}$$

In view of (130) and (131), this is equivalent to the following inequality:

$$J^* < (1-p\rho\eta)x^*f(x^*). \tag{178}$$

Assume that f is the distribution described by (154) in the first numerical example. Let the parameters be as follows:

$$p = .8 \tag{179}$$

$$V = 1 \tag{180}$$

$$L = .25 \tag{181}$$

$$\alpha = .6 \tag{182}$$

$$\tau = .4 . \tag{183}$$

Equation (156) yields:

$$x^* = .607 . \tag{184}$$

We obtain:

$$J^* = .493 \tag{185}$$

and

$$(1-p\rho\eta)x^*f(x^*) = .549. \tag{186}$$

Obviously, (178) is satisfied. No strong evolutionarily stable strategy c* exists in this numerical case.

10. Parameter Influ nces

The model does not determine a unique equilibrium investment level, but a whole range of strong evolutionarily stable strategies, if such a strategy exists at all. Consider a situation where an investment level $c*$ in the interior of the range of strong evolutionarily stable strategies is chosen by all members of both species and imagine a small change of the environment which can be described as an increase or decrease of one of the parameters. If the change is sufficiently small, the range shifts, but $c*$ remains in the range. This means that a small environmental change does not necessarily exert any adaptive pressure.

If one wants to ask the question how parameter changes influence resources offered to pollinators, one must focus on the potential borders of the range c_1^*, c_2^*, and c_3^* rather than $c*$. Unfortunately, the attempt to determine the signs of parameter influences on c_1^*, c_2^*, and c_3^* was not very successful. More can be said on ratios between these variables. We shall look at c_2^*/c_1^* and on c_3^*/c_1^*.

Existence of strong evolutionarily stable strategies requires $c_3^* > c_1^*$. Therefore, the ratio c_3^*/c_1^* is of special interest. A parameter change which increases c_3^*/c_1^* from below 1 to above 1 makes strong evolutionarly stable strategies possible. In this sense, the ratio c_3^*/c_1^* is an *existence indicator*. Parameter changes which increase c_3^*/c_1^* enhance the chances for existence.

If strong evolutionarily stable strategies exist, then c_1^* is the lower border of their range and the minimum of c_2^* and c_3^* is the upper border; c_2^*/c_1^* and c_3^*/c_1^* are upper bounds for the ratio of upper border and lower border. The length of the range divided by the lower border is the *relative size* of the range. A parameter change which increases both ratios c_2^*, c_1^* and c_3^*, c_1^* increases the relative size of the range.

Equations (130), (131), and (132) define c_1^*, c_2^*, and c_3^*. We have

$$\frac{c_2^*}{c_1^*} = \frac{1}{1-p\rho\eta} \tag{187}$$

$$\frac{c_3^*}{c_1^*} = \frac{1}{1-p\rho\eta} \cdot \frac{J*}{x*f(x*)} \ . \tag{188}$$

The common factor $\rho U/P*$ of c_1^*, c_2^*, and c_3^* cancels out in these ratios. If one wants to determine influences on c_1^*, c_2^*, and c_3^*, one has to look at this common factor. In view of (110) we have

$$\frac{\rho U}{P*} = \frac{1}{\frac{1}{2\rho U} + J*} \ . \tag{189}$$

Equation (189) shows that influences on $\rho U/P*$ are determined by influences on $1/2\rho U$ and on $J*$. Therefore, we shall look more closely at $1/2\rho U$. With the help of (101) one obtains:

$$\frac{1}{2\rho U} = \frac{1-\rho\eta(p+q\rho\eta J*)}{\rho\tau(1+q\rho\eta J*)} . \qquad (190)$$

Equivalent transformations yield:

$$\frac{1}{2\rho U} = \frac{1+q\rho\eta}{\rho\tau(1+q\rho\eta J*)} - \frac{\eta}{\tau} \qquad (191)$$

$$\frac{1}{2\rho U} = \frac{\eta}{\tau} \cdot \frac{qF*}{1+q\rho\eta J*} + \frac{1}{\rho\tau} - \frac{1}{\tau} + 1 \qquad (192)$$

$$\frac{1}{2\rho U} = \frac{qF*}{\frac{\tau}{\eta} + q\rho\tau J*} + \frac{1}{\tau}(\frac{1}{\rho} - 1). \qquad (193)$$

For fixed q, τ, and ρ an increase of $x*$ exerts a positive influence $1/2\rho U$, since $F*$ is increased and $J*$ is decreased. Therefore, the joint effect on $\rho U/P*$ cannot be easily determined. For this reason, influences on c_1^*, c_2^* and c_3^*, which involve $x*$ as an intervening variable, will not be examined. $x*$ depends on V, L, and p, but not on α and τ. Equation (193) shows that an increase of ρ or τ results in a decrease of $1/2\rho U$ and therefore in an increase of $\rho U/P*$.

Table 2 shows the signs of the derivatives of the row variables with respect to the column parameters. The influences of V, L, p, and a mean preserving spread have already been discussed in Section 5. The lack of any influence of τ and α on $x*$ is indicated by the entry "0" in the corresponding fields of Table 2.

In view of $\rho = 2\alpha-1$, an increase of α results in an increase of ρ. The parameters α and τ have positive influences on $\rho U/P*$ and, therefore, also on c_2^* and c_3^*. Since $1-\rho\rho\eta$ is also increased by an increase of τ, the parameter τ has a positive influence on c_1^*. The influence of α on c_1^* cannot be determined in this way since an increase of α decreases $1-\rho\rho\eta$. It follows by (187) and (188) that an increase of τ results in a decrease of c_2^*/c_1^* and c_3^*/c_1^* whereas an increase of α results in an increase of c_2^*/c_1^* and c_3^*/c_1^*.

The ratio c_2^*/c_1^* does not depend on $x*$ and, therefore, is not influenced by V, L, or a mean preserving spread. (187) shows that an increase of p increases c_2^*/c_1^*. The ratio c_3^*/c_1^* depends on $x*$. The expression $x*f(x*)$ is non-decreasing in $x*$ if and only if we have:

$$\frac{\partial x f(x)}{\partial x}\bigg|_{x*} = x*f'(x*) + f(x*) \geq 0. \qquad (194)$$

In view of $f(x^*) > 0$, this inequality can be divided by $f(x^*)$. With the help of (141) the resulting inequality can be expressed as follows:

$$\varepsilon(x^*) \geq -1. \qquad (195)$$

	τ	α	V	L	p	mean preserving spread
x^*	0	0	+	−	+	+
$\dfrac{c_2^*}{c_1^*}$	−	+	0	0	+	0
$\dfrac{c_3^*}{c_1^*}$	−	+	$\begin{array}{c}-\\ \varepsilon^*\geq-1\end{array}$	$\begin{array}{c}+\\ \varepsilon^*\geq-1\end{array}$		
c_3^*	+	+				
c_2^*	+	+				
c_1^*	+					

Table 2: Signs of parameter influences. The entry $\varepsilon^* \geq -1$ indicates that the sign is correct if this condition is satisfied; ε^* stands for $\varepsilon(x^*)$.

Explanation of symbols:

- τ - pollen transfer rate
- α - probability to encounter the same type of flower in near search
- V   - time for visiting and handling one flower, including local travel in near search
- L   - travel time for far search
- p   - probability of unchanged microhabitat conditions in near search
- c_1^* - lower bound for the equilibrium investment level c^*
- c_2^*, c_3^* - upper bounds for the equilibrium investment level c^*.

For the sake of shortness, we write ε^* instead of $\varepsilon(x^*)$. Inequality (195) is stronger than condition (E) which only requires $\varepsilon^* > -2$. Under the assumption $\varepsilon^* \geq -1$ the quotient $J^*/x^*f(x^*)$ is decreased by a sufficiently small increase of x^*. This, together with (188), shows that for $\varepsilon^* \geq -1$, the ratio c_3^*/c_1^* has a negative partial derivative with respect to V and a positive one with respect to L.

Interpretation: An increase of the transfer rate τ results in an increase of c_1^*, c_2^*, and c_3^*. The range of strong evolutionarily stable strategies is shifted in the direction of more resources offered to pollinators, but at the same time the ratios c_2^*/c_1^* and c_3^*/c_1^* are decreased. An increase of τ results in new local equilibrium conditions with a higher minimum investment c_1^*, but the higher the equilibrium investment c^* is, the greater is the saving of investment costs obtained by a mutant who does not offer resources to pollinators. From a certain point onwards, the high investment requirements imposed by local conditions may destroy the validity of global conditions.

The parameter α is related to spatial correlation of local species density. It is easily understandable that an increase of α results in an increase of c_1^*, c_2^*, and c_3^*. The higher the parameter α is, the more important it is to induce near search. However, why does an increase of α increase the ratios c_3^*/c_1^* and c_2^*/c_1^* whereas an increase of τ has the opposite effect? An answer to this question can be found with the help of (127). If α is increased, the kink of $E(c,c^*)$ at c^* becomes sharper; this kink becomes less sharp if τ is increased. The kink is due to the fact that a slight change of c at c^* has only a *short run effect* in the case of an increase and *long run effect* in the case of a decrease. The difference between the short run effect and the long run effect is decreased by an increase of τ. The more pollen is transferred in the short run, the less important is the long run. Contrary to this, an increase of α increases the difference between the short run effect and the long run effect; a higher spatial correlation of species abundance increases the advantage of near over far search. A sharper kink is connected to a wider gap between c_1^* and c_2^*.

The derviation of the necessary conditions in Section 8 shows that in the case $c^* = c_1^*$ the right derivative of $E(c,c^*)$ at $c = c^*$ is zero. Therefore, a sharper kink at c^* in the case of $c^* = c_1^*$ is connected to a steeper slope of $E(c,c^*)$ to the left of c^*. Maybe it is this feature of the shape of $E(c,c^*)$ which creates a connection between the sharpness of the kink and the condition that $E(c^*,c^*)$ is greater than $E(0,c^*)$.

246

11. Discussion

The aim of the model presented here is to try to understand why plants offer resources to pollinators. At first glance, this question may sound trivial. The viability of plant-pollinator interaction systems requires resources offered to pollinators. However, this is not a satisfactory explanation of the phenomenon. It is necessary to explain why it is in the interest of an individual plant to offer resources to pollinators.

Usually, pollinators do not observe the yield of a flower before a visit. The yield is known only after a visit. This opens an opportunity for cheating. A mutant plant may save energy for other purposes by not offering resources to pollinators. What are the adverse consequences for the mutant's fitness? Why is it evolutionarily stable to offer resources to pollinators? Our modelling efforts are devoted to this basic issue.

A great variety of different ecological systems of plant-pollinator interactions are observed in nature. One cannot expect that the reasons for the production of resources offered to pollinators are the same in every case. This paper has explored only one of the possibilities.

The model presented here describes an idealized system of one pollinator species interacting with two plant species. It is assumed that both plant species are equal in all relevant aspects such as flower shape and color, overall frequency, spatial correlation of local densities and resource utilization capabilities. Pollinators do not distinguish between both types of flowers. The resource offered is not replenishable and one plant has only one exploitable flower at each given point of time. The number of pollinators per exploitable flower is constant over time. These simplifying assumptions enhance analytical tractability, but they also focus attention on salient features of the proposed explanation of the phenomenon of resources offered to pollinators.

Our explanation does require the presence of several plant species visited by the same pollinators, but it does not rely on essential differences between plant species. Multiple flowers at one plant, blooming at the same time, may provide additional incentives for offering resources to pollinators, but our explanation already works with only one flower at each plant. In our model the environmental pattern of a mosaic of microhabitats is the source of spatial correlation of yield levels. Positive spatial correlation of yield levels is necessary for the type of foraging behavior entering our explanation. If the microhabitat quality of plant sites were randomly distributed without any spatial correlation, the pollinators would have no decision problem between near search and far search. It would always be optimal to engage in near search.

Since we assume non-replenishability, our model needs an exogenous source of positive spatial correlation. If resources were replenishable, the search behavior of pollinators would be an endogenous source of positive spatial correlation. The mathematical difficulties arising from an unknown spatial yield distribution determined by an unknown optimally adapted search policy are avoided here. Our model shows that the environmental pattern of a mosaic of microhabitats alone, without replenishability of resources, is sufficient to generate the positive spatial correlation which creates the pollinator's decision problem between near search and far search.

In our model, the amount of resources offered to pollinators depends multiplicatively on microhabitat quality and investment. The level of investment can be interpreted as the percentage of total available energy devoted to resources offered to pollinators. Investment is the strategic variable of the plant. The level of investment is chosen before a response to microhabitat conditions is possible.

In this paper, attention is concentrated on strong evolutionarily stable investment strategies. We look at equilibrium situations in which every plant chooses the same level of investment.

Results on Optimal Pollinator Foraging

The derivation of results on optimal pollinator behavior is based on the assumption that all plants choose the same investment level. It is shown that the optimal foraging policy is characterized by a critical yield level. The yield of the last visited flower is compared with the critical level. Near search is chosen above the critical level and far search is chosen below the critical level.

The critical yield level corresponds to critical microhabitat quality. At a microhabitat with this quality a plant with the common investment level produces exactly the critical yield. The analysis shows that the critical microhabitat quality x^* does not depend on the common investment level (see Theorem 2).

Even if no explicit formula for the critical microhabitat quality has been obtained, Eq. (57) permits the determination of parameter influences (see Table 1). The probability p of staying in the same microhabitat after the choice of near search has a positive influence on the critical microhabitat quality x^*. The higher p is, the stronger is the positive spatial correlation of yield levels. The positive influence of p on x^* shows that a stronger positive correlation of yield levels leads to a wider range of microhabitat qualities which induce far search. However, a higher p also results in longer stretches of near search.

As one would expect, an increase of the far travel time L narrows the range of microhabitat qualities which induce far search; similarly, an increase of the time V of handling and local travel decreases the range of microhabitat qualities followed by near search.

Table 1 also describes the effect of a *mean preserving spread*. Roughly speaking, a mean preserving spread is a flattening of the distribution of microhabitat qualities, subject to certain non-parametric restrictions. A mean preserving spread increases critical microhabitat quality. The range of microhabitat qualities inducing far search is widened. The effect has nothing to do with an increased risk inherent in a flatter distribution, but rather with the greater chances of finding an extraordinarily good microhabitat offered by a greater variability of microhabitat quality.

Investment and Pollen Flow

Pollen flow is defined as the total expected amount of pollen transferred from one flower to other flowers of the same species. A plant's male fitness is its pollen flow divided by the average pollen flow in the population. Up to a constant total fitness is male fitness minus cost of investment. Female fitness is not explicitly modelled.

The assumptions on pollen transfer are based on the idea that a pollinator carries a constant *transferable load* of pollen. At each flower visit a fraction τ of the transferable load is lost and replaced by pollen of this flower. The probability that the next flower visited is of the same species as the last one depends on the search behavior of the pollinator. This probability is $\alpha > 1/2$ for near search and equal to $1/2$ for far search. Local densities of both species exhibit positive spatial correlation.

On the basis of the results on optimal pollinator foraging, we have examined the pollen flow of a mutant as a function of its investment level. An explicit expression has been obtained with Eq. (109). A higher investment increases the pollen flow, since it extends the range of microhabitat qualities connected to a yield above the critical level; therefore, a higher investment leads to a higher probability of near search after a visit. Since near search is connected to a higher probability for the next flower being of the same species, near search after a visit is more advantageous for the pollen flow than far search.

The Pollen Flow Kink

It is an important result of our analysis that the pollen flow of a mutant as a function of its investment has a kink at the equilibrium investment level. A small increase of investment beyond this level has a weaker effect on pollen flow than a small decrease by the same amount.

The pollen flow kink can be interpreted as an informational phenomenon. An investment which differs from the equilibrium level transmits "misinformation" on the microhabitat quality. In anthropomorphic terms one may say that the pollinator "overestimates" the microhabitat quality in the case of an investment above the equilibrium level and "underestimates" it in the opposite case. Overestimation may result in near search where normally far search would be indicated and thereby enhance pollen flow. However, if the microhabitat stays the same, the wrong impression is quickly corrected by the next flower. Contrary to this, if underestimation leads to far search, where normally near search is indicated, the wrong impression is never corrected.

The informational asymmetry between underinvestment and overinvestment creates the pollen flow kink. At the equilibrium level of investment, the advantage of inducing overestimation by a small increase of investment is less pronounced than the disadvantage of inducing underestimation by an investment decrease of the same size.

The Range of Strong Evolutionarily Stable Investment Strategies

With the help of the pollen flow function (109) necessary conditions for a strong evolutionarily stable investment strategy have been derived (Theorem 3). Moreover, two sets of sufficient conditions have been presented (Theorems 4 and 5). Under the assumptions of Theorem 5, the fitness of a mutant plant as a function of its investment level has the shape shown in Figure 6. With increasing investment, fitness is first decreasing, then increasing up to the kink at the equilibrium investment level, and then decreasing again. The kink is due to the pollen flow kink. If investment is too low, there may be no chance to produce the critical yield, even in the best microhabitat; nevertheless, investment costs are incurred. This explains the initial fitness decrease in Figure 6.

Apart from degenerate cases, the model admits a whole range of strong evolutionarily stable strategies if such strategies exist at all. A lower bound c_1^* and an upper bound c_2^* for the range are connected to the right-hand and the left-land slope at the kink. An additional upper bound c_3^* is due to the condition that zero investment should be less advantageous. The additional conditions listed in either Theorem 4 or Theorem 5 have the consequence that all investment levels within these bounds (with the exception of a border point at c_3^*) are strong evolutionarily stable strategies.

The emergence of a range of strong evolutionarily stable strategies is due to the kink at the equilibrium investment level. In view of the kink, local optimality requires that two inequalities are satisfied for the two slopes on both sides. Without the kink, the necessary condition for local optimality would be just one equation. This equation would uniquely determine the equilibrium investment level.

A strong evolutionarily stable strategy in the interior of the range remains stable after a sufficiently small change of the parameters of the model. This robustness property is an additional consequence of the presence of a kink.

Parameter Influences on the Range of Strong Evolutionarily Stable Investment Strategies

A change of the parameters of the model may shift the lower bound c_1^* and the potential upper bounds c_2^* and c_3^* of the range. We have also looked at influences on the quotients c_2^*/c_1^* and c_3^*/c_1^*. Those results which could be obtained are summarized by Table 2.

Strong evolutionarily stable strategies do not exist if c_3^*/c_1^* is below 1. Therefore, one may say that a positive influence on this quotient enhances the chances for existence of strong evolutionarily stable investment strategies.

The transfer rate τ can be interpreted as a parameter which measures the speed of pollen transfer. The expected number of flowers until a pollen grain is lost is $1/\tau$. Table 2 shows that τ has a positive influence on all three bounds of the range. Therefore, higher values of τ can be expected to be associated with higher investment levels. However, the quotients c_2^*/c_1^* amd c_3^*/c_1^* are diminished by an increase of τ. A decrease of c_3^*/c_1^* to a value below 1 results in the breakdown of the conditions for a strong evolutionarily stable investment strategy connected with positive amounts of resources offered to pollinators. Consequently, the positive influence of τ on the investment level may stop at a value of τ where stability breaks down.

The probability α of the next flower visited in near search being of the same species is a measure of the extent of positive spatial correlation of local species densities. Table 2 shows that α has a positive influence on c_2^*, c_3^* and c_2^*/c_1^* as well as c_3^*/c_1^*. Obviously, an increase of α leads to a more favorable situation for the existence of a strong evolutionarily stable strategy. Moreover, the situation after an increase of α permits higher levels of equilibrium investments.

A plausible interpretation of the influences of α suggests itself. A higher α results in a more pronounced difference between near search and far search as far as the pollen flow is concerned. This augments the plant's incentive to induce more near search by more investment.

It is important for our interpretation of the influences of τ and α that the conditions (A) to (F), which appear in the sets of sufficient conditions given by Theorems 4 and 5, do not depend on these parameters. Only the necessary conditions (G) are influenced by α and τ.

The influences of V, L, p, and a mean preserving spread on c_1^*, c_2^*, and c_3^* remain unclear. For further details, see Section 10.

Alternative Explanations

The amount of pollen transferred from a flower to a pollinator may be an increasing function of handling time, and handling time may be an increasing function of the yield level. It is doubtful whether these causal relationships hold in many cases, but wherever they are valid they provide an easy *handling time explanation* for the phenomenon of resources offered to pollinators. Even without any microhabitat heterogeneity and without the danger of losing pollen on a different species, pollen flow would be an increasing function of investment.

Replenishability of resources offered to pollinators is a wide-spread phenomenon in nectar producing flowers. Usually, nectar is refilled after exploitation. If resources are replenishable, a positive spatial correlation of yield levels is likely to result from the foraging behavior of the pollinators. The construction of a exact analytical model for a *replenishment explanation* meets the difficulty of a complex interdependence of the spatial yield distribution and pollinator behavior. Both must be determined simultaneously.

The simultaneous presence of many flowers at the same plant is the rule rather than the exception. A plant with more than one flower has an incentive to offer resources to pollinators in order to induce them to visit further flowers after a visit of the first one. Several papers by G.H. Pyke (1978, 1984) present an empirically based explanation which combines multiple flowers on a vertical inflorescence with a replenishable resource (nectar). The theory has been successfully applied to hummingbird pollination of *Ipomomopsis aggregata*. Empirical observations are used in order to determine the yield distribution. On the basis of the empirically obtained distribution a sequence of critical yields is computed depending on the number of flowers already visited on the inflorescence.

The relationship between the number of flowers visited on an inflorescence and the plant's pollination success is obtained by ingenious empirical methods. The final results are in agreement with the hypothesis of evolutionarily stable nectar production. Undoubtedly, Pyke's studies of concrete plant-pollinator systems are of great importance. It is not a valid objection against his work that he does not offer a theoretical solution for the problem of interdependence between optimal foraging and yield distribution. His approach remains theoretically incomplete. However, in the analysis of field data, it is fully justified and maybe even preferable to combine empirically obtained distributions and functional relationships with partial theoretical derivations.

In this paper, no attempt has been made to confront our model with data. The aim of our efforts has been a more modest one, namely, the complete theoretical investigation of an idealized case. In a similar approach to a model of replenishment with multiple flowers on a vertical inflorescence, one would have to face the problem posed by the interdependence of optimal foraging and yield distribution.

A game model proposed in the literature (Bell 1986) assumes only two possible strategies for plants: they can either produce a fixed amount of nectar or no nectar at all. Pollinators also have only two possible strategies: they can either distinguish both types of flowers (this is costly) or neglect to do so. In view of the impossibility of genuinely mixed strong evolutionarily stable strategies in asymmetric games (Selten 1980), the model yields cyclic fluctuations. As Bell points out himself, the prediction of a substantial fraction of flowers without nectar does not agree with field observations. In order to obtain reasonable results, it sems to be better to treat the amount of resources offered to pollinators as a continuous variable.

For mass flowering trees or bushes it may be a justifiable approximation to assume that the search behavior of pollinators leads to an ideal free distribution of pollinators over plants. This amounts to a direct proportionality between nectar production and the number of visits. An *ideal free distribution explanation* along these lines has the virtue of simplicity. The authors are involved in a project of constructing a formal model based on this approach (Ellner, Selten and Shmida, 1989, in preparation).

Concluding Remark

The model of pollinator foraging and flower competition presented in this paper describes an idealized special case. Ecological systems of plant-pollinator interaction are very complicated. A logically stringent and theoretically complete explanation of the phenomenon of resources offered to pollinators must be based on an analytically tractable formal model. Simplifying assumptions are the price for analytical tractability.

Our model is sufficiently simple to permit an integrated treatment of the interconnected decision problems of plant and pollinator. This makes it possible to obtain results like the kink in the pollen flow function and the emergence of a whole range of strong evolutionarily stable investments in resources offered to pollinators. An integrated treatment of both decision problems is also a necessary condition for a meaningful discussion of parameter influences on equilibrium levels of variables. A model which is sufficiently complex to be directly applied to empirical data is likely to be too difficult to permit the same depth of analysis.

In spite of many simplifying assumptions the explanation of resources offered to pollinators emerging from the analysis of this paper seems to capture important features of real plant-pollinator systems. Our description of the pollinator's foraging behaviour focusses on the decision between near search and far search. It is a well documented fact that search distances of pollinators decrease with increasing yield levels (Waddington 1983). This suggests that not only in our model but also in the field yield levels transmit information on the favorability of the immediate neighborhood.

As has been pointed out in the introduction, in plant species which offer resources to pollinators usually all individuals do this. An equilibrium in which all individuals choose the same investment level agrees with this observed pattern.

The distribution of many plant species over space tends to be patchy. Not only in our model, but also in the field this fact must be expected to make near search after a visit advantageous from the point of view of the plant. Therefore, the informational effect of the yield level creates an incentive to supply resources offered to pollinators.

References

Baker, H.G., and P.D. Hurd (1968). Intrafloral Ecology. Ann. Rev. Entomol. **13**: 385-414.

Barth, F.G (1985). Insects and Flowers - The Biology of a Partnership. Princeton: Princeton University Press.

Bell, G. (1986). The Evolution of Empty Flowers. J. Theor. Biol. **118**: 253-258.

Bently, B.L., and T.S. Elias (1983). The Biology of Nectaries. New York: Columbia University Press.

Bertsch, A. (1987). Flowers' Food Sources and the Cost of Outcrossing. In: E.D. Schulze, and H. Zwölfer (eds.), Ecological Studies, Vol. **61**, p. 277-293, Berlin: Springer.

Boyden, T.C. (1982). The Pollination Biology of Calypso Bulbosa (Orchidaceae): Initial Deception of Bumblebee Visitors. Oecologia **55**: 178-184.

Brink, D., and J.M. de Wit (1980). Interpopulation Variation in Nectar Production in Aconitum columbianum. Oecologia **74**: 160-163.

Brown, J.H., and A. Kodric-Brown (1979). Convergence, Competition and Mimicry in Temperate Community of Hummingbird Pollinated Plants. Ecology **60**: 1022-1035.

Charlesworth, D., D.W. Schemske, and V.L. Sork (1987). The Evolution of Plant Reproductive Characters; Sexual Versus Natural Selection. In: B.C. Stearns (ed.), The Evolution of Sex and its Consequences. Basel: Birkhäuser, pp. 317-335.

Charnov, E.L., and J.J. Bull (1986). Sex Allocation, Attraction and Fruit Dispersal in Cosexual Plants. J. Theor. Biol. **118**: 321-325.

Cochran, M.E. (1986). Consequences of Pollination by Chance in the Pink Lady's-slipper, Cypripedium acaule. Ph.D. Dissertation, University of Tennessee, Knoxville.

Dafni, A. (1984). Mimicry and Deception in Pollination. Ann. Rev. Ecol. Syst. **15**: 259-278.

Eastabrook, G.F., and D.C. Jespersem (1974). Strategy for Predator Encountering a Model-mimic System. Am. Nat. **108**: 443-457.

Eickwort, G.C., and H.S. Ginsberg (1980). Foraging and Mating in Apoidea. Ann. Rev. Entomol. **25**: 421-446.

Faergi, K., and L. Van der Pijl (1966). The Principle of Pollination Ecology. London: Pergamon Press.

Feinsinger, P. (1978). Ecological Interactions Between Plant and Hummingbirds in a Successional Tropical Community. Ecol. Month. **48**: 269-287.

Harstadt, R.M., and A. Postlewaite (1981). Expected Utility-Maximizing Price Search with Learning. Manag. Sci. **27**: 75-80.

Heinrich, B. (1979a). "Majoring" and "Minoring" by Foraging Bumblebees, Bombus vagans - An Experimental Analysis. Ecology **60**: 245-255.

Heinrich, B. (1979b). Resource Heterogeneity and Patterns of Movement in Foraging Bumblebees. Oecologia **40**: 235-245.

Heinrich, B. (1983). Insect Foraging Energetics. In: C.E. Jones and, R.J. Little (eds.), Handbook of Experimental Pollination Biology. New York: Van Nostrand Reinhold, pp. 187-214.

Heinrich, B., and P. Raven (1972). Energetics and Pollination Ecology. Science **176**: 597-602.

Houston, A.I., and J. McNamara (1980). The Application of Statistical Decision Theory to Animal Behavior. J. Theor. Biol. **85**: 673-690.

Houston, A.I., A. Kacelnik, and J.McNamara (1982). Some Learning Rules for Acquiring Information. In: D.J. McFarland (ed.), Functional Ontogeny. Boston: Pitman, pp. 140-191.

Kacelnik, A., A.I. Houston and P. Schmid-Hempel (1986). Central Place Foraging in Honeybees. Behav. Ecol. Sociobiol. **19**: 19-24.

Kacelnik, A., J.R. Krebs, and B. Ens (1987). Foraging in a Changing Environment: An Experiment with Starlings. In: M. Commons, A. Kacelnik, and S. Shettleworth (eds.), Quantitative Analyses of Behavior. Vol. **6**: Foraging. Lawrence Erlbaum Ass.

Kacelnik, A., and I.C. Cuthill (1987). Starling and Optimal Foraging Theory: Modelling in Practical World. In: A. Kamil, J.R. Krebs, and R. Pulliam (eds.), Foraging Behavior. New York: Plenum Press.

Kohn, M.G., and S. Shavell (1974). The Theory of Search. J. Econ. Th. **9**: 93-124.

Knuth, P. (1906). Handbook of Flower Pollination, 3 Vols. Oxford: Clarendon Press.

Laverty, L.M. (1980). The Flower-visiting Behavior of Bumblebees: Floral Complexity and Learning. Can. J. Zool. **58**: 1324-1335.

Linsley, E.G. (1958). The Ecology of Solitary Bees. Hilgardia **27**: 543-601.

Lippman, S.A., and J.J. McCall (eds.) (1979). Studies in the Economics of Search. Amsterdam: North Holland Publ.

Maynard Smith, J. (1982). Evolution and the Theory of Games. Cambridge: Cambridge University Press.

Maynard-Smith, J., and G.R. Price (1973). The Logic of Animal Conflict. Nature **246**: 15-18.

McNamara, J., and A. Houston (1985). A Simple Model of Information Use in the Exploitation of Patchly Distributed Food. Anim. Behav. **33**: 553-560.

Nilsson, L.A. (1983). On Flower Deception Between Orchid an Campanula. Nature **305**: 779.

Pleasants, J.M., and S.J. Chaplin (1983). Nectar Production Rates in Asclepas Quadrifolia: Causes and Consequences of Individual Variation. Oecologia **59**: 232-238.

Plowright, R.C., and T.M. Laverty (1984). The Ecology and Sociobiology of Bumble Bees. Ann. Rev. Entomol. **29**: 175-199.

Proctor, M., and P. Yeo (1973). The Pollination of Flowers. London: Collins, St. James Place.

Pyke, G.H. (1978). Optimal Foraging in Bumble-Bees and Coevolution with their Plants. Oecologia **36**: 281-293.

Pyke, G.H., 1984. Optimal Foraging Theory: A Critical Review. Ann. Rev. Ecol. Syst. **15**: 523-575.

Raven, P.H., R.F. Everts, and Curtis (1980). Biology of Plants. New York: North Holland Publ. Inc.

Real, L. (ed.) (1983). Pollination Biology. Orlando: Academic Press.

Rothschild, M. (1974). Searching for the Lowest Price, when the Distribution is Unknown. J. Polit. Econ. **82**: 689-711.

Rothschild, M., and J. Stiglitz (1970). Increasing Risks, 1. A Definition. J. Econ. Th. **2**: 225-243.

Schmid-Hempel, P., A. Kacelnik, and A.I. Houston (1985). Honeybees Maximize Efficiency by not Filling their Crop. Behav. Ecol. Sociobiol. **17**: 61-66.

Selten, R. (1980). A Note on Evolutionarily Stable Strategies in Asymmetric Animal Conflicts. J. Theor. Biol. **84**: 93-101.

Staddon, J.E.R. (1987). Adaptation to Reward. In: A. Kamil, J.R. Krebs, and R. Pulliam (eds.), Foraging Behavior. New York: Plenum Press.

Stebbins, G.L. (1974). Flowering Plants - Evolution Above the Species Level. Cambridge, Mass.: The Belknap Press.

Stephenson, A.G., and R.I. Bertin (1983). Male Competition, Female Choice and Sexual Selection in Plants. In: L. Real (ed.), Pollination Biology. Orlando: Academic Press, pp. 109-149.

Stuart, C. (1979). Search and the Spatial Organization of Trading. In: S.A. Lippman, and J.J. McCall (eds.), Studies in the Economics of Search. Amsterdam: North Holland Publ., pp. 17-33.

Terborgh, J. (1973). On the Notion of Favorableness in Plant Ecology. Am. Nat. **107**: 481-501.

Thomson, J.D. (1983). Component Analysis of Community-level Interactions in Pollination Systems. In: C.E. Jones, and R.J. Little (eds.), Handbook of Experimental Biolog. Scientific and Academic Editions. New York: Van Nostrand Reinhold, pp. 451-46.

Thomson, J.D. (1986). Pollen Transport and Deposition by Bumblebees in Erythronium: Influences of Floral Nectar and Bee Grooming. J. Ecol.

Thomson, J.D., W.P. Maddison, and R.C. Plowright (1982). Behavior of Bumblebee Pollinators of Aralia hispida. Oecologia **5**: 326-336.

Thomson, J.D., and R.C. Plowright (1980). Pollen Carryover, Nectar Reward and Pollinator Behavior with Special Reference to Diervilla lonicera. Oecologia **46**: 68-74.

Thomson, J.D., M.V. Price, N.M. Wasser, and D.A. Stratton (1986). Comparative Studies of Pollen and Fluorescent Dye Transport by Bumblebees Visiting Eryhtronium grandiflorum. Oecologia **69**: 561-566.

Vogel, S. (1983). Ecophysiology of Zoophilic Pollination. In: O.L. Lange, P.S. Nobel, C.B. Osmond, and H. Ziegler (eds.), Physiological Plant Ecology. Vol. **3**, 559-624.

Waddington, K.D. (1983). Foraging Behavior of Pollinators. In: L. Real (ed.), Pollinator Biology. London: Academic Press, pp. 213-239.

Waddington, K.D., and L.R. Holden (1979). Optimal Foraging: On Flower Selection by Bees. Am. Nat. **114**: 179-195.

Whittacker, R.H., and S.H. Levin (1977). The Role of Mosaic Phenomena in Natural Communities. Theor. Pop. Biol. **12**: 117-139.

To trade, or not to trade; that is the question

James W. Friedman

and

Peter Hammerstein

Abstract

The black hamlet is a simultaneous hermaphrodite fish that does not fertilize its own eggs and whose reproductive success is strongly related to the number of eggs of other fish that it fertilizes. Its own eggs are an inducement to other fish to let the fish fertilize their eggs. Consequently, the hamlet engages in *egg trading*, a mating process in which one fish lets another fertilize its eggs in exchange for allowing it to fertilize the eggs of the other fish.

It would be in a fish's interest if the partner would spawn first allowing the fish to fertilize the eggs of the other fish while keeping its own eggs as a bargaining chip to make the same arrangement with another fish. Such behavior might not be viable in the long run, as the "gullible" fish would suffer low reproduction rates as compared with the "sophisticated, selfish" fish. Eventually, the latter would have only its own sort to deal with, which might even hamper the mating process. In addition, if male fish could participate in such a process, then male specialization might be more profitable than being an hermaphrodite.

Instead of the preceding scenario, the hamlets egg trade under circumstances in which it is important to use time well. The egg trading process involves two fish spawing alternately, perhaps five or six times each. As a result the fish that spawns first has little at risk. The process is carried out slowly enough that neither fish has an incentive to "kiss and run."

The purpose of this paper is to model the hamlets' mating as a game in which their chosen behavior is an equilibrium. We investigate two issues in particular: (1) Is it in the interest of an hermaphrodite fish to behave in a way that sustains the behavior that they follow? (2) If it were possible for hamlets that were specialized as males to deceive hermaphrodites into egg trading, would a population be sustainable in which such males were a significant fraction of the total?

1. Introduction

The black hamlet, *Hypoplectrus nigricans*, is a simultaneous hermaphrodite that invariably mates with other fish, rather than by selfing. When mating takes place between a pair of fish, each fish alternates several times between male and female roles in an elaborate process that is called a *spawning bout*. During the process of fertilization (a *spawning act*), the two fish are in a virtual embrace during which one fish is releasing eggs and the other is simultaneously fertilizing them. Thus one spawning bout is made up of many spawning acts with the egg clutch of each fish divided into several parcels. Our purpose in this paper is to model this behavior and explain why it persists.

The hamlet has been studied by Fischer (1980, 1981) who suggests a game-theoretic explanation for the observed behavior. He suggests that eggs are relatively few and very expensive compared with sperm, and that, therefore, it is easy for a hamlet to get its own eggs fertilized by another fish. Consequently, as its own eggs will certainly be fertilized, its (comparative) reproductive success is largely determined by the number of eggs from other fish that it fertilizes. Thus a fish can use the eggs in its possession as an inducement to other fish to allow their eggs to be fertilized. That is, fish A will allow fish B to fertilize a parcel of its eggs on the condition that fish B will reciprocate.

Hamlets form pairs only for mating on a given day. This contrasts with the behavior of the harlequin bass, *Serranus tigrinus*, which form permanent pairs that sleep, breed and forage together. The harlequin bass spawns using a two round process under which each fish releases all of its eggs in a single spawning act. See Fisher (1986) and Fisher and Peterson (1987). In the hamlet this two round process would seem vulnerable to "cheating." To see this suppose, for example, that fish are accustomed to some random means of determining which fish releases eggs first. Now introduce into the population a few fish that will only release eggs first when the other fish has definitively refused to do so and that swim away after a spawning act in which the partner released eggs first. These fish, when paired with a member of the original stock, will frequently fertiliz the eggs of the other fish and still possess their own eggs as an inducement to additional fish to mate with them.[1] Such fish would, at first, have far greater reproductive success than would members of the original stock. If, in addition, this new fish would release only a modest fraction of its eggs when going first, it would have some protection against being left in the lurch. Such a fish would probably grow in number over time until it were a large, perhaps majority part of the population.

In broad outline the fish spend a significant amount of time in search and display before they begin their alternate spawning acts (called *egg trading*). This prior search and display process is an expensive investment of time that greatly raises the cost to a fish if it deserts its partner to look for another because it would have to be repeated with a new partner. Releasing the eggs in several small parcels ensures that the advantage from fertilizing the eggs of another fish, followed by abandoning that fish, brings only modest gain. Thus a large price is payed for a modest reward if a fish that initially plays the male role fertilizes its partner's eggs and then deserts its partner. Both fish have a strong incentive to continue the reciprocal process until on fish has no more eggs.

Axelrod and Hamilton (1981) have introduced the repeated prisoner's dilemma as a means of explainin certain animal interactions where one partner appears to act generously towards another; however, the hamlets sequential process is not a repeated prisoner's dilemma and the behavior followed by the fish is different from tit-for-tat. Fisher (1988) discusses the behavior of the hamlets as a repeated prisoner's dilemma and introduces tit-for-tat strategies. Our approach differs from his in several ways. He defines the individual iterations in a way that makes them structurally interdependent; whereas in a repeated prisoner's dilemma the iterations are not interdependent. Also he does not incorporate the search and display phases into his model. We think these are crucial to providing the fish with the incentive to egg-trade.

In the remaining sections below, we sketch the mating behavior of the hamlet, drawing on Fischer, in §2. In §3 we present a game theoretic model that captures the main features of the reported behavior, in §4 some consequences of this behavior are explored, in §5 we examine whether hamlets can specialize on male function, and in §6 are summary comments and suggestions for further research.

2. A Summary of the Field Observation of the Hamlet

Fischer (1980, 1981) is the principal source of field observation of the black hamlet, *Hypoplectrus nigricans* (Serranidae). It is a coral reef inhabitant near Panama. Its eggs are buoyant when released and float off, so there is no parental care, which means that there is no parental cost after spawning. In general, each fish has its own foraging territory; however, mating generally occurs at a specific location that is apart from the foraging territory of the fish. Mating is confined to the two hours before sunset each day and is not seasonal in the sense that some fish have eggs on any given day throughout the year. Those fish having eggs on a given day will head for the common mating ground at the appointed time, form into pairs, and mate. The time of mating appears advantageous for several reasons. First, the hamlets sleep when it is dark. Second, the eggs will be dispersed before nighttime predators have appeared. And third, the daytime predators pose a smaller threat than they would be earlier in the day.

Mating takes the form of *egg trading*. That is, one fish releases 15% or so of its eggs, which are fertilized by the other fish, then the second fish releases a similar quantity of its eggs, then the first fish releases a similar quantity of eggs, etc. The process is very time consuming, and its slowness is part of what makes it work. Some minutes, perhaps ten or so, are typically needed to find a fish to pair with. When two fish first pair, they go through a display process, taking roughly ten minutes, that looks as if it allows each fish to inspect the belly of the other fish and get an idea of its stock of eggs. Then, if they proceed to egg trading, they spend two or three minutes displaying before each later spawning act. The time taken to release eggs in a single spawning act is short compared to the display time between spawning acts. If it becomes the turn of a fish to release eggs and it does not do so (perhaps because it has no more), the other fish will usually swim away. Consequently, nearly an hour can go by from the time a fish reaches the mating ground to the time when it finishes a spawning bout with another fish.

Fish lacking eggs do not go to the mating ground, suggesting that they cannot fool the other fish. If one fish of a pair swims away because its partner ceases to release eggs it will seek another partner if time permits and may well have another spawning bout. The timing described above is an average of what Fischer observed and he also observed that all processes started off more slowly. Then, as the two hour period draws to an end, the pace tends to quicken. The patterns described here are not invariable and iron-clad, but they are typical. The egg clutch possessed by a fish must be used within one day. In general, the egg production of a fish is likely to be closely related to the size of the fish.

Thus the egg trading ritual has incentives that make it in the selfish interest of an individual fish to adhere to them. A fish without eggs cannot easily fool a fish that has eggs. Mating takes place at a special time in a common location distinct from the foraging territories of the fish. Consequently only fish with eggs to trade will meet at the mating location. Then, the fish have every incentive to remain in the pairs that they form until one fish has released its last eggs. This is because the amounts released in a single spawning act are relatively small and the ritual behavior required to get to the point of spawning with a fish is time consuming. To "kiss and run" a fish taking first the male role will get to fertilize only a small quantity of eggs and will then have to use a large amount of valuable time to be able to spawn with a different fish. The fish will, in fact, be better off playing by the common rules.

To compare the hamlets' game with a repeated prisoner's dilemma, think of the hamlet's game as being divided into a number of successive time periods. In the first period all egg-carrying fish search to find partners and in the second, those fish with partners display to one another. In the third period, the fish that have displayed can choose to leave their partners or engage in a spawning act. The fourth period is the same as the third for fish that spawned or displayed in period three. In the repeated prisoner's dilemma, each period is an ordinary prisoner's dilemma and is consequently identical to every other period. The immediate payoff to a player in a single period is always higher if he "cheats" than if he "cooperates." The next period is the same irrespective of behavior in the earlier periods.

Now look again at the hamlets' game. During search and display periods the immediate payoff is zero. After display or after a spawning act a fish receives zero if it swims away (the analog of "cheat") and receives a positive payoff if both fish proceed to spawning (the analog of "cooperates"), although the fish in the female role gets a lesser (though positive) payoff than the fish in the male role. The latter has expended no eggs. Unlike the prisoner's dilemma, one period often differs from another, the payoff matrix for a period depends on the prior behavior of the fish, and search and display periods are required before spawning can fruitfully take place.

There is another fundamental difference between the repeated prisoner's dilemma and the hamlets' game. In the former a pair of players, say Klaus and Anne, will play a number of successive, identical prisoner's dilemma games. The number could be randomly determined. In the hamlet's game, two fish may display together and then spawn together; however, at any time either fish can quit the pairing and seek a new partner.

Finally consider the applicability of tit-for-tat behavior in the hamlets' game. This strategy dictates that fish A will choose in any period the same behavior exhibited by fish B in the immediately preceding period. And vice versa. Thus under tit-for-tat, if they both display in one period, they will both display in the next. Even if we take a looser interpretation of tit-for-tat, broadly using it to identify "cooperate" with "proceed to spawning after display and switch roles after each spawning act," then it is true that a fish generally cooperates after each period of cooperation however, if fish A cheats on fish B, then they no longer have any association. Fish B has no opportunity to cheat fish A in the next period, nor would fish B necessarily cheat some other fish with which it is next paired.

3. A Simplified Model of Mating Behavior

An idealized model of black hamlet mating behavior is presented below that we believe captures the essentials. Suppose the hamlets have a mating period of 70 minutes per day, coming just near sunset, and that the stock of eggs held by a fish will be good on that day only. Also assume that a fish has exactly one of only three possible quantities of eggs: 0, 1, z (where z > 1). The quantity "1" does not mean one egg, but denotes a way of normalizing the units; a clutch typically contains several hundred or more eggs.

The 70 minute mating period can be thought of as 7 time periods; for we assume that search takes 10 minutes, display takes 10 minutes, and egg exchange takes 30 minutes. We also assume that egg exchange will never start unless at least 30 minutes remain in the mating period. This assumption makes our model more tractable, but we do not believe it changes the nature of the results. When two fish exchange, the amount of eggs each gives up is the same; therefore, it equals the stock of the fish carrying the lesser number.

Figure 1 shows the model is schematic fashion. In the figure the horizontal axis measures time elapsed from the start of the start of the mating period. A broken line indicates time during the mating period that cannot be put to use. For example, a fish that searches for ten minutes, displays for ten minutes, and mates for

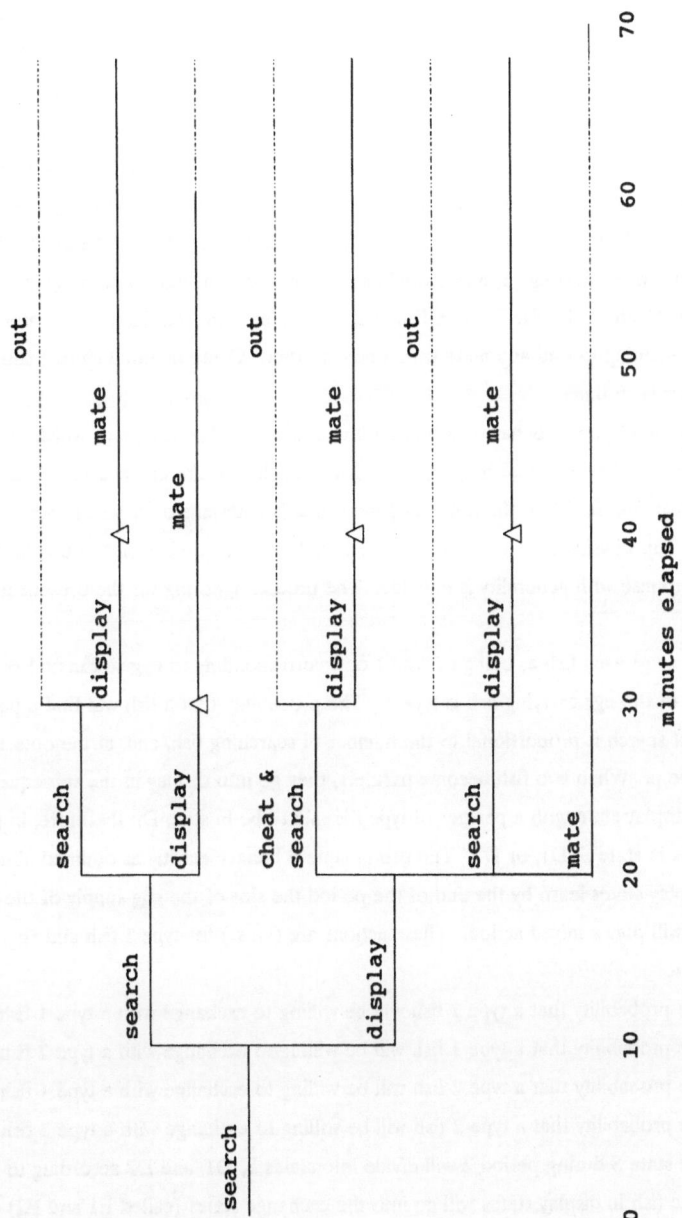

Figure 1

thirty minutes has twenty minutes at the end of the mating period when nothing pertaining to mating can be done. Likewise, a fish that fails in three successive periods (thirty minutes) to find a partner has reached a point when there is not sufficient time to search further, display, and mate. Consequently, such a fish is noted as being "out" at that point and this is indicated by a broken line. The figure does not show all conceivable actions a fish could take; it is restricted to those action that are potentially interesting. This is particularly relevant to "kiss and run" behavior under which a fish possessing eggs fertilizes a parcel of eggs from a partner and then fails to reciprocate. This can happen at any time during mating if the fish in the male role still has eggs; however, it is not always interesting. If a fish would gain from this behavior, then the gain is largest when the partner releases its first parcel. Furthermore, a fish cannot possibly gain from kissing and running unless there is sufficient time remaining to find and mate with a new partner. Consequently Figure 1 shows the kiss and run possibility only for such times.

Our analysis of equilibrium fish behavior is divided into two parts. The first part, which immediately follows, analyzes equilibrium behavior when the parcel size is too small to make kiss and run behavior beneficial to the fish engaging in it. Then the range of parcel sizes is determined for which fish would never gain from this behavior. This division is carried out because it makes the analysis easier to do and to understand, but no compromise with generality is entailed. The process, ignoring for the present the issue of parcel size, proceeds thus:

Period 1: We may regard each fish as being of type 1 or 2 corresponding to egg stocks of 1 or z. A fraction q of the egg-carrying fish is type 1. The probability that a fish will find a partner during a period of search is proportional to the number of searching fish, and, at the outset, this is taken to be p. When two fish become partners, they go into display in the subsequent period.

Period 2: A fish in display state with a partner of type i is said to be in state Di; therefore, in period 2 each fish is in state S, D1, or D2. The fish in state S behave exactly as outlined already. The fish in display states learn by the end of the period the size of the egg supply of the partner. Each fish will play a mixed action. These actions are (s_1, s_2) for type 1 fish and (σ_1, σ_2) for type 2 fish where

s_1 = the probability that a type 1 fish will be willing to exchange with a type 1 fish
s_2 = the probability that a type 1 fish will be willing to exchange with a type 2 fish
σ_1 = the probability that a type 2 fish will be willing to exchange with a type 1 fish
σ_2 = the probability that a type 2 fish will be willing to exchange with a type 2 fish

Period 3: The fish in state S during period 2 will divide into states S, D1, and D2 according to rules given above. The fish in display states will go into the exchange states (called E1 and E2) or into state S. Going into state an exchange state requires willingness on the part of both members of a pair.

The fish that go into exchange at period 3 are engaged for periods 3, 4, and 5. When they are done, there is not enough time to have another round of exchange, as they would need a minimum of one search and one display period before beginning another exchange. This would eat up the two remaining periods. Fish that are in display at time 2 and spurn (or are spurned by) their partners can search in period 3, display in period 4, and exchange in 5, 6, and 7. Thus it may be reasonable for a type 2 fish to spurn a type 1 fish in period 2. But fish in display during period 3 will always want to exchange, because they cannot have another opportunity. To spurn a partner in period 3 means to search in 4 and display no earlier than 5 which does not leave enough time for the 3 period exchange session.[2] Likewise, fish in display in time 4 will always exchange.

Given the distribution of egg-carrying fish between types 1 and 2 and given the probability of finding a partner, at time 2 the fish are distributed among states as shown in Table 1. And at time 3 they are distributed among states as shown in Table 2. The payoffs to the two types of fish are shown in equations (5) and (6). The payoff to a fish is measured by the sum of its own fertilized eggs plus the number of eggs of other fish that it fertilizes. The details are developed in the next several pages.

Distribution of Fish Among States at Time 2

	────────────── State ──────────────		
	S	D1	D2
Type			
1	$(1-p)q$	q^2p	$(1-q)qp$
2	$(1-q)(1-p)$	$(1-q)qp$	$(1-q)^2p$

Table 1

Distribution of Fish Among States at Time 3

		──────────── State ────────────			
	S	D1	D2	E1	E2
Type					
1	p'	$q^2p(1-p)^2$	$q(1-q)p(1-p)^2$	$q^2ps_1^2$	$q(1-q)ps_2\sigma_1$
2	q'	$q(1-q)p(1-p)^2$	$(1-q)^2p(1-p)^2$	$q(1-q)ps_2\sigma_1$	$(1-q)^2p\sigma_2^2$

Table 2

To calculate the fish distribution at time 3 recall that the probability of finding a partner is proportional to the number of fish in state S, and this was initially p. At time 2 the number of fish in the state S is 1 - p times the original number; therefore, the probability of finding a partner is p(1 - p). The fraction of the original total population of egg carrying fish that are searching in time 2 and that fail to find partners is

$q(1 - p)(1 - p + p^2)$ of type 1 and $(1 - q)(1 - p)(1 - p + p^2)$ of type 2. Consequently the proportion of searchers of type 1 and type 2, respectively, at time 3 are:

$$p' = q[(1 - p)(1 - p + p^2) + qp(1 - s_1^2) + (1 - q)p(1 - s_2\sigma_1)] \tag{1}$$

$$q' = (1 - q)[(1 - p)(1 - p + p^2) + (1 - q)p(1 - \sigma_2^2) + qp(1 - s_2\sigma_1)] \tag{2}$$

Now the payoff function of an individual fish is constructed. Let the fish type be i (= 1, 2) and suppose a fish of type i plays a mixed action (t_1^i, t_2^i) where t_j^i is the probability that this fish will consent to exchange with a fish of type j. The population, apart from this individual, is assumed to be choosing according to $(s_1, s_2, \sigma_1, \sigma_2)$. The probability that the fish is in any of various states during any of several periods is given below in Table 3. That is, an entry such as the time 3 entry in the D1 column for type 1 is the probability as of the start

The Probability of a Fish Being in Various States

Time	Type	S	D1	D2	E1	E2
1	both	1	0	0	0	0
2	both	1-p	pq	p(1-q)	0	0
3	1	Q_1	$(1-p)^2 pq$	$(1-p)^2 p(1-q)$	$pqs_1 t_1^1$	$p(1-q)\sigma_1 t_2^1$
3	2	Q_2	$(1-p)^2 pq$	$(1-p)^2 p(1-q)$	$pqs_2 t_1^2$	$p(1-q)\sigma_2 t_2^2$
4	1		$Q_1 pq$	$Q_1 p(1-q)$	$(1-p)pq$	$(1-p)p(1-q)$
4	2		$Q_2 pq$	$Q_2 p(1-q)$	$(1-p)pq$	$(1-p)p(1-q)$

Table 3

of the mating process (at the beginning of time 1) that this fish will be in state D1 when time 3 comes. Let

$$Q_1 = (1 - p)(1 - p + p^2) + pq(1 - s_1 t_1^1) + p(1 - q)(1 - \sigma_1 t_2^1) \tag{3}$$

$$Q_2 = (1 - p)(1 - p + p^2) + pq(1 - s_2 t_1^2) + p(1 - q)(1 - \sigma_2 t_2^2) \tag{4}$$

In filling out the table, the probabilities for exchange states are given only for the first period in the state. This means that the probability given for state E1 or state E2 at time 4 is the probability of being in the first of the three periods of exchange at that time.

Transition probabilities are displayed in Table 4. Each line of the table shows a path that a fish can take over time between states. For example, in the third row, the succession of states is S, D1, S, D1, E1. This indicates that a fish starts in state S, displays in time 2 with another type 1 fish, returns to state S at time 3, then at time 4 it displays with a type 1 fish, and in time 5 it is in state E1. All fish are in state S at time 1 and all the relevant possibilities are represented in the various rows. In a row, between a pair of state designations, the probability is written of going from the state on the left to the state on the right, given that the fish is already in the state on the left. For example, in the third row the entry between D1 (in time 2) and S (in time 3) is $1 - s_1 t_1^1$, which is the probability that a pair of fish in display, one of type i and one of type 1, will next be in the search state.

The product of the probabilities in a specific row is the probability, as of the start of time 1, that a fish will proceed to egg exchange by the path specified in that row. As there are eight such paths, the total probability of proceeding to egg exchange by any route is the sum of the eight row-products. The payoff to a fish is equal to its own egg production (1 or z) plus eight probability-egg products. Each of these products corresponds to a row of Table 4 and is the product of the probabilities in the row multiplied by z if both fish in the exchange state are type 2 and multiplied by 1 otherwise.

The payoff functions are

$$F_1 = pqs_1 t_1^1 + p(1 - q)\sigma_1 t_2^1 + p^2 q(1 - s_1 t_1^1)Q_1 + p^2(1 - q)(1 - \sigma_1 t_2^1)Q_1 + p(1 - p)^2 + 1 \tag{5}$$

$$F_2 = pq[s_2 t_1^2 + pqQ_2(1 - s_2 t_1^2) + p(1 - q)Q_2(1 - \sigma_2 t_2^2) + (1 - p)^2]$$

$$+ p(1 - q)[\sigma_2 t_2^2 + pqQ_2(1 - s_2 t_1^2) + p(1 - q)Q_2(1 - \sigma_2 t_2^2) + (1 - p)^2]z + z \tag{6}$$

These payoff functions measure the expected number of eggs fertilized by the fish, including both the fish's own eggs and those of other fish.

4. Optimal Behavior and Equilibrium

At a population equilibrium, there will be values $(s_1^*, s_2^*, \sigma_1^*, \sigma_2^*)$ such that when they are being used by all other fish, a type 1 fish will maximize its payoff by choosing $(t_1^1, t_2^1) = (s_1^*, s_2^*)$ and a type 2 fish by choosing $(t_1^2, t_2^2) = (\sigma_1^*, \sigma_2^*)$. The first analytical question to ask is what are the values of (t_1^1, t_2^1) that maximize the payoff of the fish. This must be asked both for type 1 and type 2 fish. The second question to which we turn is whether a wholly different way of behaving would be in the interests of a fish when the rest of the population is following the general behavior outlined in §§2, 3. In particular, would "kiss and run" behavior be profitable? That is, behavior under which a fish fertilizes a parcel of another fish's eggs and then goes off to find another partner. Turning to the first question, the derivatives of the F_1 with respect to t_1^1 and t_2^1 are

$$\frac{\partial F_1}{\partial t_1^1} = F_1^1 = pqs_1(1 - pQ_1) \tag{7}$$

$$\frac{\partial F_1}{\partial t_2^1} = F_1^2 = p(1 - q)\sigma_1(1 - pQ_1) \tag{8}$$

$$\frac{\partial F_2}{\partial t_1^2} = F_2^1 = pqs_2[1 - pqQ_2 - p(1 - q)Q_2z] \tag{9}$$

$$\frac{\partial F_2}{\partial t_2^2} = F_2^2 = p(1 - q)\sigma_2[z - pqQ_2 - p(1 - q)Q_2z] \tag{10}$$

Transition Probabilities for a Fish of Type i

			Time					
1			2		3		4	5
S	pq	D1	$s_i t_1^i$	E1				
S	p(1-q)	D2	$\sigma_i t_2^i$	E2				
S	pq	D1	$1-s_i t_1^i$	S	$Q_i pq$	D1	1	E1
S	pq	D1	$1-s_i t_1^i$	S	$Q_i p(1-q)$	D2	1	E2
S	p(1-q)	D2	$1-\sigma_i t_2^i$	S	$Q_i pq$	D1	1	E1
S	p(1-q)	D2	$1-\sigma_i t_2^i$	S	$Q_i p(1-q)$	D2	1	E2
S	1-p	S	pq(1-p)	D1	1		E1	
S	1-p	S	p(1-p)(1-q)	D2	1		E2	

Table 4

For a type 1 fish, $t_1^1 = 1$ and $t_2^1 = 1$ are always optimal. If $s_1 > 0$, then $F_1^1 > 0$ and the optimal value of t_1^1 is the largest value, $t_1^1 = 1$. If $\sigma_1 > 0$ then only $t_2^1 = 1$ is optimal. The type 1 fish jointly raise their payoffs by selecting positive s_1, and the only individually optimal value is one. So the interesting case here is $s_1 = t_1^1 = 1$. Regarding t_2^1 and s_2, behavior depends on what the type 2 fish do. If $\sigma_1 = 0$, then a (1, 2) pairing for exchange in period 3 cannot occur and the choice of t_2^1 or s_2 will not change this; however, it can never hurt the type 1 fish to choose $s_2 = t_2^1 = 1$. That is, a perfect equilibrium (Selten, 1975) requires $s_2 = 1$. Therefore, $s_2 = t_2^1 = 1$ will be used below.

Looking now at the type 2 fish, it is immediate that the arguments applied above to t_1^1 and s_1 apply to the choice of t_2^2 and σ_2. That is, if $\sigma_2 > 0$, the sign of F_2^2 is the same as that of $z - pqQ_2 - p(1 - q)Q_2z = z(1 - Q_2) + pqQ_2(z - 1)$, which is clearly positive. So, here too the perfect equilibrium requires $\sigma_2 = 1$. To determine the optimal value of t_1^2, note first that $F_2^1 = 0$ unless $s_2 > 0$. Supposing $s_2 > 0$, the sign of F_2^1 is the same as the sign of $1 - pqQ_2 - p(1 - q)Q_2z$. This, in turn, is positive if

$$z < 1 + \frac{1 - pQ_2}{pQ_2(1 - q)} = z^* \tag{11}$$

The right hand side of equation (11) is a critical value for z, denoted z^*. If $z < z^*$, then it is optimal for a type 2 fish to mate with a type 1 fish when such a pair is displaying in period 2, but if $z > z^*$, it is better to spurn the type 1 fish and search again in the hope of finding a type 2 fish. It is clear that z^* increases as q increases and falls as p increases. That is, the rarer are the type 2 fish, the less selective they will be and the easier it is to find a partner during a search phase, the more selective they will be. If type 2 fish were 70% of the population (i.e., $q = .3$) and the probability of pairing were initially .7, then $z^* = 2.38$. These conditions favor a low critical value, but $z^* = 2.38$ is high in the sense that one would expect few members of the egg carrying fish population to vary this much. The reason is that clutch size is probably about proportional to body mass and the adult hamlets show little size variation. According to Fischer (1986) the mean length of fish he observed is 83.3mm and the standard deviation of length about 5mm.

Thus we have values $(s_1^*, s_2^*, \sigma_1^*, \sigma_2^*)$ that give equilibrium behavior as long as no fish playing first the male role has an incentive to search for a new partner immediately after fertilizing the first parcel of its current partner's eggs. Whether this incentive is present depends upon the parcel size. We investigate this question next by finding the parcel sizes such that no fish would gain fitness from such behavior.

The incentive to fertilize a few of someone else's eggs and then swim away can be dealt with as follows: Suppose that the first parcel of eggs in an egg trading session is released so fast that the fish in the male role can fertilize them and get into the search mode quickly enough that it has effectively the whole period to search. Doing this can only be valuable in period 3. Any later time is too late to search, display, and trade. To make "kiss and run" behavior unprofitable, the value of an individual parcel of eggs must be less than the difference between the payoff as of period 3 to being in exchange and to being in search.

Let the egg quantity in a single parcel be λ, so that each type 1 fish will lay $1/\lambda$ parcels.[3] Suppose it is the start of time 3 and that we are examining the situation of a fish that has just entered state E. The payoff to a type 1 fish that does not fertilize and run is 2 and the (expected) payoff if it does fertilize and run is $pQ_1 + 1 + \lambda$. The former is larger than the latter if

$$\lambda < 1 - p[(1 - p)(1 - p + p^2) + p(1 - q)(1 - \sigma_1)] = \lambda_1^* \tag{12}$$

The right hand side of equation (12) gives the critical parcel size for a type 1 fish, λ_1^*, and it can be analyzed for the two extreme cases of (a) $q = 0$, $\sigma_1 = 0$ and (b) $\sigma_1 = 1$. For the former case

$$\lambda_1^* = 1 - p + p^2 - 2p^3 + p^4 \tag{13}$$

$$\frac{\partial \lambda_1^*}{\partial p} = - (1 - p)^2 - p^2(5 - 4p) < 0 \tag{14}$$

From equation (14) it is clear that equation (13) reaches a minimum when $p = 1$ and that minimum is $\lambda_1^* = 0$. For case (b),

$$\lambda_1^* = 1 - p + 2p^2 - 2p^3 + p^4 \tag{15}$$

$$\frac{\partial \lambda_1^*}{\partial p} = - 4(p - \tfrac{1}{2})^2 + 4p^2(p - \tfrac{1}{2}) \tag{16}$$

The second derivative of equation (15) is postive for all values of p between zero and one, so equation (15) has a unique minimum at $p = 1/2$, where equation (16) is zero. The corresponding value of λ_1^* is 13/16. Note, too, that, for any value of p, λ_1^* rises as $(1 - q)(1 - \sigma_1)$ falls. Therefore, there is no smallest parcel size in the extreme case (a), but otherwise there is. If $0 \le q \le .5$, there is a smallest parcel size that is never less than q, and the smallest value occurs when $p = 1$. Denote the critical parcel size for type 1 fish by $\lambda_1^* = \Lambda_1(p, q, \sigma_1)$.

For a type 2 fish that enters the exchange state with another type 2 fish, the payoff is 2z when it does not "kiss and run" and the (expected) payoff if it does is $pQ_2z + z + \lambda$. The former is larger than the latter if

$$\lambda < z - p[(1 - p)(1 - p + p^2) + pq(1 - \sigma_1)]z = \lambda_2^* \tag{17}$$

It is clear from inspection of equation (17) that many characteristics of the critical parcel size, which may be denoted $\lambda_2^* = \Lambda_2(p, q, \sigma_1)$, can be inferred from the analysis of $\Lambda_1(p, 1 - q, \sigma_1)$. In particular for any (p, q, σ_1), $\Lambda_2(p, q, \sigma_1) = z\Lambda_1(p, 1 - q, \sigma_1)$.

It is easy to calculate sample values for λ_1^* and λ_2^* using plausible numbers for the parameters. Suppose $z = 1.5$. As long as z^* exceeds z, the equilibrium value of $(s_1^*, s_2^*, \sigma_1^*, \sigma_2^*)$ is (1, 1, 1, 1) and the size of q does not affect either λ_1^* or λ_2^*. Then λ_1^* never falls below .8 and λ_2^* never falls below 1.2 as p varies between .1 and .9. The minimum values occur at $p = .5$. These calculations indicate that two parcels per fish (i.e., $\lambda = \frac{1}{2}$) would be sufficient most of the time. Observed values of λ are more typically $\frac{1}{4}$ and smaller, indicating four or more parcels per fish. The larger number, of course, provides considerable leeway to accommodate pairing of fish that are very disparate in size. The hamlets, then, appear to parcel their eggs to a great enough extent to effectively discourage the "kiss and run" tactic. At the same time, their degree of parceling does not go far beyond what is required and, finally, Fischer observed very little "kiss and run" behavior.

5. The Scope for Male Specialization

As the behavior and apparent perceptions of the hamlets are modeled above, fish without eggs never go to the mating area because the hamlets can tell whether a fish has eggs and they will not initiate exchange with a fish lacking eggs. In the present section that assumption is slightly modified. We assume that hermaphrodite fish without eggs never go to the mating area, but that there are pure male fish that do go and that have a chance to fool other fish into starting the egg trading process. If a pure male could do this, it would have the advantage that it could breed every day; whereas, the hermaphrodite fish breed only at intervals of some days when they have eggs. The periodicity of egg production in hamlets is not known; however, the analysis below suggests whether male specialization is viable by giving an estimate of conditions under which it would pay.

To simplify matters, we assume here that there is only one sort of hermaphrodite hamlet -- one having one unit of egg capacity. The timing process is like this: At the start of the 70 minute breeding period, all fish are in the search state S. The proportion of males is q and of hermaphrodites is 1 - q. All hermaphrodites at the breeding area possess eggs. At the end of the search process fish are paired randomly according to the rules used in earlier sections. If a male (M) is paired with an hermaphrodite (H), there is a probability r that the male has fooled the other fish and they proceed to display. After display, they go to egg trading, but after one spawning bout the male does not reciprocate and further spawning ceases. We assume this happens so fast that the whole period is spent by both fish in search. With probability 1 - r the hermaphrodite is not fooled, so the fish do not proceed to display, but, instead, return directly to the search state.

Several assumptions are made that are meant to favor the fish specialized on male function.

(1) An hermaphrodite who is not fooled will discover its partner is male at the start of the display process, rather than at the end. This permits the male to return more quickly to finding a different partner and increases his chances to breed. Of course, it helps the other fish to get more quickly into circulation too, but the potential gain to the male is greater.

(2) When a male pairs with, and fools, an hermaphrodite and they proceed to egg trading, the hermaphrodite is automatically the first to spawn. There is no reason to presume this based upon observed fish behavior; indeed it is more plausible that each fish has ½ chance of being first, and, if the male were first, the trading process would stop without any spawning.

(3) If pure males can fool egg carrying fish, then presumably an hermaphrodite without eggs could do so as well. We assume here that the eggless hermaphrodites do not attempt to breed because the additional offspring are not worth the decrease in foraging time. Being a pure male means a fish does not have the cost of female reproductive organs; however, the pure male may be better off spending more time breeding, and less foraging. The hermaphrodite lacking eggs may not find it worthwhile to curtail foraging time in favor of breeding while eggless.

The three conditions listed above are relaxed later in the section. Tables 5 to 7 show the distribution of the population at times 2, 3, and 4, respectively. The state to which fish go is, in each of the tables, broken down into categories to better understand what is happening. The top row in each table accounts for fish that failed to make a pairing. The row showing SHM accounts for fish that are paired with fish of the other type where the male did not fool the hermaphrodite. The row showing SMM accounts for males paired with males. The row labeled DHH accounts for hermaphrodite pairs and DHM refers to pairs of different types where the male fooled the hermaphrodite.

The total number of male fish searching at time 2 is $M_2 = q(1 - pr + pqr) = q(1 - Q_2)$ and the number of hermaphrodite fish is $H_2 = (1 - q)(1 - p + pq - pqr)$. Let $S_2 = H_2 + M_2$, $p_2 = pS_2$, and $q_2 = M_2/S_2$.

The total number of male fish in the search state at time 3 is $M_3 = M_2(1 - prH_2) + qQ_2$ and the number of hermaphrodite fish is $H_3 = H_2(1 - pH_2 - prM_2) + qQ_2$. Let $S_3 = H_3 + M_3$, $p_3 = pS_3$, and $q_3 = M_3/S_3$. The number of eggs fertilized by hermaphrodite fish per single hermaphrodite fish per day is

$$F_H = p(1 - q) + \frac{p}{1 - q}(H_2^2 + H_3^2) \tag{18}$$

Fish Distribution at Time 2

State at time 1	2	M	H
S	S	$(1 - p)q$	$(1 - p)(1 - q)$
S	SHM	$pq(1 - q)(1 - r)$	$pq(1 - q)(1 - r)$
S	SMM	pq^2	
S	DHH		$p(1 - q)^2$
S	DHM	$pq(1 - q)r = qQ_2$	$pq(1 - q)r = qQ_2$

Table 5

The number of eggs per male fish per day that are fertilized by males is

$$F_M = \lambda[Q_2 + prH_3 + prH_2(1 - Q_2)(1 - prH_3)] \tag{19}$$

Suppose that all offspring of an H-H mating are hermaphrodites, that the offspring of an H-M mating split equally between hermaphrodite and male, and that a male can participate in μ days of breeding activity for each single day that a hermaphrodite can participate. This means that the total population of hermaphrodite fish is $\mu(1\text{-}q)$, given a proportion of hermaphrodites to males at the breeding ground of $(1 - q)$ to q. The total number of eggs fertilized by hermaphrodites is $\mu(1 - q)F_H$, all of which yield hermaphrodite offspring, and the number of eggs fertilized by males, yielding half hermaphrodite and half male offspring, is $\mu q F_M$. If the proportion of males in the population is to grow, then the growth rate of males must exceed the growth rate of females, or

$$\frac{\mu q F_M}{2q} > \frac{\mu(1 - q)F_H + \frac{1}{2}\mu q F_M}{\mu(1 - q)} \qquad \text{or} \tag{20}$$

$$\mu > \frac{2F_H}{F_M} + \frac{q}{1 - q} = \mu^* \tag{21}$$

Values for μ^* are given in Table 8 for sample values of p and r. If the actual value of μ is beneath that shown in Table 8, then the male population would shrink over time. If the probability of pairing were initially .6 and the probability that a male could fool an hermaphrodite were .3, then there would have to be at least 31 days between egg clutches for the hermaphrodites in order that males could prosper. As q, the relative size of the male population, grows, the critical value of μ increases from 31, which holds for q = 0. At q = .3, the critical value of μ is 47, at q = .7 it is q = 56.

In the preceding analysis three assumptions were made that favor the pure male. They are 1) if an H-M pair forms during a search phase and the hermaphrodite is not fooled, then the two fish separate

Fish Distribution at Time 3

State at time 2	State at time 3	M	H
S	S	$(1 - p_2)M_2$	$(1 - p_2)H_2$
S	SHM	$pM_2H_2(1 - r)$	$pM_2H_2(1 - r)$
S	SMM	pM_2^2	
S	DHH		pH_2^2
S	DHM	pM_2H_2r	pM_2H_2r
D	S	qQ_2	qQ_2
D	E		$p(1 - q)^2$

Table 6

Fish Distribution at Time 4

State at time 3	State at time 4	M	H
S	S	$(1 - p_3)M_3$	$(1 - p_3)H_3$
S	SHM	$pM_3H_3(1 - r)$	$pM_3H_3(1 - r)$
S	SMM	pM_3^2	
S	DHH		pH_3^2
S	DHM	pM_3H_3r	pM_3H_3r
D	S	pM_2H_2r	pM_2H_2r
D	E		pH_2^2

Table 7

immediately and spend the subsequent period in the search state, 2) in an H-M pair when the hermaphrodite is fooled, the hermaphrodite always spawns, and 3) hermaphrodite fish that lack eggs never attempt to breed. These three assumptions are removed singly below. That is, while any one is removed, the other two remain in effect.

Removing the first assumption means supposing that when an H-M pair forms, they proceed through the next period in the display state. Then, at the end of the display state, the hermaphrodite is fooled with probability r and not fooled with probability 1 - r. Males that fail to fool their hermaphrodite partners return as slowly to the search state as those who succeed. F_H in equation (18) is unchanged except that $M_2 = q(1 - p + pq)$, $H_2 = (1 - p)(1 - q)$, $M_3 = (1 - pH_2)M_2 + pq(1 - q)$, and $H_3 = M_3 - pH_2^2$. F_M is changed, and is given below.

Critical Values for the Breeding Frequency
of Hermaphrodites for $\lambda = \frac{1}{6}$ and q = 0

r	.1	.2	.4	.6	.8	1
1	11.6	11.3	10.7	10.5	10.7	12
.7	16.3	15.6	14.7	14.3	14.7	17.1
.4	28.0	26.5	24.5	23.7	24.7	30
.3	37.2	35.0	32.1	31.0	32.5	40
.2	55.4	51.9	47.3	45.7	48.1	60
.1	110.1	102.6	93.1	89.8	95.0	120

Table 8

$$F_M = \lambda pr \left[1 - q + \frac{M_2H_2 + M_3H_3}{q} \right] \tag{22}$$

Table 9 repeats the information from Table 8 with the situation modified to reflect the slower discovery process. The critical values of μ rise across the board except when p = 1. For example, for p = .6 and r = .3, the value in Table 9 is 46.8 as compared with 31 in Table 8.

Suppose our assumption that H-M pairs always have a single spawn is modified to assume they have one spawn every second pairing, based on the condition that a fish is randomly selected for the female role and, when a male is supposed to take the female role first, the pairing breaks up with no spawn occurring. Then, this is equivalent to halving the value of λ and, in turn, doubles all the values in Table 8. Thus for p = .6 and r = .3 the critical value of μ becomes 62. If the hermaphrodites can have eggs about six times per year, it is too often to allow for viable males. Both assumptions (1) and (2) can be removed simultaneously; then the values in Table 9 must be doubled.

Finally, suppose that all the hermaphrodites that are eggless behave like the pure males and attempt to breed. Suppose also that an hermaphrodite with eggs treats males and eggless hermaphrodites identically. Letting q denote the fraction of the population consisting of pure males plus eggless hermaphrodites and 1-q

Critical Values for the Breeding Frequency of Hermaphrodites

for $\lambda = \frac{1}{6}$ and q = 0 and with Slow Discovery

r	p					
	.1	.2	.4	.6	.8	1
1	21.8	19.9	16.6	14.0	12.5	12
.7	31.1	28.4	23.7	20.0	17.8	17.1
.4	54.5	49.7	41.4	35.1	31.2	30
.3	72.7	66.3	55.2	46.8	41.6	40
.2	109.0	99.4	82.8	70.2	62.5	60
.1	218.0	198.8	165.7	140.3	124.9	120

Table 9

the proportion of hermaphrodites with eggs, equation (18) gives the number of fertilized eggs per hermaphrodite with eggs that results from a pairing of two egg carrying fish. Equation (19) gives the number of fertilized eggs per non-egg carrying fish that results from a pairing of an egg carrying fish with a non-egg carrying fish. Let u denote the proportion of the population consisting of pure males. Then the total number of fertilized eggs of all types is $qF_M + (1 - q)F_H$. Of these $qF_M/2$ result in male fish and $(1 - q)F_H + (q - u/2)F_M$ result in hermaphrodite fish. The fraction of male fish in the population will grow if

$$\frac{\tfrac{1}{2}uF_M}{u} > \frac{(1 - q)F_H + (q - \tfrac{1}{2}u)F_M}{1 - u} \qquad \text{or} \qquad (23)$$

$$(\tfrac{1}{2} - q)F_M > (1 - q)F_H \qquad (24)$$

Thus q, the proportion of males plus the proportion of eggless hermaphrodites cannot exceed ½ in any case. This places a severe restriction on the frequency with which hermaphrodites must be carrying eggs if a male population is to be sustained. If hermaphrodites have ripe eggs every μ days, then the ratio of eggless hermaphrodites to egg carrying hermaphrodites is μ - 1 to one. From 1 = u + μ(1 - q) it follows that μ =

$(1 - u)/(1 - q)$. An upper bound on the value of μ, given that $q < \frac{1}{2}$ and $u \geq 0$, is 2. Thus, if the fish have eggs less frequently than every second day, it will be impossible to sustain a population containing male fish.

The analysis above suggests that the chances for success as a specialized male are virtually nil. On the one hand, the form of the display process and the nonparticipation of eggless hermaphrodites strongly suggests the fish can perceive whether another fish is carrying ripe eggs. This, in turn, implies a low value for r and that means the hamlets would have to spawn extraordinarily seldom if the male specialist were to be able to compete successfully.

6. Concluding Comments

The behavior of the hamlet is sophisticated in the sense that cooperation is involved between fish and a fish avoids behaving in a myopically selfish fashion as it would if it deserted a partner as soon as it fertilized the partner's eggs. Their actual behavior is, nonetheless, selfish in the sense of being noncooperative equilibrium behavior; however, the fish have evolved strategies of play that bring about a cooperative outcome by means of noncooperative behavior. There appear to be four "enforcement mechanisms" for the hamlets' behavior. They are (1) hamlets do not form long term mating relationships, (2) the search and display time, (3) the dividing of eggs into many parcels, and (4) the one day viability of the eggs. As the observation of *Serranus tigrinus* indicates, the fish would have an incentive to alternate male and female roles with one another if they formed long term mating relationships. As the hamlets do not do this, they need a substitute means of ensuring that a fish playing the female role will have the opportunity to play the male role. Given (1), then (2), (3), and (4) all reduce the desirability of cheating. Search and display time introduces a cost to seeking a new partner, while the limited viability of the eggs intensifies this cost. The parceling of eggs reduces the gain to cheating and, in addition, means that a fish that has just spawned remains attractive because it still has eggs to continue attracting its current partner. What, then, are the essential mechanisms among these four? We believe all are essential except for the one day viability of eggs. Mechanisms (2) and (4) both pertain to switching costs, but search and display would cause a switching cost even if eggs had an indefinitely long life. This is because the time and energy used for search and display takes away from foraging time. Suppose, for example, that the fish had the ability to make eggs ready for use just seconds before spawning and that the unreadied eggs would last the lifetime of the fish. This would reduce the pressure on the fish to find partners and stick with them. However, if the display time were significantly lengthened, then the cost of switching could be increased sufficiently to provide the proper incentive.

Notes

1. If a random device were to select, with probability .5 for each fish, one fish of a pair to take the female role first, and if the selfish mutant actually releases its eggs when selected to be first, then the policy of swimming away when the partner were first selected would raise the mutant's fitness by about 50%.

2. Maybe it is too rigid to force this condition. One could say that limited exchange can take place. The simpler assumption is made because it eases calculation, leaving only one time period in which a real choice is possible.

3. Strictly speaking, the number of parcels would be the smallest integer that is no less than $1/\lambda$.

References

Axelrod, Robert and William D. Hamilton, 1981, "The Evolution of Cooperation," *Science* 211, 1390-1396.

Fischer, Eric A., 1980, "The Relationship Between Mating System and Simultaneous Hermaphroditism in the Coral Reef Fish, *Hypoplectrus Nigricans* (Serranidae)," *Animal Behavior* 28, 620-633.

Fischer, Eric A., 1981, "Sexual Allocation in a Simultaneously Hermaphroditic Coral Reef Fish," *The American Naturalist* 117, 64-82.

Fischer, Eric A., 1986, "Mating Systems of Simultaneous Hermaphroditic Serranid Fishes," in T. Uyeno, R. Arai, T. Taniuchi, and K. Matsuura, eds. *Proceedings of the Second Indo-Pacific Fish Conference*, Ichthological Society of Japan, Tokyo, 776-784.

Fischer, Eric A., 1988, "Simultaneous Hermaphroditism, Tit-for-Tat, and the Evolutionary Stability of Social Systems," *Ethology and Sociobiology* 9, 119-136.

Fischer, Eric A. and Chris W. Petersen, 1987, "The Evolution of Sexual Patterns in the Seabasses," *BioScience* 37, 482-489.

Selten, Reinhard, 1975, "Reexamination of the Perfectness Concept for Equilibrium Points in Extensive Games," *International Journal of Game Theory* 4, 25-55.

COMPETITION AVOIDANCE IN A DRAGONFLY MATING SYSTEM

by

Hans J. Poethke and Franz J. Weissing

Abstract

Mature males of the dragonfly species Aeschna cyanea regularly visit ponds or rivers where they patrol along the shoreline in search for females. Whenever two males encounter another during their visits at the water, a peculiar interaction develops which quite often results in one or even both participants leaving the mating place. In the present paper we shall analyse the question whether the tendency to leave after an intermale encounter can be interpreted as an adaptive response to local circumstances. In particular, the leaving tendency will be viewed as a strategic decision to avoid local high density situations on the basis of private information which is gained by encountering other males.

It is easy to see that an adaptive decision to leave a mating place should depend on the presumed reaction of the opponent. Accordingly, an intermale encounter will be modelled as an evolutionary normal form game where the payoffs are given in terms of mating chances which are to be expected on the basis of a male's information situation. Optimal updating of a male's private information strongly depends on the behavioural norms which are established in the population. This implies that there is a strategy—payoff feedback: On the one hand, selection induced by payoff differences leads to a change in the strategic structure of the population; on the other hand, the payoffs are not externally given and fixed, but they themselves evolve in reaction to the evolution of the population strategy.

The concept of evolutionary stability will be extended in order to cope with such strategy—payoff feedbacks which arise quite naturally in biological applications. On the basis of this concept, several evolutionary games will be analysed. In all our models, we get a unique evolutionarily stable leaving tendency that corresponds to a completely mixed strategy. Even for realistic parameter constellations, however, our models generally overestimate the leaving tendency when compared to empirical data. In the last section, we shall outline a refined model which is in better agreement with field observations.

1. Introduction

The greatest part of each day Aeschnid dragonflies spend hunting for prey. During this time they can be found flying at varying heights along forest fringes and hedges and catching insects. When hunting, dragonflies hardly react to conspecifics crossing their way. Several times per day, however, mature males visit ponds or rivers in search for females which come to the water for oviposition. In contrast to some other dragonfly species (Corbet 1980), Aeschna cyanea males will usually not occupy territories at the mating place. Instead of staying at a certain site, they patrol along the whole shoreline randomly changing the direction of their flight and often hovering on the spot (for the spatial organization at a pond see Kaiser (1976), Poethke & Kaiser (1987), and Poethke (1988)).

Females arrive at a pond at random throughout the day, but there are always more males than receptive females at the water (Kaiser 1985). Thus, for a single male the chance of encountering a receptive female is a decreasing function of the male density at the pond. The more males are competing for females at the same time, the smaller are the mating chances for each of them. Since every male has only a limited time budget to spend at the mating place (Kaiser 1974b), a male should leave the pond when male density is high, and he should try to visit the pond at those times where there is less competition.

However, males have very limited information about how many rivals actually are present at a pond. In fact, they are usually not able to notice a conspecific that is further away from them than five meters, and sometimes they do even not perceive the presence of a conspecific which is less than one meter away. As soon as a patrolling male catches sight of another dragonfly (or any other moving object of approximately the same size), he rushes at it. If it is a conspecific female, she will flee and the male will follow her and try to seize her and to copulate with her. If the moving object is another male, the two males will dart towards each other. Circling around each other and sometimes even clashing with each other they will leave the shoreline together (Kaiser 1974c).

After such an intermale encounter, usually both males will seperately return to the water. Frequently, however, one or even both of them do not return to the pond but switch to another mating place or to feeding flight. From field observations, there are no indications that intermale encounters have the character of a fight where the loser is chased away by the winner. Hence, we have to address the question whether leaving is profitable for the leaving individual itself.

An individual male should leave the mating place whenever another try at another pond appears to be more profitable to him than staying at a shoreline that is crowded by competitors. If there are large fluctuations in local male density, an individual dragonfly should use all the information he has on actual male density as a basis for his decision whether to leave or to stay. In view of the limited visual capacities of dragonflies, we assume that intermale encounters provide the main source of information an individual male may get on actual male density at the pond. Encountering another male, an individual dragonfly knows that there is at least one competitor at the pond, and he should use this knowledge to update his estimate of local male density. If the updated estimate indicates that local male density is high, he should leave the pond; otherwise, he should stay.

However, there is a game—like component in such an updating process leading to the optimal use of private information. In fact, after an encounter has taken place, the new estimate of local male density should be influenced by the presumed reaction of the competitor. The updated estimate should be higher if the competitor is expected to stay, and it should be lower if he is expected to leave. In this paper, we shall develop several evolutionary normal form games in order to model this situation. The payoffs in each of these games are given by expected mating chances, where expectations are based on a male's information situation.

We assume that any new piece of 'private' information is taken into account by means of a Bayesian updating process. It is easy to see that optimal updating should strongly depend on the behavioural norms that are established in the population. Thus, we cannot assume that the payoffs of our games are externally given and fixed. Instead, we have to build a 'closed' model, where there is a *strategy–payoff feedback:* the strategies evolve according to the payoff structure in a binary interaction — the payoffs, however, evolve themselves since they are functions of the population strategy. We believe that situations like this arise quite naturally in biological applications. In Section 5, we shall outline how the concept of evolutionary stability can be extended to those evolutionary games, where the payoffs are internally given by the strategic structure of the population.

2. The Information Situation at the Mating Place

Aeschna cyanea males visit a pond for mating, but the chance to be the first male to meet a receptive female will be the smaller the more conspecific males are patrolling at the shoreline. Let i be the male density at the water. If $g(i)$ denotes the chance that a particular male will get access to one of the receptive females present at the pond, $g(i)$ will be a decreasing function of i:

$$g(i) < g(j) \quad \text{for} \quad i > j. \tag{1}$$

This implies that it is profitable for a male to make use of the information he has on local male density. When male density is far above average, he should leave the pond and fly to another mating place or switch to another activity in order to reserve his mating–time budget for low–density situations.

The model considerations in the next two sections will be based on a somewhat simplified view of a dragonfly's informational state. A male dragonfly will usually not know how many rivals are with him at the water, since his visual sense does not allow him to have an overview over the whole mating place. Let us assume that we know how many dragonfly males are in the area and how they usually behave. Based on this knowledge, we will be able to calculate the density distribution p of males at a mating place. We shall derive this distribution in Section 4.2.

A particular male at a pond, however, will never face this 'objective' density distribution p, but rather a 'subjective' distribution which is conditional on his particular situation. If he has just arrived, he 'knows' that there is at least one dragonfly male at the water (he himself) and that there has not been time enough for him to meet a conspecific. The conditional male density distribution corresponding to this information situation will be called q^a. The situation is changed if the male gets additional information by encountering another male. Encountering a rival, he 'knows' that there are at least two males at the water. Taking into account the density dependent encounter rates, the corresponding updated density distribution q^e can be calculated.

For simplicity, the models developed in the next two sections will be based on the assumption that newly arriving at a pond on the one hand and encountering a male rival on the other are the only pieces of information on which a male may base his strategic decision. In addition, we shall implicitly assume that any new private information on male density at the mating place will decay in the course of time. If some *relaxation time* has passed after the acquisition of a new piece of information, a male dragonfly will again be confronted with q^m, the mean density distribution a male experiences during a visit at a mating place.

In order to motivate the analysis in the next two sections, let us assume that the three density distributions q^a, q^e, and q^m which correspond to different information situations may be characterized by their mean values

$$n^a := \sum_i q^a(i) \cdot i, \tag{2}$$

$$n^e := \sum_i q^e(i) \cdot i, \tag{3}$$

$$n^m := \sum_i q^m(i) \cdot i. \tag{4}$$

Newly arriving at a pond, a male 'expects' a mean density which is given by n^a. During his visit, the density he expects to be realized will change according to gain and decay of information. Information is gained whenever he meets a conspecific rival. Encountering another male, the expected density will change to n^e. However, the encounter may cause the competitor to leave the pond. If α denotes the probability of leaving, the expected density immediately after returning from an encounter will be given by

$$(1-\alpha) \cdot n^e + \alpha \cdot (n^e - 1) = n^e - \alpha. \tag{5}$$

This process of gaining and losing information will be discussed in detail in Sections 4 and 8. A preliminary scetch of it is shown in Figure 1. Figure 1a depicts the information situation of a male assuming that his informational state may be represented by his expectation n^{exp} of male density at the mating place. Without further information, a male should expect a density that is given by n^m. Accordingly, the additional information he has got by encountering other males corresponds to the difference between n^{exp} and n^m. The diagram in Figure 1a is based on the assumption that this 'private' information of a male decays at an exponential rate. The expected (instantaneous) mating success G of a male is a direct function of his expectation of male density. It is depicted in Figure 1b, assuming that $g(i)$ is inversely proportional to i.

In the following sections, the *leaving tendency* α will be considered as a strategic parameter describing a male's reaction to the change of his information situation caused by an intermale encounter. We shall develop some game models which will permit us to derive the evolutionarily stable leaving tendency α^*.

(a)

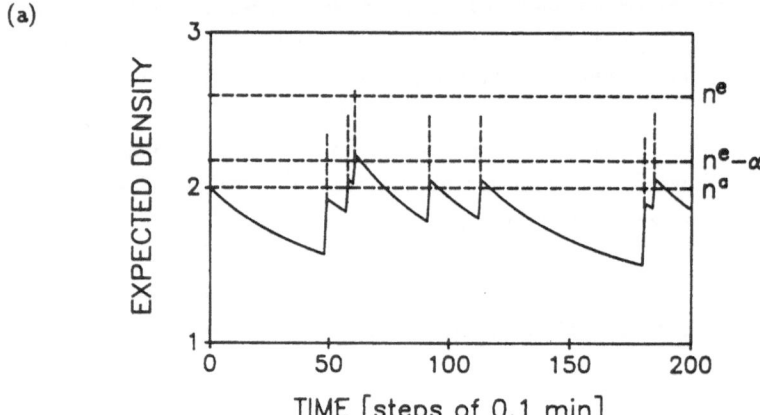

(b)

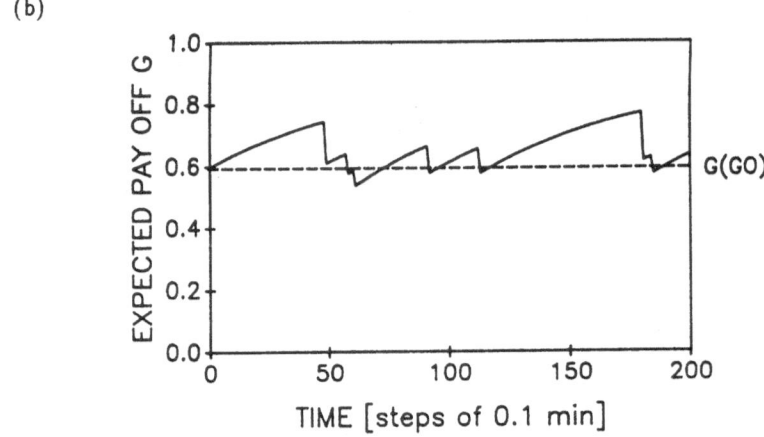

FIGURE 1: (a) Change of a male's expectation of male density during his visit at the pond.
Private Information on male density is assumed to decay at an exponential rate.

(b) Associated change in expected instantaneous mating success. Mating success is assumed to be inversely proportional to male density.

3. The Low Density Case

Let us first consider a simple world consisting of a small number of dragonfly males sharing a large number of ponds. In this world, the probability that more than two males are simultaneously at a pond is extremely small and as a first order approximation we may ignore it:

$$q^a(i) \ll 1 \quad \text{for} \quad i > 2. \tag{6}$$

We are interested in the situation where two dragonflies meet each other. Let us assume that there are two males (called X and Y) patrolling at the same pond searching for females. Suddenly X encounters Y. Both males dart towards each other (thus indicating that they are males) and leave the shoreline together. As soon as male X loses visual contact to Y, he faces the question whether to return to the pond ('STAY') or to leave it ('GO'). If he leaves, he may switch directly to another pond. We assume that X does not know whether Y returned to the pond or whether he left.

Based on our assumption on the decay of information, we shall focus our interest on a short time interval immediately after the encounter and on the instantaneous payoff (= expected mating chances per short time unit) male X will have during this interval (see Section 7). Since we ignore the possibility that there is a third dragonfly male at the water, a male which stays after an encounter will either get a payoff of $g(2)$ (if the opponent stays too) or a payoff of $g(1)$ (if the competitor leaves the pond). A male that leaves after an encounter will get another chance at another pond. Ignoring travel costs (see Section 7), his expected instantaneous payoff depends on q^a, and it is given by

$$G(q^a) = \sum_i q^a(i) \cdot g(i). \tag{7}$$

Using this notation, the conflict of interest may be characterized by a symmetric normal form game which is described by the following payoff matrix (only the payoff for individual X is shown):

X: \ Y:	STAY	GO
STAY	$g(2)$	$g(1)$
GO	$G(q^a)$	$G(q^a)$

(8)

We shall now calculate the unique symmetric Nash equilibrium point of this game. Notice that in view of (6), $G(q^a)$ may be written as

$$G(q^a) = q^a(1) \cdot g(1) + q^a(2) \cdot g(2) \tag{9}$$

which implies

$$g(1) > G(q^a) > g(2) . \tag{10}$$

These inequalities show that (8) has two asymmetric Nash equilibria in pure strategies (X stays and Y goes; X goes and Y stays). Since the labels X and Y were attached arbitrarily to the two participants of an encounter, the opponents cannot base their strategic decisions on them. Consequently, only symmetric equilibria are biologically meaningful (see Selten 1983) and we have to look for an equilibrium in mixed strategies. Every symmetric Nash equilibrium of (8) may be characterized by the leaving probability α^* it ascribes to each of two participants of an encounter. α^* will be called a *Nash equilibrium strategy* if the strategy pair (α^*, α^*) is a symmetric Nash equilibrium of (8).

Suppose that male X encounters an opponent that has the tendency α to leave the mating place. Then the expected payoff to X for his two pure strategies is given by

$$G_\alpha(\text{STAY}) = \alpha \cdot g(1) + (1-\alpha) \cdot g(2) , \tag{11}$$

$$G_\alpha(\text{GO}) = G(q^a) . \tag{12}$$

A mixed equilibrium strategy α^* has to be such that both pure strategies yield the same payoff, since only then there is no selective advantage in deviating from it. Therefore, we must have

$$G_{\alpha^*}(\text{STAY}) = G_{\alpha^*}(\text{GO}) . \tag{13}$$

In view of (9), (13) yields:

$$\alpha^* = q^a(1) , \tag{14}$$

i.e., the equilibrium leaving tendency is identical to the probability that a newly arriving male dragonfly finds himself alone at the pond.

It is easy to see that this prediction of our low density model is in sharp contrast to the behaviour of dragonfly males which is observed in the field. From field data we know that Aeschna cyanea males leave the pond after an encounter in about 15% of all cases. If the situation for this species were described correctly by our model, we would have $q^a(1) = \alpha^* = 0.15$ and, in view of (6), $q^a(2) = 1-\alpha^* = 0.85$. Notice that this means that in 85 per cent of all arrivals at the water, a newly arriving male would be confronted with a competitor already being there. Obviously, this implication of (14) is not compatible with our low density assumption. In order to get a better fit between model predictions and empirical data, we have to develop a more detailed model.

4. A General Normal Form Game

4.1 The Payoff Matrix

Even if we do not assume that male density at all ponds is extremely low, the conflict of a dragonfly male that has just encountered a competitor can still be modelled as a symmetric normal form game. The resulting payoff matrix resembles very much that obtained for the low density case, but it is complicated by the fact that now the payoffs for 'STAY' are not known exactly but have to be derived from the expected density distribution q^e.

As with the low density case, each male that leaves the mating place will have another try at another pond. Correspondingly, the expected payoff for leaving will be given by (7). If *both* males stay, they will be confronted with the density distribution q^e, and their expected payoff will be given by

$$G(q^e) = \sum_i q^e(i) \cdot g(i) . \tag{15}$$

If one of the competitors leaves the water, male density will be reduced by one, i.e. the new density distribution is described by q^e_{-1} where

$$q^e_{-1}(i) := q^e(i+1) \quad \text{for all } i . \tag{16}$$

Notice that in this case the expected payoff is given by

$$G(q^e_{-1}) = \sum_i q^e(i) \cdot g(i-1) . \tag{17}$$

Consequently, we are dealing with a game the payoff structure of which is characterized by the following payoff matrix:

	Y: STAY	GO
X: STAY	$G(q^e)$	$G(q^e_{-1})$
GO	$G(q^a)$	$G(q^a)$

(18)

Obviously, we have

$$G(q^e) < G(q^e_{-1}) . \tag{19}$$

It is easy to see that the pure strategy 'STAY' is a Nash equilibrium strategy of (18) — and therefore a candidate for being an *evolutionarily stable strategy* or *ESS* (see Section 5) — if and only if

$$G(q^e) \geq G(q^a) . \qquad (20)$$

On the other hand, the pure strategy 'GO' is a Nash equilibrium strategy of (18) if and only if

$$G(q^a) \geq G(q^e_{-1}) . \qquad (21)$$

We shall now argue that it is rather unlikely that (20) or (21) hold true for 'realistic' parameter constellations. Recall that $G(q^a)$ is the expected payoff immediately after a male's arrival at the mating place and that $G(q^e)$ denotes the expected payoff on the basis of the information that he has just encountered a conspecific male. After his arrival, a male 'knows' that at least one dragonfly male (he himself) is at the pond whereas an encounter reveals to him that at least two males are present. Thus, it is rather implausible that inequality (20) should hold. On the other hand, the payoff difference $G(q^e_{-1}) - G(q^e)$ is positive (see (19)) since it is always good for a male if one of his competitors leaves the pond. In fact, this outcome is the best conceivable for a male which decides to stay at the mating place. Accordingly, it is likely that the payoff difference $G(q^e_{-1}) - G(q^e)$ exceeds that of $G(q^a) - G(q^e)$, a fact that is not compatible with inequalities (19) and (21). For these reasons, we shall focus our further considerations on the inequalities

$$G(q^e_{-1}) > G(q^a) > G(q^e) , \qquad (22)$$

which are satisfied in all examples that will be considered below.

If (22) holds true, the game (18) has a unique mixed Nash equilibrium strategy, i.e. a leaving tendency α^*, $0 < \alpha^* < 1$, which is characterized by the fact that the expected payoffs for the two pure strategies are equal to another. As in (11) and (12), the expected payoff to an individual male X for his pure strategies are of the form

$$G_\alpha(\text{STAY}) = \alpha \cdot G(q^e_{-1}) + (1-\alpha) \cdot G(q^e) , \qquad (23)$$

$$G_\alpha(\text{GO}) = G(q^a) , \qquad (24)$$

provided that he encounters an opponent whose leaving tendency is given by α. Equating (23) and (24) for $\alpha = \alpha^*$ yields the following formula for the leaving tendency α^* at equilibrium:

$$\alpha^* = \frac{G(q^a) - G(q^e)}{G(q^e_{-1}) - G(q^e)} . \qquad (25)$$

In the following two sections, we will derive the density distributions p, q^a, q^e, and the density dependent payoffs $g(i)$. This will then enable us to calculate α^*.

4.2 The Density Distributions

Arrivals at and departures from a mating place may be viewed as a birth and death process, and we need a detailed description of this process in order to derive the expected male density distributions p, q^a, and q^e.

In developing our model of male dragonfly behaviour we have a situation in mind where there is a fixed number N of competitors who visit a fixed number M of mating places. Each male has a certain time budget – a mean of B time units per day – that he may spend searching for mates. The rest of each day has to be spent with other activities like hunting for prey. Whenever a male switches from another activity to mate searching behaviour, a 'mating bout' begins. Suppose that a typical male starts m mating bouts per day. Then the mean length T of each bout is given by $T = B/m$, where T is measured in arbitrary time units.

We shall not assume that all mating bouts are of the same length. Instead, we suppose that the termination of a mating bout is a stochastic event: each male has a certain tendency to terminate a mating bout *spontaneously* without encountering a conspecific. We shall not model the decision process leading to spontaneous departure, and we shall not discuss the function of this behaviour. We take it as externally given and incorporate it into our model by assuming that each male has a constant probability c of spontaneous departure from the mating place within one time unit.

If all mating bouts are terminated spontaneously, the expected length T of a mating bout is related to c by the following formula:

$$T = B/m = 1/c. \tag{26}$$

In addition to the spontaneous termination of a mating bout, we shall, however, also consider the strategic decision to leave a mating place after an encounter with a conspecific male. We interpret this decision as a means to allocate the available mating time budget B in a way that avoids inter–male competition for females as far as possible. It should, however, neither lengthen nor shorten the budget which is available for mate searching. In fact, if an increase in the tendency to leave the mating place after an encounter would decrease the mating time budget B, it would probably never be profitable for a male to leave, i.e., a male should always choose $\alpha = 0$.

Accordingly, the decision to leave after an encounter should not result in a termination of the mating bout. Instead, the pure strategy GO of our normal form model should be interpreted as a decision to *switch* to another pond. The mating bout continues, but there is a chance that competition is less intense at the other mating place.

Let us now focus on a specific mating place and a fixed short time interval Δt, which will represent the time unit of our model. During such an interval, males arrive at the water with a certain probability r. The males at the water have a probability c to terminate spontaneously their mating bout. Intermale encounters are random events and — as has been shown by Kaiser (1974a) — the probability $k(i)$ for such an encounter to take place within the time interval Δt depends on the density i of male dragonflies at the water. Following Poethke & Kaiser (1985), we set

$$k(i) := \frac{V}{L} \cdot \frac{i \cdot (i-1)}{2} \cdot \Delta t , \qquad (27)$$

where V is the relative speed between dragonflies and L denotes the length of the shoreline.

Based on these details, we may now describe the whole process by a flow chart (Figure 2) which summarizes the actions of a particular male dragonfly (called X), the state of the system (i.e. male density i at the water), and the resulting payoff $g(i)$ individual X can expect to get during a time interval Δt.

Let us consider a male X, whose leaving tendency after an encounter is given by α', whereas his competitors have a leaving tendency of α. Suppose that X arrives at a mating place. With probability $q^\alpha(i)$, it will find the pond in state i (i.e. with X included, there are i males patrolling at the shoreline). Within a short time interval Δt a number of events may happen, which possibly may change the state of the system:

1. With probability r, a new dragonfly may arrive at the pond. If this happens, the state of the system will change from i to $i+1$ and male X may expect a payoff of $g(i+1)$ for the next time interval. Assuming that the time interval Δt is very short, we may ignore the possibility that more than one dragonfly male do arrive at the same time.

2. With probability $1-(1-c)^{i-1} \approx (i-1) \cdot c$, at least one of the $i-1$ competitors of X will spontaneously leave the pond. If a competitor leaves, the state of the system will change from i to $i-1$ and X can expect a payoff of $g(i-1)$ for the next time interval. Again, we have assumed that Δt is small enough to neglect the spontaneous departure of more than one competitor.

3. With probability c individual X may itself spontaneously terminate his mating bout. Accordingly, he well not get any payoff for the next time unit.

4. With probability $k(i)$ two males may encounter each other. Ignoring the possibility that more than one encounter takes place at the same time, we shall distinguish two cases:

    a) With probability $2/i$ male X is involved in the encounter. With probability α' he will leave the pond and switch to another mating place. If he leaves, his expected payoff for the next time interval will be given by $G(q^\alpha)$ regardless of the decision of his opponent. If X does not leave, his expected payoff for the next time interval depends on the behaviour of his opponent.

287

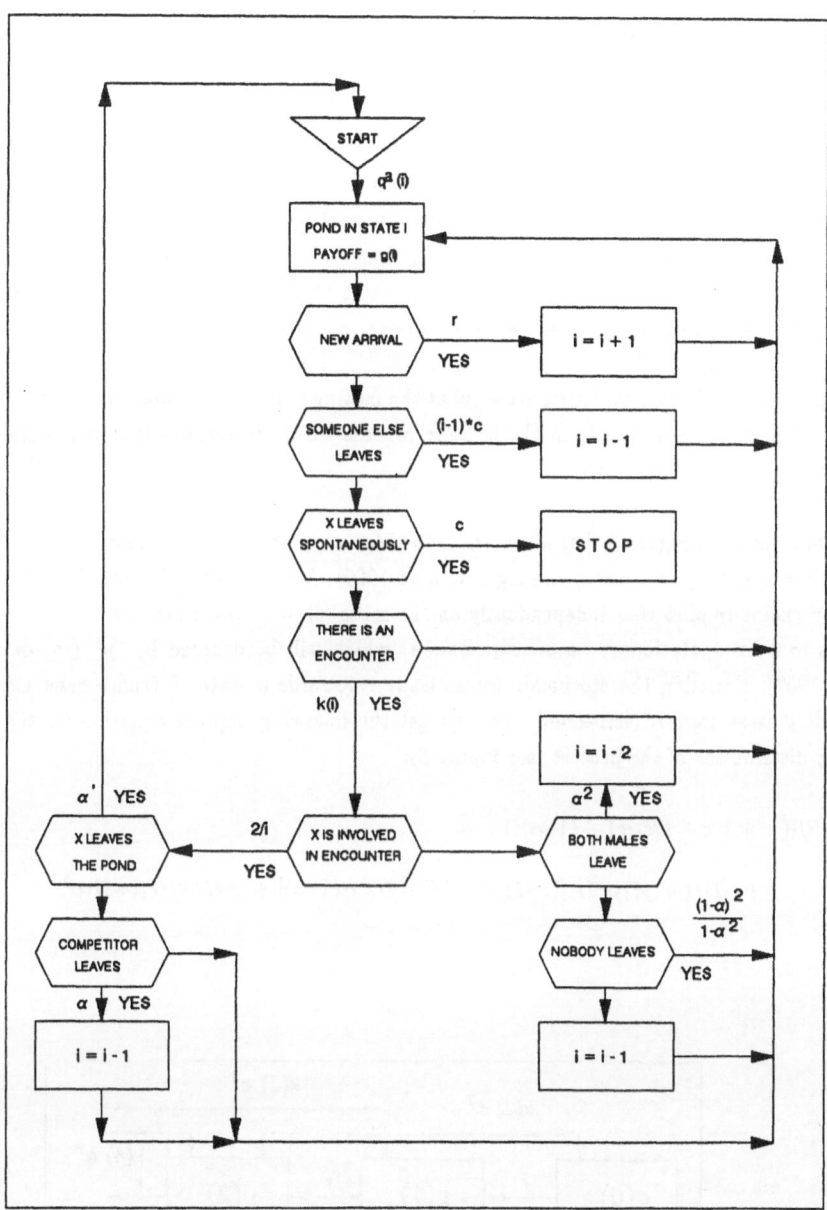

FIGURE 2: A flow chart describing the fate of an individual dragonfly male X and the transitions of the state of the system which are possible within one short time interval Δt.

If the opponent leaves (probability α), the state of the system changes from i to $i-1$, and the payoff will be given by $g(i-1)$; if the opponent does not leave (probability $1-\alpha$), the state of the system remains unchanged and the expected payoff for X will be given by $g(i)$.

b) With probability $1-2/i$ male X will not be involved in the encounter. In this case, there are three different outcomes which are relevant for X: With probability α^2 both participants of the encounter leave the pond; with probability $2 \cdot \alpha \cdot (1-\alpha)$ one of the two participants leaves; and with probability $(1-\alpha)^2$ both participants decide to stay. Accordingly, the state of the system changes to $i-2$ or to $i-1$ or it remains at i. T e expected payoff to X for the next time interval will be given by $g(i-2)$, $g(i-1)$, or $g(i)$ respectively.

To derive the male density distribution p at the mating place, we are not interested in the fate of any single individual, but only in the Markov process which determines this distribution. This process is illustrated in Figure 3.

It is obvious from Figure 3 that all states of the Markov process communicate with another, i.e. that they are 'regularly recurrent' (see e.g. Breiman 1969). A well known theorem from the theory of Markov chains implies that independently on the initial state of the system, the Markov process converges to a fixed, stationary *limit distribution* which will be denoted by p^* (for details see Breiman 1969). Equating the stochastic forces leading towards a state i (right–hand side) with those leading away from it (left–hand side), we get the following implicit equation for the unique stationary distribution of the process (see Figure 3):

$$p^*(i) \cdot (r + i \cdot c + k(i) \cdot (1 - (1-\alpha)^2)) = \tag{28}$$

$$p^*(i-1) \cdot r + p^*(i+1) \cdot [(i+1) \cdot c + k(i+1) \cdot 2\alpha \cdot (1-\alpha)] + p^*(i+2) \cdot k(i+2) \cdot \alpha^2.$$

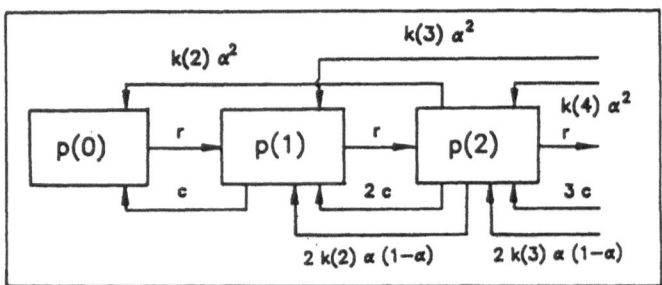

FIGURE 3: State transition probabilities for the states $p(0)$, $p(1)$, and $p(2)$ of the Markov process governing male density at the water.

Equation (28) allows us to calculate the stationary distribution p^* as a function of α, r, c, and k. As the limit distribution of the Markov process, p^* describes the density distribution our dragonfly system will reach after a sufficient time. Given that this *relaxation time* has passed, $p^*(i)$ is the probability that at any given point of time there will be exactly i male dragonflies at the water. Thus, p^* corresponds to the 'objective' density distribution p that was introduced in Section 2:

$$p(i) := p^*(i) \quad \text{for all } i. \tag{29}$$

Based on p, we can now calculate $q^m(i)$, the conditional probability that there are in all i males at the mating place given that one particular male X is also present:

$$q^m(i) := \text{Prob } \{ \ i \text{ males at the pond } | \ X \text{ is at the pond } \}. \tag{30}$$

Since there is a total number of N males in our system, Bayes' formula (see Breiman 1969) implies

$$q^m(i) = \frac{p(i) \cdot i/N}{\sum\limits_{j} p(j) \cdot j/N}. \tag{31}$$

If $\bar{n}$ denotes the mean male density at the water,

$$\bar{n} := \sum_{j} p(j) \cdot j, \tag{32}$$

the density distribution q^m is given by

$$q^m(i) = p(i) \cdot i / \bar{n}. \tag{33}$$

By similar considerations, we get the density distribution q^a immediately after the arrival of a particular dragonfly male X. It is given by

$$q^a(i) = \frac{p(i-1) \cdot 1/N}{\sum\limits_{j} p(j-1) \cdot 1/N} = p(i-1). \tag{34}$$

Let us finally calculate $q^e(i)$, the conditional probability that there are in all i males at the water given that individual X has just encountered another male. Recall that $q^m(i)$ is the probability that X is facing i–1 competitors given that it is at the water. The probability that an encounter is taking place within the time interval Δt is given by $k(i)$, and X participates in such an encounter with probability $2/i$. From this we get

$$q^e(i) = \frac{q^m(i) \cdot k(i) \cdot 2/i}{\sum\limits_{j} q^m(j) \cdot k(j) \cdot 2/j}. \tag{35}$$

which may be written as

$$q^e(i) = \frac{p(i) \cdot i \cdot (i-1)}{\sum\limits_{j} p(j) \cdot j \cdot (j-1)}. \tag{36}$$

4.3 The Density Dependent Payoffs

Our considerations on dragonfly behaviour are based on the assumption that male mating chances decrease with an increasing number of competitors. To get an analytical expression for the density dependent payoff $g(i)$, we assume that female arrivals at the water are random events. Females are a scarce resource, and their density at the mating place is always much smaller than that of the males. Depending on the conspicuousness of female behavior, there is a certain probability w that any particular male that is present at the water will find a particular female during her stay at the pond.

Since a male's search for females is independently from that of any other male, the probability that any particular female is not found at all is given by $(1-w)^i$. Correspondingly, the female will be found with probability $1-(1-w)^i$, and with probability $1/i$ it will be a particular male X who found her. Therefore, it is plausible to assume that the expected mating success of any particular male X who is confronted with $i-1$ competitors is proportional to the product of $1/i$ and $1-(1-w)^i$, and we may set

$$g(i) := \frac{1-(1-w)^i}{i}.$$ (37)

Notice that in case of maximal searching efficiency ($w = 1$), $g(i)$ is given by

$$g(i) = 1/i,$$ (38)

i.e., it is inversely proportional to the number of males present at a pond.

4.4 Strategy–Payoff Feedback

Our models are based on the assumption that the leaving tendency after an encounter, α, is a strategic parameter which evolves as an adaptive reaction to the fluctuations in local male density as experienced by individual males. Accordingly, we interpret α as a selective response to the 'subjective' density distributions q^a and q^e. Notice, however, that q^a and q^e are themselves dependent on the population strategy α. In fact, they are functions of p, and p is the limit distribution of a Markov process which is at least partly determined by α. The dependence of the distributions q^a and q^e on the strategic parameter α is illustrated in Figure 4.

Therefore, we are confronted with an evolutionary normal form game, where the 'elementary' payoffs $G(q^a)$, $G(q^e)$, and $G(q^e_{-1})$ are not *externally* given and fixed. Instead, they are changing themselves during the selection process: there is a feedback between the payoff structure (which governs the evolution of the strategic parameter α) and the population strategy (which partly determines the Markov process and thereby the elementary payoffs). In Section 5, we shall show how to analyse such a situation.

(a)

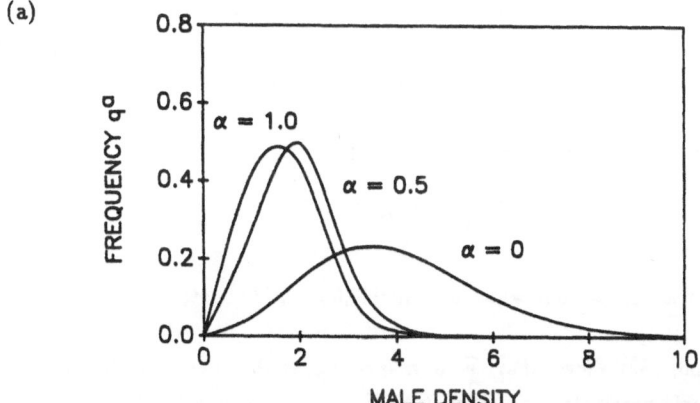

(b)

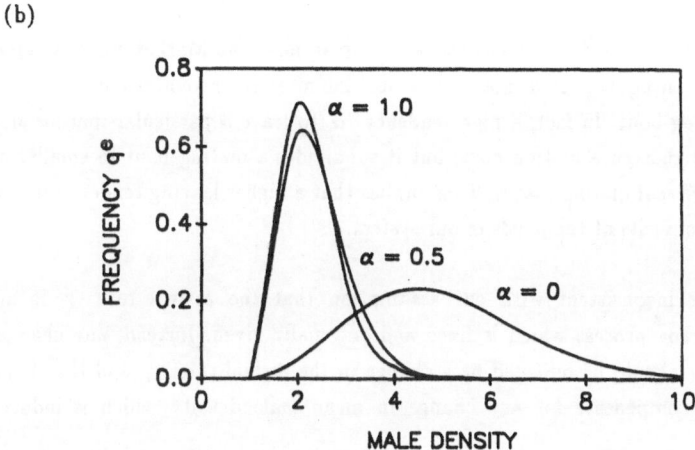

FIGURE 4:   Dependence of the 'subjective' density distributions q^a and q^e on the population strategy α.

Before doing this, we have to remove an inconsistency from our model which will provide an additional feedback between the Markov process and the strategic parameter α. In formulating our model, we had a situation in mind, where a fixed number N of male dragonflies having a certain mating time budget B visit a fixed number M of mating places. Under these assumptions, the mean male density at a specific mating place should be given by

$$\bar{n} = \frac{P \cdot N}{M},$$ (39)

where P denotes the fraction of a day which is available for mating bouts.

On the other hand, (32) shows that $\bar{n}$ is a function of the limit distribution p, and p depends on the strategic parameter α. Obviously, $\bar{n}$ is a decreasing function of α : A higher tendency to leave after an encounter shortens the mean duration of male visits at the mating place, and it results in a lower mean male density at the water.

It is easy to see the reason for this inconsistency: up to now, the Markov process described in Figure 3 does not yet distinguish between a leaving decision after an encounter and the spontaneous termination of a mating bout. In fact, a high tendency α to leave a particular mating place after an encounter does not shorten a mating bout, but it subdivides a mating bout in smaller subunits which are spent at *different* mating places. This implies that a higher leaving tendency is associated with a larger number of visits at the ponds of our system.

Of course, this is inconsistent with our assumption that the arrival rate r is an input parameter of the Markov process which is fixed and externally given. Instead, any change in the population strategy α should be reflected by a change in the arrival rate r, and the change in r should automatically compensate for any change in mean male density which is induced by a change in α.

In order to incorporate this into our model, let us consider the limit distribution p and the induced mean male density $\bar{n}$ as functions of α and r:

$$\bar{n}_{(\alpha,r)} = \sum_j p_{(\alpha,r)}(j) \cdot j.$$ (40)

For a fixed strategy α, $\bar{n}$ is strictly increasing with an increase in the arrival rate r. Therefore, there is exactly one $r = r(\alpha)$ such that the equation

$$\bar{n}_{(\alpha,r(\alpha))} = \frac{P \cdot N}{M}.$$ (41)

holds true. From now on, we shall interpret the mean male density $\bar{n}$ as being fixed and externally given by (39). Treating all other paramters of the Markov process as constants, the arrival rate r will be considered as being a function of α which is implicitly given by equation (41).

5. Evolutionary Stability in Games with Strategy–Payoff Feedback

As has been explained above, we are dealing with an evolutionary normal form game, where there is a feedback between the evolving population strategy α and the selective forces governing this process. In order to analyse the system, we have to consider *two* processes which strongly affect another: The Markov process (as depicted in Figure 3) determines the density distributions p, q^a and q^e and the resulting payoff structure of the normal form game. The selection process – which is governed by this payoff structure – determines the population strategy α which is an important input parameter for the Markov process.

Although the two processes depend on each other, we may deal with them separately by taking account of the fact that they proceed on rather different time scales:

– The Markov process is governed by the arrival rate $r(\alpha)$, the rate of spontaneous departure c, the encounter rate $k(i)$, and the rate of departure α after an encounter has taken place. Numerical values for these parameters based on field data (see (51) and (52)) show that substantial changes in the density distributions q^a and q^e may take place within minutes. Usually, it will not take longer than about half an hour until the equilibrium distribution is reached.

– Selection, on the other hand, operates on a much longer time scale. In fact, we are dealing with rather weak selection, and the changes taking place within one generation are almost imperceptibly small. Aeschna cyanea has a generation time of about two years, and it will take several generations until the composition of a population has significantly changed.

An individual Aeschnid dragonfly surviving to adulthood spends the greatest part of its life in one of the larval stages. For only about three months it lives as a winged imago outside the water. This time is much longer than the relaxation time of the Markov process but much too small to be affected by evolutionary changes. Thus, for an individual's decision, the composition of the population is effectively constant while the frequency distributions p, q^a and q^e will usually be in equilibrium.

Since selection is so much slower than the Markov process, the two processes may be considered as effectively uncoupled. Whenever a population strategy α has been realized in a population, we may perform an *adiabatic approximation* and assume that the parameters of the Markov process remain constant on its way to its equilibrium state. Any change in the population strategy α will immediately be reflected in a corresponding change in the equilibrium distributions q^a and q^e.

These considerations suggest to analyse the situation as follows: Let us focus our attention on a monomorphic population where a certain strategy α has been fixed in the course of evolution. Accordingly, all members of the population are 'α–strategists', i.e. all individuals have a tendency α to leave the mating place after an encounter with a competitor.

For a fixed α, the limit distribution $p = p(\alpha)$ of the Markov process is well defined. Based on p, the distributions $q^a(\alpha)$, $q^e(\alpha)$ and $q^e_{-1}(\alpha)$ may be calculated (see (34), (36), and (16)) as well as the payoff matrix $A = A(\alpha)$ (see (18)). This matrix characterizes the selective forces which are currently at work. In particular, A yields the expected payoffs for the two pure strategies 'STAY' and 'GO' which are given by (23) and (24) and which may now be written in the form:

$$G_\alpha(\text{STAY}) = \alpha \cdot G(q^e_{-1}(\alpha)) + (1-\alpha) \cdot G(q^e(\alpha)), \tag{42}$$

$$G_\alpha(\text{GO}) = G(q^a(\alpha)). \tag{43}$$

Notice that $G_\alpha(\text{STAY})$ and $G_\alpha(\text{GO})$ are *non-linear* functions of α which is in contrast to those evolutionary normal form games whose payoffs are given externally.

Having defined the selective forces, we are now in a position to judge whether a monomorphic population of α-strategists is *evolutionarily stable*, or whether there is a selective tendency leading to its destabilization. In order to do this, suppose that $0 < \alpha < 1$ is such that the expected payoff for 'STAY' is different from that for 'GO':

$$G_\alpha(\text{STAY}) \neq G_\alpha(\text{GO}). \tag{44}$$

We shall argue that in this case α cannot be evolutionarily stable since a population of α-strategists may be invaded by a mutant strategy $\alpha' \neq \alpha$.

First, let us consider the case $G_\alpha(\text{GO}) > G_\alpha(\text{STAY})$, a mutant strategy α' which is slightly larger than α, and a bimorphic population consisting of a mixture of α-strategists and α'-strategists. Although we have not explicitly derived the Markov process and the resulting payoff structure for a polymorphic population, it is plausible to assume that the inequality $G(\text{GO}) > G(\text{STAY})$ continues to hold if only α' is close enough to α. Since the α'-strategists have a higher leaving tendency than the α-strategists, they also get more often the higher payoff for the pure strategy 'GO'. This, however, implies that the α'-strategists have an expected payoff which is consistently higher than that of the α-strategists. Due to their selective advantage, the α'-strategists will successfully invade the α-population, and after some time strategy α will be replaced by α'.

Similarly, the inequality $G_\alpha(\text{STAY}) > G_\alpha(\text{GO})$ would result in an evolutionary trend towards strategy parameters α' which are larger than α. Correspondingly, the identity

$$G_\alpha(\text{STAY}) = G_\alpha(\text{GO}) \tag{45}$$

is a necessary condition for $0 < \alpha < 1$ to be immune against invasion by slightly different mutant strategies.

Let us now suppose that (45) holds true. Let us further suppose that $G_{\alpha'}(\text{GO}) > G_{\alpha'}(\text{STAY})$ holds for a mutant strategy α' which is slightly larger than α. In view of (45), it is plausible to assume that the inequality $G(\text{GO}) > G(\text{STAY})$ continues to hold for any bimorphic population consisting of a mixture of α–strategists and α'–strategists. Like before, this implies that after some time α will be replaced by α'. Similarly, α is not invasion–proof if $G_{\alpha'}(\text{STAY}) > G_{\alpha'}(\text{GO})$ holds true for a mutant strategy α' which is slightly smaller than α.

Taken together, these considerations motivate the following definition of evolutionary stability which is very similar in spirit to Eshel's (1983) concept of *continuous stability* :

DEFINITION: Evolutionary stability.

A monomorphic population of α^*–strategists is *evolutionarily stable*, and α^* is an evolutionarily stable strategy or *ESS*, if there exists an $\epsilon > 0$ such that for $0 \leq \alpha \leq 1$ the following two conditions are satisfied:

(a)
$$G_{\alpha}(\text{STAY}) > G_{\alpha}(\text{GO}) \quad \text{for} \quad \alpha^*-\epsilon < \alpha < \alpha^*; \tag{46}$$

(b)
$$G_{\alpha}(\text{STAY}) < G_{\alpha}(\text{GO}) \quad \text{for} \quad \alpha^* < \alpha < \alpha^*+\epsilon. \tag{47}$$

Obviously, this definition easily generalizes to the class of all evolutionary 2x2 normal form games with strategy–payoff feedback. It is easy to see that for the special case of those evolutionary games the payoff structure of which is fixed and externally given, the definition above coincides with the usual concept of evolutionary stability as introduced by Maynard Smith & Price (1973) (see also Lessard, 1990).

The expected payoffs of the two pure strategies are continuous functions of α, since the limit distribution $p(\alpha)$ depends continuously on α. As a consequence, (46) and (47) imply

$$G_{\alpha^*}(\text{STAY}) = G_{\alpha^*}(\text{GO}) \quad \text{for} \quad 0 < \alpha^* < 1, \tag{48}$$

i.e., at a completely mixed ESS α^* the payoffs for the pure strategies have to be equal to another.

For a pure strategy α^*, either condition (46) or condition (47) is empty. By continuity of payoffs in the strategic parameter α, we get the following necessary conditions for α^* being a pure ESS:

$$G_{\alpha^*}(\text{STAY}) \geq G_{\alpha^*}(\text{GO}) \quad \text{for} \quad \alpha^* = 0; \tag{49}$$
$$G_{\alpha^*}(\text{STAY}) \leq G_{\alpha^*}(\text{GO}) \quad \text{for} \quad \alpha^* = 1. \tag{50}$$

(48) to (50) are just the conditions for α^* to be a Nash equilibrium strategy. Accordingly, we get the following result:

(a)

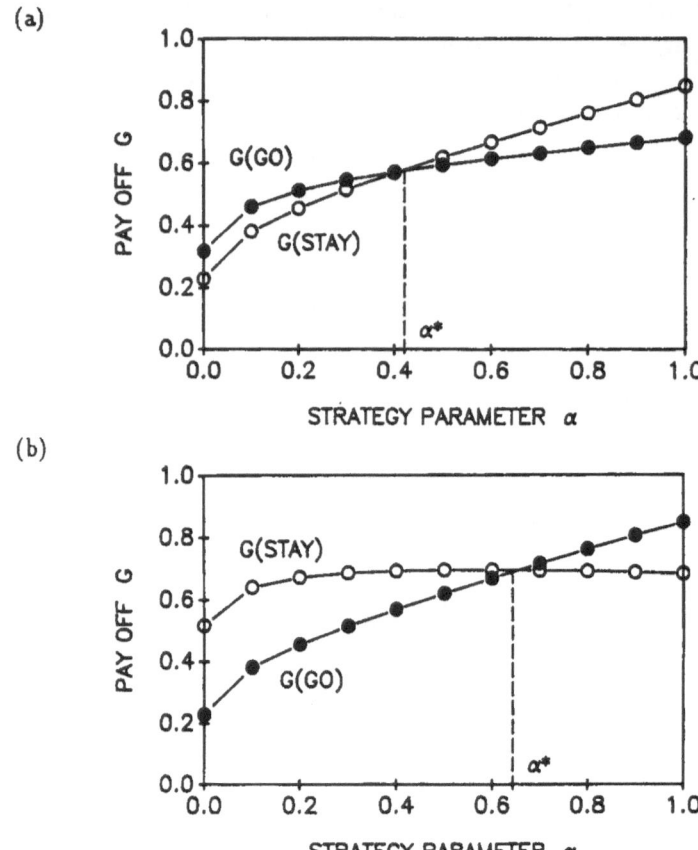

(b)

Evolutionary stability in an evolutionary game with strategy–payoff feedback.
(a) α^* is a mixed ESS.
(b) α^* is not an ESS. Instead, both pure strategies are evolutionarily stable.

See text for details.

PROPOSITION: Nash feedback of evolutionarily stable strategies.
Every evolutionarily stable strategy α^* is a Nash equilibrium strategy of the evolutionary normal form game that is induced by α^* itself.

Figure 5 illustrates that the conditions for evolutionary stability can easily be checked graphically by plotting the payoffs for pure strategies, G(STAY) and G(GO), as functions of α. In view of (48), the intersection points of G(STAY) and G(GO) correspond to those mixed strategies α^* which are candidates for being evolutionarily stable. (46) and (47) imply that α^* is indeed evolutionarily stable, if and only if $G_\alpha(\text{STAY})$ is smaller than $G_\alpha(\text{GO})$ on the left side of the intersection point and larger than $G_\alpha(\text{GO})$ on the right—hand side of α^*. Notice that this means that $G(\text{STAY})$ intersects $G(\text{GO})$ 'from below to above' (see Figure 5a). If $G(\text{STAY})$ intersects $G(\text{GO})$ 'from above to below' (see Figure 5b), α^* is not an ESS. Instead, there are two evolutionarily stable strategies which are given by the two extremal strategy parameters $\alpha = 0$ and $\alpha = 1$.

6. Numerical Results

In view of the strategy—payoff feedback discussed above, (25) is an implicit equation that is not easily solved analytically. Therefore, we shall present some numerical results which give an impression of the size of the evolutionarily stable leaving tendency α^* as well as its dependence on the parameters of the Markov process.

In Figures 6 and 7, for varying male searching efficiencies w and various mean male densities $\bar{n}$ the expected payoffs for 'STAY' and for 'GO' are depicted as functions of the strategic parameter α. Using a step length of one second,

$$\Delta t = 1 \text{ sec}, \tag{51}$$

the parameters for the numerical solution were chosen according to Poethke & Kaiser (1985):

$$V = 0.25 \text{ m/sec}; \quad L = 69 \text{ m}; \quad c = 5.5 \cdot 10^{-4}; \tag{52}$$

$$0.5 \leq \bar{n} \leq 1.5. \tag{53}$$

In all cases, $G(\text{STAY})$ intersects $G(\text{GO})$ 'from below to above' at a mixed strategy α^*. As has been explained above, this implies that α^* is an ESS. In Section 4.1, we have argued that it is rather plausible that only strategies which are completely mixed may be evolutionarily stable for the games that are considered in this paper. Accordingly, in all our simulations we have never encountered a situation where $G(\text{STAY})$ and $G(\text{GO})$ did not intersect at all or where $G(\text{STAY})$ intersected $G(\text{GO})$ 'from above to below'.

As can be seen from Figures 6a to 6c, the searching efficiency w has only a negligible effect on the point α^* where $G(\text{STAY})$ and $G(\text{GO})$ do intersect. We shall therefore concentrate on the case of maximal searching efficiency whenever we are mainly interested in the value of the evolutionarily stable leaving tendency α^*. In view of (38), the basic payoffs $g(i)$ are especially simple if we have $w = 1$.

(a)

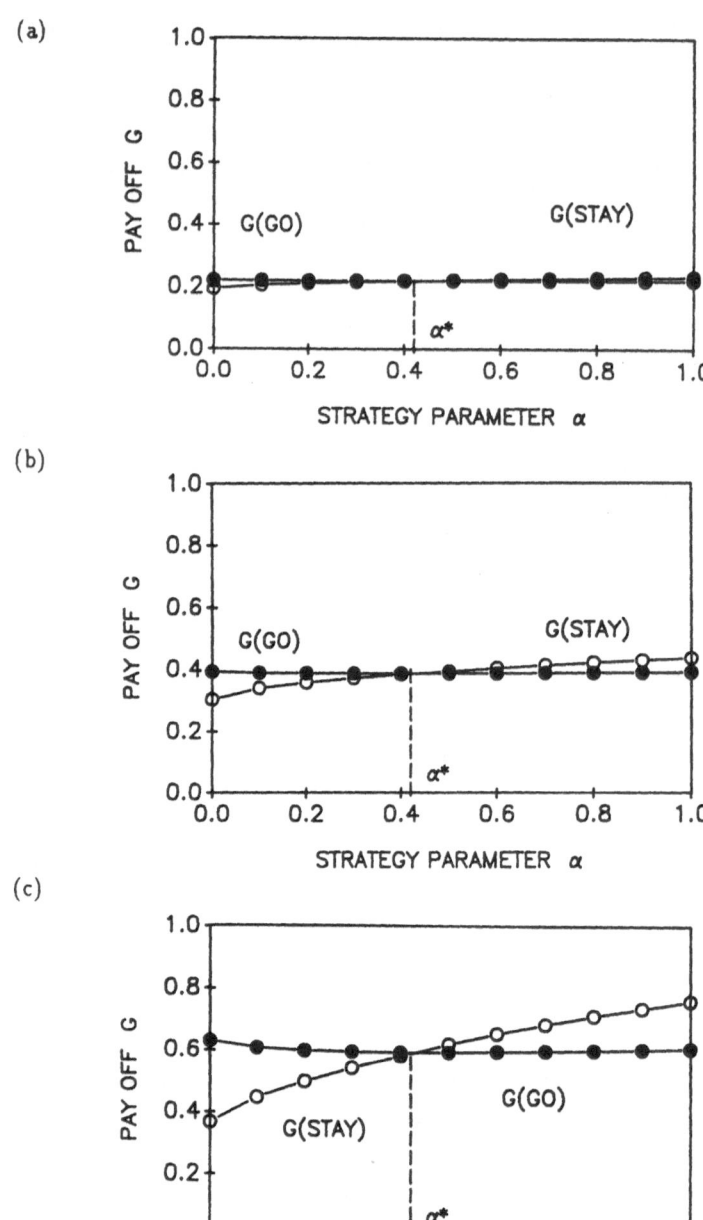

(b)

(c)

FIGURE 6:   Expected payoff of pure strategies for a mean male density $\bar{n} = 1.0$
and different male searching efficiencies w:
(a) $w = 0.25$; (b) $w = 0.5$; (c) $w = 1.0$.
Open circles: G(STAY), closed circles: G(GO).
α^* denotes the evolutionarily stable tendency to leave after an
encounter.

(a)

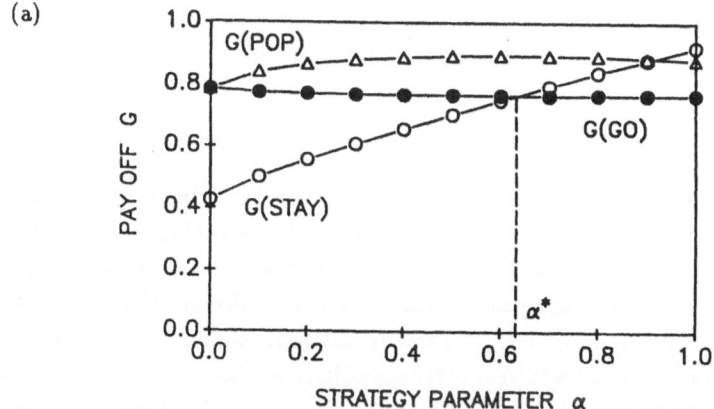

(b)

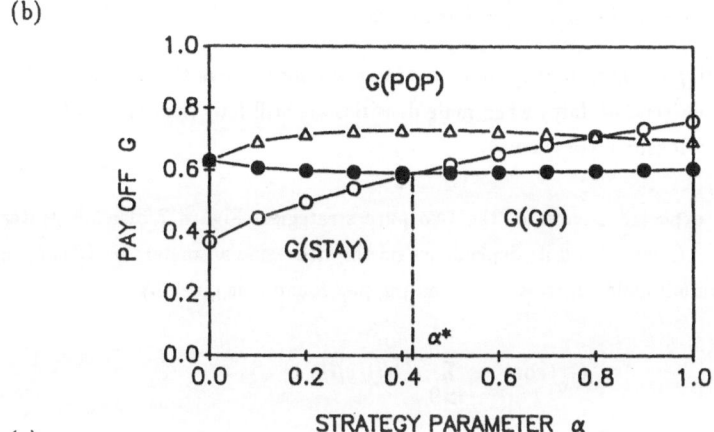

(c)

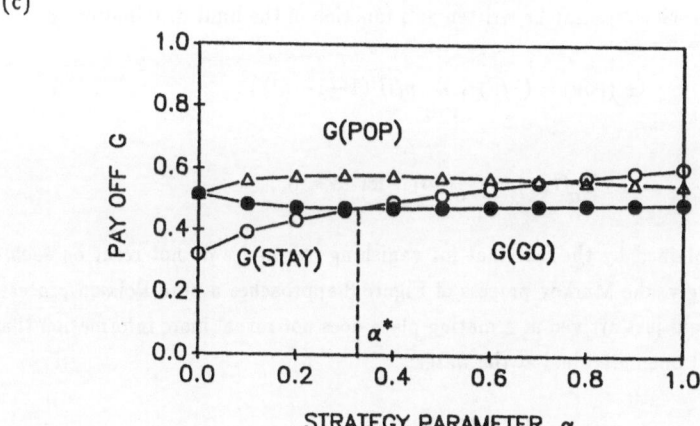

FIGURE 7: Expected payoffs of pure strategies and mean population payoff for a maximal male searching efficiency $w = 1.0$ and different mean male densities:

(a) $\bar{n} = 0.5$; (b) $\bar{n} = 1.0$; (c) $\bar{n} = 1.5$.

Open circles: G(STAY), closed circles: G(GO), triangles: G(POP).

Notice, however, that mating success is positively correlated with searching efficiency. Therefore, the parameter w has an influence on the absolute value of expected payoffs. In addition, it has a strong effect on payoff differences: the *strength of selection* is also positively correlated with the searching efficiency of each particular male.

Mean male density, on the other hand, strongly affects the value of the evolutionarily stable leaving tendency α^* as well as absolute payoffs and payoff differences (Figures 7a to 7c). It is obvious that payoffs decrease if more males compete for the same number of females, but it is interesting to notice that with increasing mean male density $\bar{n}$ males should be less willing to leave the pond after an intermale encounter. Apparently, the relative difference between n^a and n^e is largest in low–density situations. In view of the law of large numbers (see e.g. Breiman 1969), this effect is not too surprising since the relative fluctuations in male density should become smaller if the number of competitors at the mating place increases.

It would be interesting to check if dragonflies are more willing to leave the mating place at the start of the mating season (end of July) when male densities are still low, than lateron (e.g. end of August) when peak male densities are reached.

In addition to the expected payoffs of the two pure strategies, Figure 7 also illustrates the *population mean payoff*, $G(\text{POP})$, and its dependence on the strategic parameter α. $G(\text{POP})$ is the mean payoff of an individual male which is at the mating place, and it is given by

$$G_\alpha(\text{POP}) := \sum_{i>0} q^m(i) \cdot g(i) \,. \tag{54}$$

In view of (33) and (37), $G(\text{POP})$ may be written as a function of the limit distribution p:

$$G_\alpha(\text{POP}) = (1/\bar{n}) \cdot \left[\sum_{i>0} p(i) \cdot (1-(1-w)^i) \right] \,. \tag{55}$$

Notice that

$$G_\alpha(\text{POP}) = G_\alpha(\text{GO}) \quad \text{for } \alpha = 0. \tag{56}$$

This identity is explained by the fact that for vanishing α, males do not react on each other any more. Correspondingly, the Markov process of Figure 3 approaches a pure Poisson process, and the fact that an individual just arrived at a mating place does not reveal more information than the fact that there is at least one individual at the water:

$$q^a = q^m \quad \text{for } \alpha = 0. \tag{57}$$

Notice also that the mean population payoff always reaches its maximum at the ESS $\alpha = \alpha^*$. Apparently, competition avoidance on the basis of 'individual rationality' leads to a distribution of males which is optimal on the population level.

Figure 7 shows that for realistic parameter constellations our model predicts an evolutionarily stable leaving tendency α^* which lies in the range between 0.3 and 0.7. Therefore, α^* is still considerably larger than the leaving tendency observed in the field ($\alpha \approx 0.15$) for the same range of male densities ($0.5 \leq \bar{n} \leq 1.5$).In the following section, we shall discuss an extension of our model which might result in a better agreement between empirical results and model predictions.

7. A Game Model With Switching Costs

Up to now, we have completely neglected the costs which are associated with the decision to switch to another mating place. A higher tendency to leave the mating place makes the visits at the water shorter and more numerous with the consequence that more time and energy is spent for travelling. All the costs which arise from the decision to travel to another mating place will collectively be called *switching costs*. Obviously, the omission of these costs from our model has led us to overestimate the evolutionarily stable leaving tendency α^*.

The easiest approach to take account of switching costs in our normal form model consists in replacing the expected payoff $G(q^a)$ for 'GO' in (18) by $G(q^a)-C$, where the positive parameter C represents the costs of switching. In view of (25), it is easy to see that the evolutionarily stable leaving tendency α^*_{new} for this modified model results from α^*_{old}, the ESS of our old model, by means of the following formula:

$$\alpha^*_{new} = \alpha^*_{old} - \frac{C}{G(q^e_{-1})-G(q^e)}. \tag{58}$$

Notice that α^*_{new} is always smaller than α^*_{old}, so that the ESS of the modified model is probably more in accordance with the field observations than the ESS of the old model. In addition, notice that the difference between the old and the new ESS is inversely proportional to the payoff difference $G(q^e_{-1})-G(q^e)$, which – like $G(q^a)-G(q^e)$ – might be interpreted as an indicator for the *strength of selection*. This is hardly surprising: the costs of switching should matter most in those situations, where the benefits of switching are small. Since the selection differentials are negatively correlated with $\bar{n}$ and positively correlated with w (see Figures 6 and 7), the neglect of switching costs should be most important if male density is high or if mate searching effectivity is low.

The approach outlined above is, however, not very convincing. The cost parameter C is hardly commensurable with the fitness components $G(q^a)$, $G(q^e)$, and $G(q^e_{-1})$, and it is not obvious how a term like $G(q^a)-C$ may be given a meaningful interpretation. In fact, $G(q^a)$ denotes the expected *instantaneous* mating success ($\cong$ mating chance per unit of time) of a male which has just arrived at a new mating place. C, on the other hand, corresponds to a *period* of time which is lost for mate searching behaviour due to travelling from one place to another. In order to compare the costs and benefits of switching, both have to be measured on a comparable scale.

On the one hand, the the instantaneous payoffs of the leaving decision after an encounter might be *accumulated* over a certain period of time. The accumulated benefit of switching might then be compared to the accumulated loss in mating chances due to travelling from one mating place to another. Here, we shall take the opposite approach and try to translate the costs of switching into an *instantaneous* loss in mating chances.

Let us suppose that the main costs of switching arise due to loss in time and that on average a switching flight consumes S time units of the precious mating time budget T (see (26)) of a mating bout. The mating time budget T should not be changed by a leaving decision, but part of it should be allocated to travelling instead of mate searching. In order to keep our model consistent, we shall therefore assume that also during switching flight there is a positive tendency c to spontaneously terminate the mating bout (see Section 4.2).

What are the consequences of this assumption? Spontaneous termination of a mating bout is modelled in this paper as a Poisson process where the past does not matter for the future. Accordingly, as soon as a new mating place is reached, it does not matter whether it was reached after a long or after a short switching flight. However, there is a difference in the moment where the decision is made since the probability *whether* a new mating place will be reached strongly depends on S, the mean duration of a switching flight. In fact, the probability that the mating bout will not be terminated within a time period S is given by $(1-c)^S$. This implies that — evaluated at the moment where the decision has to be made — the expected payoff for leaving should be given by the expected payoff $G(q^a)$ at the arrival at a new mating place multiplied with $(1-c)^S$, the probability that a new mating place is reached, and we get:

$$G_\alpha(\text{GO}) = (1-c)^S \cdot G(q^a). \tag{59}$$

Correspondingly, the switching costs are represented by *discounting* the benefits of switching. Notice that (58) still gives us the new ESS α^*_{new}, if we set

$$C := (1 - (1-c)^S) \cdot G(q^a). \tag{60}$$

We close this chapter by noting that there is still one problem left: If male dragonflies switch frequently within a mating bout, a considerable fraction of their mating time budget could be used up for switching flight. Accordingly, the fraction P of each day that they spend at the water (see Section 4.4) cannot longer be assumed to be fixed, but it is a decreasing function of the leaving tendency α. In view of (38), a higher leaving tendency is associated with a smaller mean male density at the water but it is difficult to quantify this relationship. Therefore, equation (39) for the determination of the arrival rate $r(\alpha)$ only makes sense if switching time is small enough to neglect its influence on mean male density.

9. References

Breiman, L. (1969): Probability and Stochastic Processes. With a View Towards Applications. Boston: Houghton Mifflin Company.

Corbet, P.S. (1980): Biology of Odonata. Ann. Rev. Entomol. 25: 189–217.

Eshel, I. (198?): Evolutionary and Continuous Stability. J. theor. Biol. ???

Kaiser, H. (1974a): Die Regelung der Individuendichte bei Libellenmännchen (Aeschna cyanea, Odonata). Oecologia 14: 53–74.

Kaiser, H. (1974b): Die tägliche Dauer der Paarungsbereitschaft in Abhängigkeit von der Populationsdichte bei den Männchen der Libelle Aeschna cyanea (Odonata). Oecologia 14: 375–387.

Kaiser, H. (1974c): Verhaltensgefüge und Temporialverhalten der Libelle Aeschna cyanea (Odonata). Zeitschr. Tierpsychol. 34: 398–429.

Kaiser, H. (1976): Quantitative Description and Simulation of Stochastic Behaviour in Dragonflies (Aeschna cyanea, Odonata). Acta Biotheoret. 25: 163–210.

Kaiser, H. (1985): Availability of Receptive Females at the Mating Place and Mating Chances of Males in the Dragonfly Aeschna cyanea. Behav. Ecol. Sociobiol. 18: 1–7.

Lessard, S. (1990): Evolutionary Stability: One Concept, Several Meanings. Theor. Pop. Biol. 37: 159–170.

Maynard Smith, J. (1982): Evolution and the Theory of Games. Cambridge: Cambridge University Press.

Poethke, H.J. (1988): Density–dependent Behaviour in Aeschna cyanea (Müller) Males at the Mating Place (Anisoptera: Aeschnidae). Odonatologica 17: 205–212

Poethke, H.J. & H. Kaiser (1985): A Simulation Approach to Evolutionary Game Theory: The Evolution of Time–sharing in a Dragonfly Mating System. Behav. Ecol. Sociobiol. 18: 155–163.

Poethke, H.J. & H. Kaiser (1987): The Territoriality Threshold: A Model for Mutual Avoidance in Dragonfly Mating Systems. Behav. Ecol. Sociobiol. 20: 11–19.

Selten, R. (1983): Evolutionary Stability in Extensive Two–Person Games. Mathem. Social Sciences 5: 269–363.

ON THE   EVOLUTION OF GROUP-BASED ALTRUISM

David M. Messick

1.Introduction
   Whether or not altruism may have evolved in humans through a
process of natural selection is a question that lies at the heart of
Campbell's (1975) Presidential address to the American Psychological
Association and that is central to the contemporary view of human
nature. Campbell argued that the nature of the moral responsibilities
of psychologists (and other citizens) are deeply linked to the answer
on gives to the questions of whether altruism could have evolved in
our species, and if so, what form it might have taken. In this paper,
I will present a case for the evolution of an ethnocentric or group-
based altruism. If the position presented here is correct, it has
important implications for Campbell's conclusions as well as for our
understanding of the roots of human sociability. A summary of
Campbell's argument follows to set the stage for the ideas to be
developed subsequently.
   Campbell's (1975) position is that biological evolution has been
inadequate to create the high level of self-sacrificial altruism that
is required to maintain the tight social interdependence of complex
urban societies. As a result, a process of social evolution has
occurred in which social or cultural institutions have evolved that
urge self-sacrifice and and helpfulness to others. These social
structures include religious institutions, patterns of moral belief,
and political or social processes that reward self-sacrificial
behavior performed to aid others. Thus, Campbell argues, humans live
in a chronic conflict between a genetic heritage that skews them
toward selfishness, and cultural processes that implore them to be
altruistic. This conflict between biology and society, argues
Campbell, must be understood and respected if psychological practice
and teaching is to be beneficial rather than harmful to our species in
the long run.
   In developing this argument, Campbell (1975) reviews the basic
problem for the evolution of altruism which may be stated as follows:
If a group contains altruists and nonaltruists, the random or

indiscriminant dispersal of altruistic benefits will cause the altruists to be s l cted against. An altruistic act, in evolutionary theory, is an act that reduces the fitness of the individual that possesses it, while at the same time increasing the fitness of other conspecifics. The altruist pays a cost in fitness, but, by so doing, the altruist also provides a benefit to one or more others. The basic problem or fundamental principle states that the uniform or random dispersal of the benefits to other conspecifics will result in the selection against and the ultimate displacement of altruists from the population. When the benefits are spread evenly among altruists and nonaltruists alike, and when only the altruists pay the costs, altrusits will be at a fitness disadvantage and altruism will not evolve.

The task of evolutionary theory, then, is to find mechanisms that lead to the non-random or non-uniform dispersal of benefits in order to compensate the altruists for their costs. Sufficient compensation may render the altruists more fit than their non-altruistic counterparts and thus provide a process by means of which altruistic traits could be selected for. Three such general mechanisms have been described so far: kin selection, reciprocal altruism, and group selection. In the next section, I will describe these three processes and introduce a variant of reciprocal altruism that I call group-based altruism. I will then note an apparent inconsistency among the fitness relationships of universal altruists, nonaltruists, and group-based altruists. I will present a simple model for the analysis of this inconsistency and show how the inconsistency may be resolved and also how group-based altruism may have evolved. I will explore the consequences two types of errors in the recognition of altruists, and will finally speculate about possible implications of group-based altruism for human social behavior.

2. Mechanisms for the evolution of altruism

Altruism will evolve only if the fitness of those individuals with the altruistic phenotype is greater than those with a nonaltruistic phenotype. The paradox arises because altruism, by definition, reduces the fitness of those it characterizes. Thus a theory of the evolution of altruism **must** be a theory that demonstrates how the altruist or altruistic trait is ultimately compensated. Three distinct processes have been described that accomplish this.

The first of these processes is <u>kin selection</u> (Hamilton, 1964). Kin selection involves the dispersal of altruistic benefits to genetically related individuals. Genetically related conspecifics are more likely than average to be altruistic themselves because of their genetic linkage to the altruist. Thus the dispersal of altruistic benefits to kin skews the distribution of benefits so that altruists receive more of the benefits than nonaltruists, on the average. The cost to the altruist of producing the benefits, however, must be modest with respect to the magnitude of the benefit. Moreover, the distribution of the benefits must be such that if it is highly concentrated in a few individuals, they must be, like offspring, genetically close to the altruist. If a larger number of individuals enjoy the benefit, the degree of genetic intimacy may be looser. Thus, while kin selection explains parental sacrifice for children (children of parents who decline to sacrifice for the children are less likely themselves to become parents to pass on the selfish gene), it also explains less obvious altruistic behaviors like alarm calling in ground squirrels (Sherman, 1977).

While kin selection is based on genetic proximity, the notion of <u>reciprocal altruism</u> (Trivers, 1971) is based solely on phenotypical characteristics. Trivers has shown that if two conspecifics "agree" to exchange altruistic benefits, those individuals will be more fit than individuals who decline to make the exchange. This idea, like that of kin selection, makes the assumption that the cost of the altruistic act is less than the benefit to be bestowed. If not, reciprocal altruism will be selected against. Trivers intended this model for species in which pairs of individuals might interact over a considerable period of time and for whom the identification of specific individuals would be possible. Thus the kinds of exchanges that Trivers seems to have had in mind are those that might be involved in more or less long term relationships between acquainted individuals. Reciprocal altruism overcomes the fundamental principle against the evolution of altruism because one is reimbursed for the altruistic cost by the reciprocated benefit. Boyd & Richerson (1988) have studied reciprocation of this sort in large groups in which repeated pairwise interactions may be infrequent.

Finally, Cohen & Eschel (1976), Matessi & Jaykar (1976), and D.S.Wilson (1975, 1980) have argued that a type of <u>group selection</u>, which depends on the segregation of a population into subgroups that are heterogeneous with regard to the altruistic trait, can also lead

to th selection of altruism. Wilson (1980) has shown, for instance,
that if a population is divided into two subpopulations, one
containing many altruists and one containing only few, then the fact
that th nonaltruists are more fit than the altruists in both
subpopulations does not imply that nonaltruists are more fit than
altruists in the total population. The altruist-dense group will enjoy
greater benefits than the altruist-scarce group by virtue of the
higher concentration of altruists. The dispersal of benefits with a
subgroup may be random, but across the population it will not be since
the altruistic clusters, by definition, are dense with altruists and
rich in benefits. What makes this form of group selection work is that
the between group benefits enjoyed by the altruistic subgroups are
larger then the within group benefits enjoyed by the nonaltruists.
Messick & van de Geer (1981) have shown that the paradox embodied in
this idea also occurs in statistical theory and decision making.

The idea that is to be developed here, group-based altruism, is
similar to the notion of reciprocal altruism. The core of the idea is
that altruists of this stripe will exchange altruistic benefits with
other altruists, but not with nonaltruists. In this sense they are
like reciprocators. Unlike reciprocators, however, I do not assume
that this form of helpfulness emerges from a history of interaction.
The temporal dimension, with the possibility for repeated interaction
that makes reciprocation so effective (Axelrod, 1984), and the
individuation that are important for Trivers' idea are not essential
elements for this one. It is sufficient for the group-based altruist
to be able to discriminate one who is an altruist from one who is not.
The proximal mechanism that supports this discrimination is, for the
moment, irrelevant. (Later it will be very imporant.) Like the
reciprocal altruists, if the cost of the altruistic act is less than
the benefit provided, it is easily demonstrated that group-based
altruists will be more fit than nonaltruists.

Here now is the puzzle. From the fundamental principle we know
that indiscriminate or universal altruists (U) will be less fit than
nonaltruists (N). We also know the group-based altruists (G) will be
more fit than nonaltruists. However, if we imagine a dichotomous
population of U's and G's, their fitness will be identical since they
differ behaviorally only in the presence of the N's. Thus we have a
weak intransitivity in the fitness of these three types since, $f(G) >
f(N)$, $f(N) > f(U)$, but $f(G) = f(U)$. What is called for is an analysis

of trichotomous popluations that contain all three types. In the next
section, I analyze a model of such a population.

3.The Model

3.1. <u>Static features</u>. We assume a population that contains P(U)
U's or universal altruists, P(G) G's or group based altruists, and
P(N) = 1-P(U)-P(G) N's or nonaltruists. We assume that the individuals
in this population are subjected to some prototypical threat--
predation, illness, or environmental hazard--such that an individual
alone when so threatened suffers a fitness loss of b units. A second
individual may help the threatened one, but only at a cost c to
itself. By providing this help, the fitness of the first individual is
restored to the base level. Thus, b is the benefit provided and c is
the cost of providing it, and -b > -c. We treat both b and c as costs
or negative quantities.

When a U encounters a threatened individual, the U will always
pay the cost to help; when an N encounters such an individual, it will
never pay the cost; and when a G meets an individual in need, it will
pay the cost to help another G or a U, those who would in turn help
the G, but not an N.
The consequences of the various possible encounters are summarized in
the payoff matrix given in Table 1. In this table, we differentiate

Table 1.

Matrix of consequences for Universal Altruists (U), Group-based
Altruists (G), and Non-altruists (N). The cost c is the decrease
(from zero) in payoff assumed by the donor of the altruistic act,
and b is the benefit to the recipient. Both b, and c are less than
zero. Payoffs to the donor are on the left of each pair and
outcomes for the recipient are on the right.

		Recipient Type		
		U	G	N
	U	c,0	c,0	c,0
Donor Type	G	c,0	c,0	0,b
	N	0,b	0,b	0,b

the two roles that an individual may occupy, that of the potential recipient of the aid, and that of the potential donor. As donors, U's always pay the cost to restore the recipient's fitness to zero. The N's never pay it. The G's pay the cost for other U's and G's but not for N's.

We assume that one individual is as likely as another to become threatened, and that one individual is as likely as another be the potential donor for one who is threatened. Finally, we assume that an individual has an equal chance of being a donor or a recipient in an interaction.

From these asssumptions we may now compute the expected payoffs to the three different types. Conditional expectations, conditioned on being a donor or a recipient, are calculated with respect to the population proportions. The unconditional expected payoff is simply the average of the two conditional ones. The payoffs for the three types are:

$$2f(U) = c + bP(N), \tag{1}$$

$$2f(G) = c(1 - P(N)) + bP(N), \tag{2}$$

$$2f(N) = b(1 - P(U)). \tag{3}$$

We next compute the differences between each of the three pairs of fitnesses to get the relative fitnesses. Simple algebra yields the following results:

$$2D(G,U) = 2[f(G) - f(U)] = -cP(N) \tag{4}$$

$$2D(U,N) = 2[f(U) - f(N)] = c - bP(G) \tag{5}$$

$$2D(G,N) = 2[f(G) - f(N)] = c[1 - P(N)] - bP(G) \tag{6}$$

An examination of these differences is revealing. Recalling that $c, b < 0$, it is clear from (4) that U's can never be more fit than the G's. The two types will be equally fit only if the altruistic act costs nothing ($c = 0$) or if there are no nonaltruists in the population ($P(N) = 0$).

Equation (5), however, is somewhat surprising in that it reveals that not only can the N's be more fit than the U's, as is implied by the fundamental principle, but that the reverse can also be true. Which of the two types is more fit depends on the relationship between the proportion of G's in the population and the cost-benefit ratio. Specifically,

$$D(U,N) > 0 \text{ iff } P(G) > c/b. \tag{7}$$

If the fraction of G's in the population exceeds the cost-benefit ratio, then U's will be more fit than N's. Each G that is added to the population, adds an increment of fitness to the U's, but not to the

N's. Thus the G's act as insulators or buffers that protect the U's from the N's. In a dichotomous population without G's, P(G) = 0, the N's will always be more fit than the U's.

Turning now to (6), we can see that the only difference between it and (5) is that the G's, instead of paying c in all encounters in which they are potential donors, only do so in encounters with other altruists. The second term on the right side of (6) is the expected benefit provided by the G's that is not extended to the N's. If the population consisted only of G's and N's, the fitness differential between the G's and N's would always be positive since, by assumption, (c-b) > 0. Thus the introduction of U's into the population reduces the relative fitness of the G's over the N's because the U's are indiscriminately altruistic to both G's and N's, but they require a cost from the G's that the N's decline to pay. it is easily shown that

$$D(G,N) > 0 \text{ iff } P(U)/P(G) < (b/c)-1. \qquad (8)$$

Thus, depending on the ratio of U's to G's in the population and on the cost-benefit ratio, G's may be either more or less fit than N's.

From (4), (7) and (8) it becomes clear that there are three distinct rank orderings of fitnesses that are possible. They are $f(G) > f(U) > f(N)$, $f(G) > f(N) > f(U)$, and $f(N) > f(G) > f(U)$. The relationships among these orderings and the distribution of the three types in the population may be graphically displayed by conditions (7) and (8) in a two dimensional simplex using triangular coordinates. This has been done in Figure 1 for b = 2c.

Using triangular coordinates it is possible to plot a three dimension probability distribution in two dimensions because the sum of the probabilites must equal unity. Each point in the equilateral triangle represents a distinct distribution of [P(U),P(G),P(N)] where the perpendicular distance from each apex to the opposite side is normed to unity and P() is the perpendicular distance from the point to an edge. For any point, the sum of the perpendicular distances to the three edges will equal one. In Figure 1, the lower right apex is P(U) = 1, i.e., [1,0,0], a homogeneous population of universal altruists; the upper apex is P(G) = 1, i.e, [0,1,0], a homogeneous population of group-based altruists; and the lower left apex is P(N) = 1, ·i.e., [0,0,1], a population of all nonaltruists. A point on the bottom edge, for instance, represents a distribution in which P(G) = 0, i.e., [x,0,1-x], a dichotomous population with P(U) = x and P(N) = 1-x.

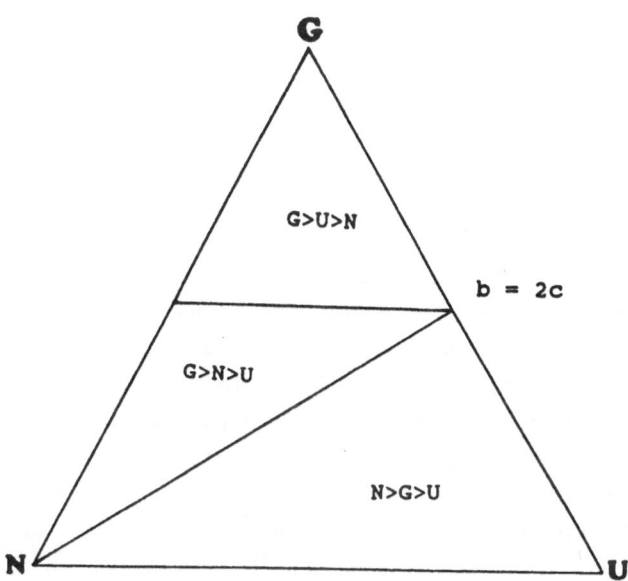

Figure 1. Basic Partition of Population Distributions into Three
Fitness Regions. Benefit/Cost Ratio is 2.

In Figure 1 the horizontal line represents P(G) = c/b = .5. For
population distributions above this line (for the given cost-benefit
ratio, of course), f(U) > f(N). The inequality is reversed for points
below the line, and f(U) = f(N) for all points on the line.

The diagonal line extending from P(N) = 1 to [.5,.5,0] on the
right hand edge is the set of all distributions for which P(U)/P(G) =
1 (= b/c - 1) which represents the boundary defined in (8). For
distributions above this line, f(G) > f(N), while the inequality is
reversed for those below. The three regions that are defined by these
two boundaries are labelled according to the appropriate fitness
ranking. Recall from (4) that f(G) = f(U) only on the right hand edge
where P(N) = 0.

The structure of the regions in Figure 1 changes in a very simple
way as a function of the cost-benefit ratio. It is easily shown that
the two boundaqry lines will always intersect at a point on the right
hand edge. As the cost-benefit ratio increases, making the altruistic
act more costly in comparison to the benefit, the point of
intersection on the right hand edge moves up and approaches the upper

apex. In the limit, naturally, when c = b, altruism no longer pays and
the N's are verywhere more fit than the altruists. As c/b grows
small, on the other hand, and approaches zero, the point of
intersection approaches the lower right apex, and the N's become the
l ast fit throughout the space. The greater the benefit per unit cost,
th less fit the nonaltruists become relative to the two types of
altruists. This is true both in terms of the absolute differences in
the fitnesses as is evidenced in (4) and (5), and also in terms of th
population distributions that favor nonaltruists. Figure 2 displays
the three subregions of the space for a rich (b = 5c) and an
impoverished (b = 1.1c) cost-benefit ratio.

We may, at this juncture, return to the weak intransitivity that
was part of the original motivation for this analysis. The three
dichotomous populations lie on the edges of the two dimensional
simplex. On the bottom edge, we see that f(N) > f(U), and on the left
hand edge we further see that f(G) > f(N). Finally, on the right hand
 dge where P(N) = 0, it is clear that f(G) = f(U). This analysis not
only reconfirms the principles summarized earlier in this paper, it
also illuminates the more general structure of which these
superficially paradoxical principles are a part.

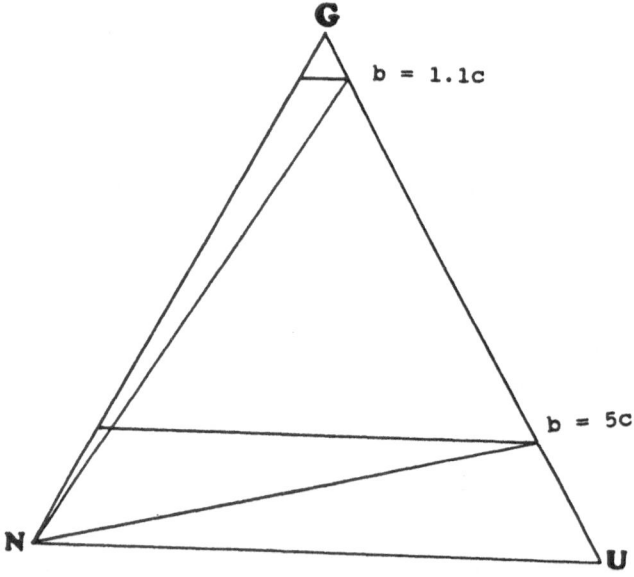

Figure 2. Partitions with Benefit/Cost Ratios of 5 (Lower
Partition) and 1.1 (Upper Partition).

3.2. <u>Dynamic Features</u>. The questions to be answered in this
section are, first, how will the population numbers change as a
function of their relative fitness, and, second, what can be said
about asymptotic or equilibrium distributions. In order to answer
these questions, let us denote the most, next most, and least fit
types by T^1, T^2, and T^3, respectively. T^1 may be either G or N
depending on the cost-benefit ratio and the population numbers, for
instance. Now for any arbitrary point or population distribution, we
assume that the relative frequency of the three types in a succeeding
generation will satisfy two conditions. We assume that the ratio of
the most fit type to the second most fit type will be at least as
large in a succeeeding generation as it was in the starting
distribution; and that the ratio of the second most fit type to the
least fit will be at least as large as it was in initial distribution.
In other words, we simply assume that the most fit will not decrease
in frequency relative to the second most fit, and that the second most
fit type will not decrease in frequency relative to the least fit type
in the population.

For any point $z = [P(T^1), P(T^2), P(T^3)]$ we can define an
arbitrary successor distribution $z' = [P'(T^1), P'(T^2), P'(T^3)]$ that
satisfies the conditions above. The set $S(z)$ of all such successor
distributions are those that satisfy the following conditions:

$$S(z) = \{z' \mid P'(T^1)/P'(T^2) > P(T^1)/P(T^2) \text{ and}$$
$$P'(T^2)/P'(T^3) > P(T^2)/P(T^3)\} \qquad (9)$$

This set $S(z)$ is the set of all possible successor population
distributions of the of the population distribution, z, with a
specified ranking of fitnesses.

This set of successor distributions is displayed in Figure 3 for
one point in each of the three distinct regions. For the point P^1, in
the N > G > U region, the set of points that satisfy (9) lie in the
shad d area to the left of the point. For a point in the G > N > U
region where the point P^2 lies, the successor points will lie above
and to the left. Finally, a point in the uppermost region, e.g., P^3,
will give rise to future populations whose distributions lie above and
to the right as indicated. From the principles underlying the
construction of these regions, it is a simple matter to deduce th
evolutionary changes that will occur for any trichotomous population.
First, if a the population begins in the upper region, it will become
ever richer in G's. Most importantly, the population composition

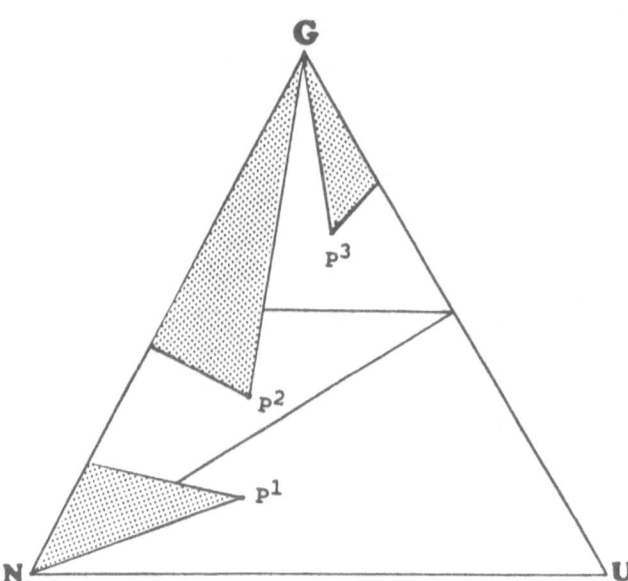

Figure 3. Graphical Display of Successor Sets for Arbitrary Points
in the Three Fitness Regions.

cannot change in such a way that it crosses into another region with a
different fitness structure. Thus, the population must asymptote on
the right hand edge or at the upper apex.

   If a population begins in the middle region, it must always cross
into the upper one. As the G's increase relative to N's, and the U's
decrease relative to N's, the population may become dichotomous with
only G's and N's, that is it may migrate to the left hand edge, or it
may cross the boundary while it is still trichotomous. If the
population moves to the left hand edge, it will migrate up to the
upper apex since the G's are more fit than the N's along this edge. If
it crosses while still trichotomous, then its direction of movement
becomes that of a point in the upper region and it will move up and to
th right. Thus a point in the middle region, like one in the region
above, will move either to the upper apex or the right hand edge.

   Th bottom region is the most interesting because there are two
possible fates for a population that begins with the N's being the
most fit. The first possibility is that it may cross the boundary into
the the region above, in which case it will move inexorably upward,
becoming more and more densely G. It may also move to the lower left
apex, to a homogeneously N population, without ever crossing this
boundary. So a point that begins in the lower region may ultimately

become homogeneously N, or, like points in the other two regions, homogeneously G or a mix of G's and U's.

These conclusions about the dynamics of a trichotomous population are easily visualized as movement in the two-dimensional simplex. The two paths that a population in the lower region may follow are graphed in Figure 4. If the population never crosses the boundary into the middle region, which is to say if the G's never become more fit than the N's, then the population will move to the all N apex. If the G's do become more fit than the N's then the population begins its upward migration and the G's will increase in frequency, eventually to the point where U's are also more fit than the N's. Ultimately, N's will disappear from the population.

It is important at this point to consider the stability of the three equilibria we have identified. The lower left apex, the all N distribution, is unstable because it is vulnerable to invasion by G's. That is to say that if a few G's were to enter an all N population, they would be selected for, and the population would begin moving up along the left hand edge where it would stop at the upper apex, the all G distribution. All N is an equilibrium, but N is not an evolutionary stable strategy (ESS) because it can be exploited by the G's.

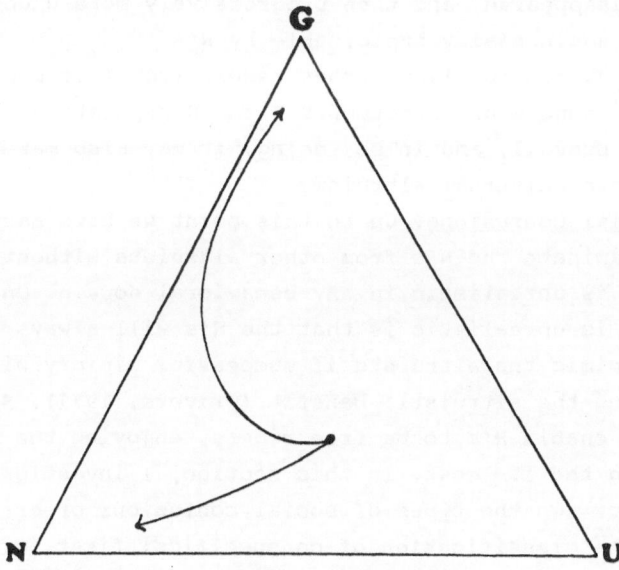

Figure 4. Two Evolutionary Paths for Populations Originating in Lower Fitness Region.

The all G equilibrium, on the other hand, cannot be exploited by N's, but there are no pressures that promote or prevent the invason of the population by U's. Thus an all G population could drift into one containing both G's and U's. While periodic invasions of such populations by N's would tend to drive the distribution upward toward all G, there are other factors that might favor U's in such dichotomous populations. The most conspicuous of these is that the G's must possess a discriminatory apparatus that is more complicated than that of the U's. U's respond uniformily to all conspicifics, whereas G's must differentiate N's from others. If this discriminatory capacity comes at a fitness cost to the G's, if it takes time and or energy from activities that enhance fitness, then U's, in an environment without N's, might have a fitness edge over the G's. The G's are, in a sense, the warriors who protect the U's from the N's, but in a world where the N's are no longer a threat, the Gs' equipment might well be a handicap. If the handicap were great, it is possible that the dichotomous population of G's and U's could move down far enough along the right edge to place the population in the bottom fitness region. An invasion by N's might then lead to a new equilibrium, namely all N. It might also result in a continuous cycle in which the population becomes ever more G until the N's have effectively disappeared, and then progressively more U until the population is again easily exploitable by N's.

The most stable equilibria that result from this model, however, are the homogeneous G or dichotomous G and U populations. Group-based altruism will prevail, and in so doing, it may also make a secure place for a more universal altruism.

3.3. <u>Social Confusions</u>. Up to this point we have assumed that the G's can discriminate the N's from other altruists without error. Such an assumption is unrealistic in any behavioral domain. One important reason why it is unrealistic is that the N's will always have an incentive to mimic the altruists if successful mimicry will result in their receiving the altruistic benefit (Trivers, 1971). Successful cheating will enable N's to be free riders, enjoying the benefit without paying the its cost. In this section, I investigate the consequences of two the types of social confusions or errors that G's can make in the classification of conspecifics: first, erroneously withholding aid to another altruist, and, second, erroneously providing the benefit to an N.

In an encounter in which a G is the potential donor, we define s
as the probability that the G mistakenly classifies another altruist,
either a U or another G, as an N and refuses aid. We define t as the
probability that the aid is erroneously given when the potential
recipient is actually an N. We next rewrite the row for G in the
payoff matrix in Table 1 to reflect the change in expected payoffs.
For instance, the expected payoff to G when encountering a U or
another G is now (1-s)c instead of c, and the expected outcome for the
victims is sb instead of 0. The expected payoff to G when encountering
an N is now tc rather than 0, and the expected payoff to the N is (1-
t)b. We compute the expected payoffs to the three types using this
modified payoff matrix and find the two crucial payoff differences to
be as follows:

$$2D(U,N) = c - bP(G)(1-s-t), \text{ and} \tag{10}$$

$$2D(G,N) = ct + (1-s-t)[P(G)(c-b) + cP(U)]. \tag{11}$$

From these differences we find the appropriate payoff boundaries.
They are as follows:

$$f(U)>f(N) \text{ iff } P(G)>c/[b(1-s-t)], \text{ and} \tag{12}$$

$$f(G)>f(N) \text{ iff } P(G)>-ct/[(c-b)(1-s-t)]-cP(U)/(c-b). \tag{13}$$

An inspection of these functions reveals several features. First,
both kinds of errors reduce the fitness of the U's relative to the
N's. The difference between (12) and (7) is that the benefit term in
the denominator of (12) is multiplied by (1-s-t). Functionally, it is
as if the errors reduce the magnitude of the benefit provided by the
altruists. Furthermore, the two types of errors have identical
effects. Relative to the N's, the U's suffer just as much from one
type of error as they do from the other.

This is not the case with regard to the relation between the G's
and the N's however. From (13) we can see that if t = 0, then (13) and
(8) are identical. In other words, if the G's never make errors of
generosity, giving aid to N's, then erroneously withholding help from
other altruists has no impact on their fitness relative to the N's. If
t is greater than zero, on the other hand, then both types of errors
are costly to the G's. The G's therefore have more to lose by making
mistakes about N's than by making mistakes about other altruists.
Hence we would expect that the G's would display a bias in the
direction of withholding, rather than extending, altruistic acts when
there is doubt about classification of the recipient.

There is an additional implication of (12) and (13) that is less easily inferred from the inequalities themselves. In (13) we see that th critical proportion of G's is a linear function of the proportion of U's in the population. The slope of that function does not vary as a function of the error rates, but the intercept does. Thus, if P(U) = 0, if the population contains only G's and N's, then the number of G's will have to exceed this critical proportion if they are to be more fit than the N's. It will no longer be the case, in other words, that the G's will always be more fit in a dichotomous population of G's and N's. This in turn implies that homogeneous populations of N's may indeed be stable in that they will be able to reestablish themselves after invasions by small numbers of G's.

The effects of the two types of errors on the fitness regions are displayed in Figures 5 through 7. In Figure 5, we set t = 0 and vary only s (in this Figure we let b = 3c). Clearly the upper boundary, the one defining whether the U's or the N's are second most fit, changes with varying levels of s, but the lower boundary, that differentiating the regions in which the G's and N's are the most fit, does not change. Consequently, the entire left hand edge lies in a region in which the G's are more fit than the N's.

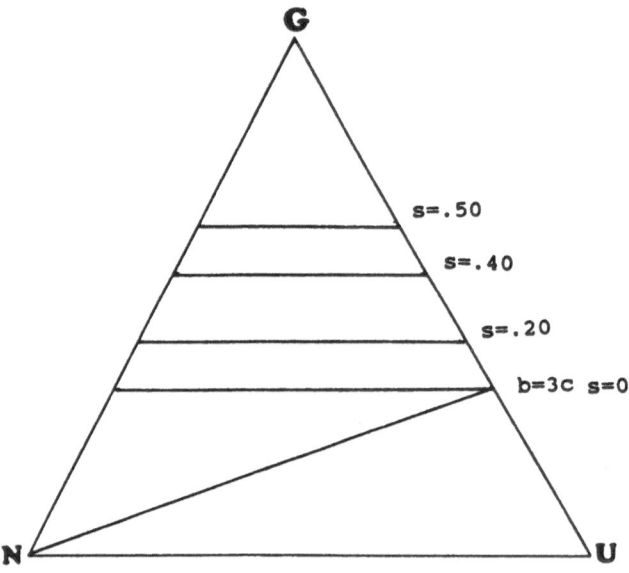

Figure 5. Effects on Fitness Regions of Varying s, the Probability of Erroneous Withholding Aid from an Altruist, when t = 0.

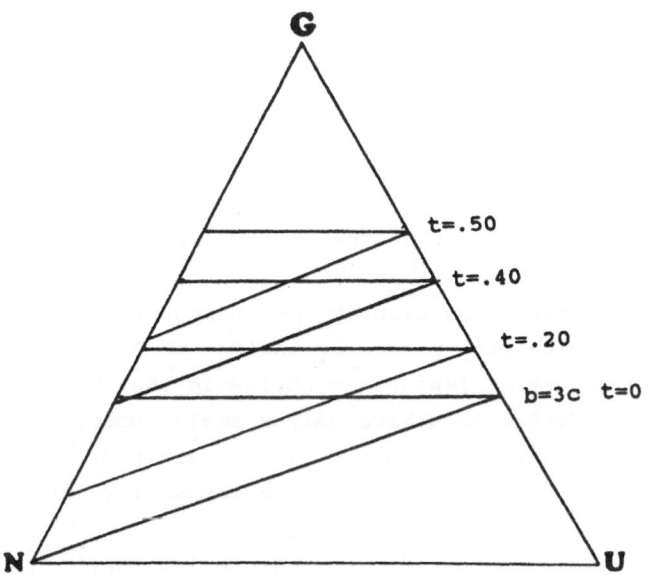

Figure 6. Effects on Fitness Regions of Varying t, the Probability
of Erroneously Providing Aid to a Non-altruist, when s = 0.

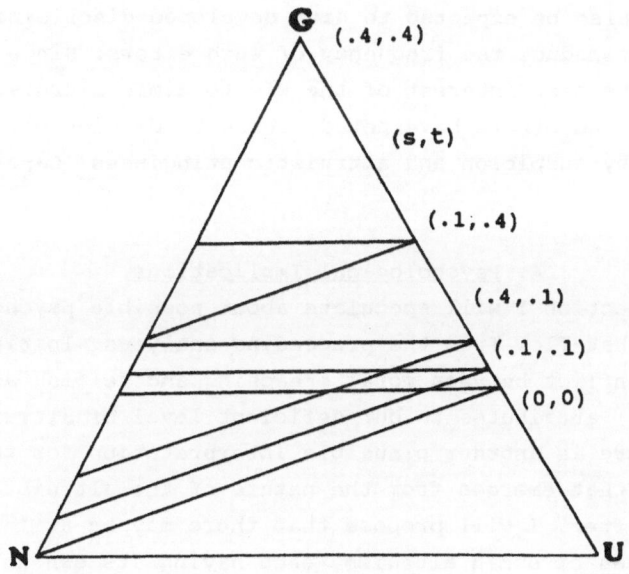

Figur 7. Effects on Fitness Regions of Simultaneously Varying s
and t.

Contrast this situation to the one displayed in Figure 6 in which
s = 0, and t is allowed to vary. Both boundaries are affected in this
case. The tendency to misidentify N's as other altruists and to aid
them increases the fitness of the N's against both G's and U's. It is
also obvious from Figure 6 that the G's are no longer more fit than
the N's everywhere along the left edge of the simplex. When t > 0, we
can now find values of t so that the N's are more fit than the G's in
dichotomous distributions.

Finally, in Figure 7 we display the consequences of varying s and
t simultaneously, with each parameter taking the values of .1 and .4.
Again, the most important feature to notice is that when the value of
t is small (.1), changes in s have only a small impact. Changes in t
have a much greater influence on the fitness structures both directly,
and indirectly, in that the larger t, the larger the impact of changes
in s.

There is a clear and unambiguous conclusion that can be drawn
from this examination of the consequences of the two types of
confusions that the G's can make. Errors of inclusion, mistakenly
including N's in the group that is aided, is a much more serious
threat to the G's (and to the U's) fitness than mistakenly depriving
an altruist of aid. Any species in which a group-based altruism has
evolved could also be expected to have developed discriminatory
tendencies that reduce the frequency of such errors. Since it will
always be in the best interest of the N's to mimic altruists, as
Trivers (1971) and others have noted, these tendencies may well be
characterized by suspicion and altruistic stinginess. Careless G's
will be doomed.

4. Psychological Implications.

In this section I will speculate about possible psychological
implications that flow from the preceeding analyses. Initially, I will
discuss the conflict between moral preaching and selfish acts that
Campbell (1975) attributes to our deficient level of altruism. I will
argue that there is another plausible interpretation for this
conflict, one that emerges from the nature of the altruistic encounter
itself. Thereafter, I will propose that there may be at least three
distinct sources of human altruism, each having its own unique
features. Some of the points I make were outlined earlier (Messick,
1976) in a different context. I will dwell at some length on group-

based altruism in this discussion and relate the idea to current
research in social psychology.

4.1. <u>Institutional Morality and Individual Selfishness</u>. Campbell
argues that moral institutions have evolved to compensate for the
inevitable selfishness that biological evolution has created in
humans. The conflict that results is between mainly selfish biological
impulses and cultural preachings that stress socially beneficial self-
sacrifice. The position that I want to offer accepts the existence of
a conflict between selfish and collective interests as commonplace
fact of social life, but views this conflict as a consequence of
social interdependence itself, not of a disparity between the demands
of biology and those of society. The heart of the conflict is evident
in Table 1. Each individual in the population may be found in the role
of the recipient or the donor. The interests of an individual in these
different roles are markedly different. As a potential recipient of
altruistic benefits, it is in one's interest to be surrounded by
altruists. One's welfare is maximized, in this role, when all of the
individuals in the social environment are willing to help. Therefore,
one should be more than willing to invest time and effort to convince
others that they (i.e., the others) should be considerate of the
welfare of their fellow humans, in particular, one's self. Support for
moralistic preaching should be widespread to the extent that such
preaching makes others more helpful.

When one is in the role of the donor, however, things are quite
different. Here one would like to evade the costs and obligations of
altruism, and one should resist the altruistic urgings that one would
endorse as a recipient. In the role of donor, one would be expected to
argue for one's right to ignore others' demands that one make
sacrifices for another's welfare. From this perspective the conflict
between selfishness and altruistic urgings has nothing to do with a
conflict between biology and culture, but has everything to do with
the fact that each individual may be either the donor or the recipient
of altruism. It is the interests of these two roles that are
incompatible and the ambivalence derives from the uncertainty about
which role one will occupy.

What emerges from this view is an illumination of the foundations
of hypocrisy, the tendency to endorse an ethical principle and to
behave according to its opposite. Our analysis clearly demonstrates
how the incentives for the endorsement and the behavior diverge. All
three types in our hypothetical population should espouse moral

principl s urging others to be altruistic, but only the behavior of
the U's would be consistent with those universally expressed
principles. The incentive to espouse moralistic principles would be
greater yet if their public avowal were taken as evidence of actually
being an altruist. The public proclamation of self-sacrificial
morality by the N's might deceive the G's about the N's true identity.
And the deceit might be even more effective if the N's first convinced
themselves that they were altruists. This insight that has led some
scholars to argue that self-deception is an evolutionary profitable
characteristic (Martin, 1986; Trivers, 1985). There is certainly
evidence that in some domains at least, self-decepton is a
psychologically beneficial quality (see, for instance, the review by
Taylor & Brown, 1988).

The hypothesis that socio-cultural evolution has developed
institutions to promote selfless, socially beneficial behavior <u>because</u>
biological evolution failed to do so, seems far from the simplest idea
available to explain the conflict between selfish behavior and
cooperative institutions.

4.2.<u>Human Altruism</u>. What do theories of the evolution of altruism
have to tell us about altruism in our own species? Hypotheses about
our behavioral past will be necessarily speculative and it is
obviously impossible to conduct experiments on our evolutionary
ancestors to test the hypotheses. However if we assume that patterns
of human helpfulness are consistent with evolutionary principles
regarding altruism, then it is possible to think about the constraints
these principles place on psychological theories. The different models
for the evolution of altruism appear to have different, but not
necessarily contradictory implications.

It is inconceivable that our species could have evolved without
the process of kin selection. Child bearing is a costly activity for
women and child rearing can be costly for men as well. Individuals in
our evolutionary past who declined to pay these costs may have had
longer life expectancies than their fertile neighbors, but we are
clearly not the descendants of the former. At least at the level of
child bearing and child caring, ours is an altruistic species.

Having said that ours is an altruistic species with regard to
offspring does not explain the proximal mechanisms that are involved.
With regard to child bearing and rearing, these mechanisms include the
pursuit of sexual pleasure, resulting in pregnancy, the elimination of
the discomfort of distended breasts by nursing an infant, and escaping

the acutely aversive sound of an infant crying by tending to its needs. The emotional bonding that we refer to as love, with its attendent concern for the welfare of the loved one, also supports the process.

It is also very likely that the benefits that derive from the reciprocation of favors were discovered by our evolutionary ancestors. Kin selection may explain the great importance placed on kin structures by virtually all human societies, but Gouldner (1960) has also argued that a "norm of reciprocity" is a universal characteristic of human society. Roughly speaking, this norm of reciprocity refers to rules that define harmful behavior an inappropriate response to behavior that was beneficial. In other words, according to Gouldner, human societies everywhere adopt a weak version of tit-for-tat (Axelrod, 1984).

It is important that the reciprocal exchange of favors is beneficial to the participants even if they are not genetically related. It is the phenotype, not the genotype that is benefitted in this case. The payoffs to those who are unable or unwilling to abide by rules of reciprocity will simply be lower than those who do accept such rules. A genetic connection is no longer needed. In an earlier paper, I hinted at possible ways in which reciprocation might be related to kin selection. I also suggested ways in which group selection might have been involved in human evolution.

What I want to focus on in this final section is group-based altruism. By that I mean self-sacrificial behavior toward another altruist (1) to whom one is not genetically related, (2) with whom one may not have a temporal relationship of the sort required by reciprocal altruism, and (3) who may not live in a population segregated into demes or sub-populations as is required by conceptions of group selection (Cohen & Eschel, 1976; Wilson, 1975). In the first part of this paper I showed that such altruism can be favored by natural selection. Here I propose that the proximal mechanisms that support this type of altruism may be those that have been referred to as ethnocentrism (Sumner, 1906; Brewer, 1986) or outgroup discrimination (Tajfel, 1971). The idea is similar to one developed by Campbell (1965) in a paper with the intriguing title "Ethnocentrism and other forms of altruistic behavior". Campbell (1972) later repudiated the concept because it appeared to depend on group selection. The model developed in this paper clearly demonstrates that "altruistic ethnocentrism" can evolve solely through individual selection.

In this model, the only thing that makes the G's distinctive is the fact that they would be willing to aid each other as well as the U's who aid everyone. The N's aid no one, not even themselves. I made the point at the beginning of this paper that the model does not rely on an anticipation of repeated interaction for the G's to have a fitness advantage over the N's. Thus in any situation in which one could be help giver or help receiver, and in which there are others who will either help indiscriminately or help not at all, there is a potential benefit to be gained by those who perceive themselves to be members of a common group, who, by virtue of their common group membership, will mutually aid one another. Put somewhat differently, those individuals who (1) draw a distinction between an ingroup, to which the individual belongs, and an outgroup, and (2) are willing to make sacrifices for the ingroup but not the outgroup, are functional G's. Our proposal, specifically, is that the tendency to partition social aggregates into ingroups and outgroups and to behave more favorably toward the former than the latter, may represent the cognitive and behavioral bedrock that supports group-based altruism. I will presently summarize some of the growing evidence for these cognitive and behavioral processes. A more complete review is given by Messick & Mackie (1989).

Central to this proposal is the idea that the individuals may not be directly aware of their prosocial orientations toward one another. They may be most aware of being members of a common group. It is the creation of the group boundary itself that defines set of individuals who would, if needed, mutually aid one another. The willingness to sacrifice for one another is a consequence of the creation of the group. The group does not happen to contain those who are altruistic; it is the creation of the group that *makes* the members altruistic. In principle, any set of individuals for whom a common group identity could be highlighted , would be equivalent to any other group. Naturally, in the actual social ecology of our species, the most salient groups will be those that share obvious features (e.g., language, nationality, gender, race, religion, age) that differentiate the members from others. However, experimental research has clearly demonstrated that groups can form around profoundly trivial distinctions as well (Sherif, 1967; Turner, 1987). And while aid is shared among the group members, it is withheld from, or is less freely

given to thos not in the group. Failure to help potential outgroup
helpers constitutes an error of overexclusion and, as I showed in a
previous section, this type of error burdens the G's with very little
cost. Alternatively, if exclusion from the group reduces the
generosity of those excluded, they become functional N's.

As a result of the fact that the expected payoff to G's is not
dependent on repeated interaction, the success of the G's does not
presume that the other G's are the same from one situation to another.
There may be multiple threats in the environment, for instance, and it
may be beneficial to be a G with regard to each of these threats. That
benefit is not based on the assumption that the other G's from one
situation are also the G's in another. A "family" ingroup may have no
overlap with a "coworker" or "classmate" ingroup. A particular
situation may accentuate one ingroup over another and slightly
different situations may elicit different salient groupings. Moreover,
the very diversity and fluidity of what constitutes an ingroup may be
the quality that permits this ethnocentric process to be studied in
so-called "minimal group" paradigm in which "groups" are formed on the
basis of essentially meaningless characteristics. The ethnocentric
process is "content free" in that it will be evoked by virtually any
social categorization.

While I have stressed that the benefits enjoyed by the G's do not
depend on the repeated interaction of the individuals, the possibility
of temporal reciprocal aid only increases the benefits of the G's (as
well as the U's) relative to the N's. If a G aids another simply
because the other was also a G in a given situation, the other may
reciprocate that aid in a different situation in which they are not
both G's. (Obviously, such reciprocation would be redundant in an
environment in which the group of G's was fixed. Help would always be
reciprocated by virtue of the common membership.) Thus a group-based
altruistic act may initiate the temporal exchange of benefits
supported by non-group-based processes.

It remains now to mention that there is ample evidence that
humans do tend to dichotomize social groupings into ingroups and
outgroups, and they do tend to respond more favorably to the former
than to the latter. Recent research on these and related issues has
been reviewed by Messick & Mackie (1989) so I will mention only a
couple of phenomena that seem central to the point of this paper.
First, there is good evidence that the cognitive representations of
ingroups and outgroups are different (see, for instance, Judd & Park,

1988; Quattrone, 1986). Outgroups tend to be seen as simpler, more homogeneous, and represented in t rms of higher level categories then ingroups. (As an illustration of this latter phenomenon, when meeting academic colleagues at international meetings, it seems a common pattern to hear one ask the institutional affiliation of someone from one's own country, but to not do so with colleagues from other countries. An American might respond to the statement "I am from Germany" with the remark, "Oh yes, I've been there." "I'm from the U. S." would more likely evoke, "From where in the U.S.?" Categorizing at the national level is sufficient for outgroups, but insufficient for ingroups.) Outgroups are thought to contain fewer subtypes of individuals and to be less complex than ingroups. The explanations for these differences are less clear. We are more familiar with ingroups than outgroups as a rule, so we may have been exposed to a greater variety of ingroup than outgroup members. However, there is evidence that the difference in perceived homogeniety is found with "minimal", that is to say artificially formed, experimental groups with no past not history as well as with natural groups. Ingroups may also be perceived as more complex than outgroups because one's self is a member of the ingroup by definition, and the self is a distinctive individual. Whatever the ultimate explanation(s), there is a growing body of evidence that people process information about outgroups differently from ingroups.

Second, there is copious evidence that people behave differently toward ingroup and outgroup members. Since the pioneering experiment by Tajfel, Flament, Billig, & Bundy (1971), an experiment that demonstrated that a history of intergroup contact was not a necessary condition for the creation of intergroup discrimination, there have been scores of experiments that have replicated the basic finding that subjects behave in a more generous or favorable manner to ingroup than to outgroup members. This bias persists even when the basis for categorizing subjects into groups is trivial and when subjects do not know the identities of the other individuals who will be affected by their decisions. Recent reviews of this work on intergroup bias have been published by Brewer (1979), Brewer & Kramer (1985), and Messick & Mackie (1989).

Finally, after reviewing the extant experimental evidence, Brewer (1979) concluded that the intergroup bias seemed to stem mainly from the over-reward of ingroup members and not from the punishment of outgroupers. While the current paper makes no claims about differences in aggression toward ingroup and outgroup targets, Brewer's conclusion reinforces the point that is implicit in this paper, namely that ethnocentrism or intergroup discrimination may exist simply because benefits are afforded or withheld to others as a function of the others' group membership. Hostility toward the outgroup need not be involved.

Whether intergroup biases occur at "natural" levels, involving nations, religions, and or language groups, where the stakes are often mortal, or at the level of trivial, experimental distinctions where the stakes are typically insignificant, such biases in favor of our ingroups appear to be fundamental to our social nature. Recent volumes have suggested that these ethnocentric tendencies may have an evolutionary origin (Crawford, Smith, & Krebs, 1987; Reynolds, 1987). In this paper I have formalized the argument and shown that a group-based or ethnocentric altruism is not only evolutionarily plausible but also consistent with what is known about human social behavior.

References

Axelrod, R. (1984). *The Evolution of Cooperation*. New York: Basic Books.

Boyd, R. & Richardson, P.J. (1988). The evolution of reciprocity in sizable groups. *Journal of Theoretical Biology, 132,* 337-356.

Brewer, M. B. (1979). In-group bias in the minimal intergroup situation: A cognitive-motivational analysis. *Psycholgical Bulletin, 86,* 307-324.

Brewer, M. B. (1986). The role of ethnocentrism in intergroup conflict. In S. Worchel & G. A. Austin (Eds.) *Psychology of Intergroup Relations.* Chicago: Nelson Hall.

Brewer, M. B. & Kramer, R. M. (1985). The psychology of intergroup attitudes and behavior. *Annual Review of Psychology, 36,* 219-243.

Campbell, D. T. (1965). Ethnocentric and other altruistic motives. In D. Levine (Ed.), *Nebraska Symposium on Motivation,* Vol 13, Lincoln: University of Nebraska Press.

Campbell, D.T. (1972). On the genetics of altruism and the counter-hedonic components of human culture. *Journal of Social Issues, 28,* 21-37.

Campbell, D.T. (1975). On the conflict between biological and social evolution and between psychology and moral tradition. *American Psycholgist, 30,* 1103-1126.

Cohen, D. & Eschel, I. (1976). On the founder effect and the evolution of altruistic traits. *Theoretical Population Biology, 10,* 276-302.

Crawford, C., Smith, M., & Krebs, D. (1987). *Sociobiology and Psychology.* Hillsdale, NJ: L. Erlbaum.

Gouldner, A. W. (1960). The norm of reciprocity: A preliminary statement. *American Sociological Review, 25,* 161-178.

Judd, C. M. & Park, B. (1988). Outgroup homogeniety: Judgments of variability at the group and individual levels. *Journal of Personality and Social Psychology, 54,* 778-88.

Martin, M. (1985). *Self-Deception and Self-Understanding*. Lawrence, KA: University of Kansas Press.

Matessi, C. & Jayakar, S. D. (1976). A model for the evolution of altrusitic behavior. *Genetics, 74,* s 174.

Messick, D. M. & van de Geer, J. P. (1981). A reversal paradox. *Psychological Bulletin, 90,* 582-593.

Messick, D. M. (1976). Comment. *American Psychologist, 31,* 366-69.

Messick, D.M. & Mackie, D.M. (1989). Intergroup relations. *Annual Review of Psychology, 40,* 45-81.

Quatrone, G. A. (1986). On the perception of a group's variability. In S. Worchel & W. G. Austin (Eds.) *Psychology of Intergroup Relations*. Chicago: Nelson Hall.

Reynolds, V., Falger, V., & Vine, I. (1987). *The Sociobiology of Ethnocentrism*. London: Croom Helm.

Sherif, M. (1967). *Group Conflict and Cooperation*. London: Routledge & Kegan Paul.

Sherman, P. W. (1977). Nepotism and the evolution of alarm calls. *Science, 197,* 1246-1253.

Sumner, W. G. (1906). *Folkways*. New York: Ginn.

Tajfel, H., Billig, M.C., Bundy,, R.P., & Flament, C. (1971). Social categorization and intergroup behavior. *European Journal of Social Psychology, 1,* 111-140.

Taylor, S. E. & Brown, J. D. (1988). Illusion and well-being: A social psychological perspective on mental health. *Psychological Bulletin, 103,* 193-210.

Trivers, R. (1971). The evolution of reciprocal altruism. *Quarterly Review of Biology, 46,* 35-57.

Trivers, R. (1985). *Social Evolution*. New York: Benjamin Cummings.

Turner, J.C. (1987). *Rediscovering the Social Group: A Self-Categorization Theory*. New York: Basil Blackwell.

Wilson, D. S. (1975) A theory of group selection. *Proceedings of the National Academy of Sciences, 72,* 143-46.

Wilson, D. S. (1980). *The Selection of Populations and Communities*. Menlo Park, CA: Benjamin Cummings.

International Journal of Game Theory

Title No. 182   ISSN 0020-7276

The **International Journal of Game Theory** is the leading international periodical devoted exclusively to game theoretical developments. Distinguished experts from around the world here present fundamental research contributions on all aspects of game theory.

Some of the interesting papers which appeared in 1990 were:

Pool-Listing Service

A listing service is offered to announce preprints of research memoranda and discussion papers in the field of game theory on a quarterly basis.

Fields of Interest

Mathematics, economics, politics, social sciences, management, operations research, the life sciences, military strategy, peace studies, theoretical biology.

Subscription Information

1991, Vol. 20 (4 issues) DM 378,– plus carriage charges or US $257.00 total

Physica-Verlag Heidelberg

Please order through your bookseller
or from Physica-Verlag, c/o Springer GmbH & Co., Auslieferungs-Gesellschaft,
Haberstr. 7, W-6900 Heidelberg, F.R.Germany

E. van Damme

Stability and Perfection of Nash Equilibria

2nd rev. and enl. ed. 1991. XVII, 345 pp. 105 figs. Softcover DM 65,–
ISBN 3-540-53800-3

This book discusses the main shortcoming of the classical solution concept from noncooperative game theory (that of Nash equilibria) and provides a comprehensive study of the more refined concepts (such as sequential, perfect, proper and stable equilibria) that have been introduced to overcome these drawbacks. The plausibility of the assumptions underlying each such concept are discussed, desirable properties as well as deficiencies are illustrated, characterizations are derived and the relationships between the various concepts are studied.

The first six chapters provide an informal discussion with many examples as well as a comprehensive overview for normal form games. The new material focuses on games in extensive form and considers such topics as: noncooperative implementation of cooperative concepts (e.g. the Rubinstein bargaining model that yields Nash's solution), repeated games (the Folk Theorem), evolutionarily stable strategies (the relevance of refinements for the biological branch of game theory), and stable equilibria (in the sense of Kohlberg and Mertens) and the adequacy of the normal form for rational decision making.

Springer-Verlag
Berlin
Heidelberg
New York
London
Paris
Tokyo
Hong Kong
Barcelona
Budapest